ISBN 978-1-334-01453-6
PIBN 10668610

MATHEMATICAL INSTITUTIONS.

In Three PARTS.

I. *CLAVIS*		The	KEY,
II. *JANUA*			GATE,
III. *ANCILLA*			HAND-MAID,

TO THE

Mathematical Sciences.

WHEREIN,

The Doctrine of *Plain* and *Spherical* TRIANGLES, is Succinctly Handled, *Geometrically Demonstrated*, Arithmetically, Geometrically, Instrumentally Performed;

And Practically Apply'd to

GEOMETRY,	SCIOGRAPHIA,
COSMOGRAPHY,	NAVIGATION,
GEOGRAPHY,	And THEORIES
ASTRONOMY,	of the PLANETS.

By WILL. LEYBOURN, Philom.

Ingredere ut Proficias.

LONDON,
Printed for *Rob. Billing*, at the *White Swan* in St. *Paul's* Church-Yard and *Geo. Briant*, at the *King's Head* near the *Royal-Exchange* in *Cornhil*. MDCCIV.

TO THE

Young STUDENT

IN THE

MATHEMATICKS.

IN the *Study* (as well as in the *Teaching*) of any *Art* or *Science*, a Gradual Proceeding therein, ought First, and Principally, to be considered: For, It is one thing to *Teach*; and another *How* to *Teach*: And, according to the Adage, *Qui bene distinguit, bene Docet*.

Now, the Design of this *Book*, being for the *Directing* and *Instructing* of such as would apply themselves to the Study of any *Mathematical Science*, so as to become such a *Proficient* therein, as to give a *Demonstrable Account* of what he does; must not attempt, at the first Onset, to fall directly upon that *Part* he principally aims at, but first acquaint himself well with such *Elements* as concern, and are subservient to, them all.

And to that end, The ſeveral *Parts* and *Secti-ons* of this *Book*, are diſpoſed in ſuch *Order*; That, Beginning with the *Firſt Part* (which is as a KEY to give you entrance into the GATE; which is the *Second Part*; Through which, having perfectly paſſed, you may confidently proceed to fall upon the *Study* (or *Practice*) of any *Mathematical Science*; And not only on thoſe that I have here (by my HAND-MAID, in the *Third Part*) directed you to; but to any other you ſhall attempt: TRIGONOMETRIA being the *Baſis*, *Foundation*; nay, the very *Primum Mobile*, of all the reſt.

For Illuſtration whereof, I have applyed the *Doctrine* of *Triangles* to *Practice* (in the ſeveral *Sections* of the *Third Part*) in the Solution of ſuch Problems, as are of frequent uſe in ſeveral Parts of the Mathematicks. So in the Firſt, Which teacheth how to take all manner of *Heights* and *Diſtances*, acceſſible or inacceſſible: And in the Sixth, which treateth of *Navigation* both by the *Plain Sea-Chart*, and that called *Mercators*; in both which, I have firſt ſhewed how *Geometrically*, by *Scale* and *Compaſſes*, to lay down (upon Paper) a *Figure* anſwerable to the *Queſtion* Propounded: In which *Figure*, ſo laid down, you will have conſtituted a *Right-Lined Triangle* or *Triangles*, of ſome kind or other, *i. e.*

either

either *Right* or *Oblique-Angled*; And in it discovered what Parts thereof are Given, by the demand in the *Question*; and then the *Part* or *Parts Unknown* will be such (as being found) will answer the *Question* demanded: And herein the Excellency of the *Doctrine* of *Plain Triangles* is in part made manifest.

Likewise, In the 1st. 2d. 3d. 4th. and 5th. Sections, which treateth of *Cosmography*; and therein of *Geography*, *Astronomy* and *Dialling*: I do first declare, and shew how the Operation of the *Question* or *Proportion*, is to be performed upon the *Terrestrial* or *Celestial Globe* (according as the thing proposed does require.) Which *Question* being resolved upon the *Globe*; and the *Globe* resting in the same *Position* it was when the *Question* was resolved thereupon: You will upon the *Body* of the *Globe*, (by the great *Circles* thereon described, and those appendant to it) apparently discover a *Spherical Triangle* or *Triangles* Constituted; in which (according to the Tenor of the *Question*) you will plainly perceive what *Parts* thereof are *Given*; and then, the other *Parts* being found, must necessarily Answer the *Question* Propounded: And herein, in some measure, is the *Doctrine* of *Spherical Triangles* made applicable to all the above-named Practices. And Moreover, by Projecting of the Sphere, as is done for the *Geographical Problems* in *Part* III. *Section* the *Second*. And for *Astronomical Problems*, as is shewed in PART II. Section II. Of Projection of the Sphere in Plano, *&c.*

Also in the VIIth. *Section*, Which treats of the *Theories*, and finding the *Places* of the *Planets*, *&c.* there the *Doctrine* of *Triangles* both *Plain* and *Spherical*, are joyntly concerned: So that, let your *Mathematical Practice* or *Study* be, *Opticks*, *Perspective*, *Fortification*, *Gunnery*, or any other *Mathematical Art*: The most *Problems* relating to any of them, must be beholding to *Trigonometry* for their Solutions.

And now I have but one thing more to advertise the Reader of: That, Whereas in the performances of all the *Operations* in the several *Parts* of this Book, the First (which treats of *Practical Geometry* only) excepted, there is continual Use to be made of a CANON, TRIGONOMETRICAL; or TABLES of Artificial SINES, TANGENTS and LOGARITHMS; So that the Reader cannot expect but that such TABLES should have been *Added* unto these *Precepts*. And, indeed, it was so intended, but that almost in every Book that hath any thing of the *Mensuration* of *Triangles* in it, there are such *Tables Printed*; and that one or other of them may be in every Man's Hands; and because all the *Problems* in this Book

may

may be ſolved by (almoſt) any of thoſe *Canons* or *Tables*; was the occaſion of the omiſſion of ſuch Tables here at preſent. But chiefly, becauſe there is intended (in ſome ſhort time) to be Printed,) A CANON of Artificial SINES and TANGENTS, both **Sexagenary** and **Centecimal**; differing from any yet Extant; it ſupplying both ways of *Numeration* at one *view*, and yet in the ſame *Room*: And to this CANON ſhall be Added Ten CHILIADS of LOGARITHMS commodiouſly Contracted; and the Uſe thereof in Logarithmical Arithmetick.

Thus having given the Reader a *General Account* in this *Preface*, I refer him to a more *Particular* in the following *Contents*, and ſo leave him to the Practice of them: In which I wiſh him good ſuceeſs; and till he hears farther from me, bid him Farewell.

WILL. LEYBOURN.

ERRATA.

Page	Line	For	Read
9	20	the given Point	the given Point P.
11	16	fame	ſame
13	28	from N to O,	from M to O,
20	19	*M ans* 8 is	*Means* 8 in 8 is
21	5	to D E.	to A E.
52	14	E C, intercepted	E F, intercepted
52	16	E and D;	E and F;
52	17	*Arch* E D is;	Arch E F is;
59	30	The *Angle* B P Q	The *Angle* P B G
63	1	*Distinct*	*Disjunct*.
69	17	contrived :	continued :
95	29	½ Sun < s :	½ Sum < s :
110	37	Meaſure is ⊙ O,	Meaſure is Q ♉,
111	15	T ♉,	Q ♉
111	16	and T,	and Q
142	23	of R, S, P, T.)	of R, A, P, T.)
144	29	draw a Line L,	draw a Line L *b*,
ibid	*ib.*	alſo b, an	alſo an
148	21	*Angle* at F,	*Angle* at D,
151	14	From A G to	From G to
164	1	of the	The
196	*ult*	Plain Triangles	Spherical Triangles
279	23	*Point* B :	*Point* in B :
281	22	Hour	Hours
282	2	26 de.	36 de.
282	16	23.25	23.35
287	2	Declination	Reclination
ibid	21	Reclination 120	Reclination 20
297	8	is to be moſt	is moſt
308	37	*Problems* and	Problems in
327	*ult*	6.50 deg.	5.50 deg.
328	32	M P O of A B,	M P of A B,
330	2	6.50 de.	5.50 de.
347	8	B T E.	B F E.

Figures in the Margins Omitted, or Miſplaced.

Page 7. Againſt *Probl.* I. put *Fig.* VIII. and againſt *Prob.* II. put *Fig.* IX. in the Margin.
Page 9. Againſt *Probl.* V. put *Fig.* XIII. in the Margin.
Page 20. For *Fig.* XXXIV. put *Fig.* XXXVII. in the Margin.
Page 50. For *Fig.* VI. put *Fig.* XII. in the Margin.
Page 89, 90. 91. *Fig.* XXXI, in the Margin, muſt be *Fig.* XXXII.
Page 274. *Fig.* LIII. in the Margin, is omitted
Page 285, and 286. for *Fig.* LIX. read *Fig.* LVIII. in the Margin.
Page 53. The three laſt Lines are thus to be read :

The Triangle { B A C / B D C / C D E } hath { One Right Angle, and Two Acute. / Two Obtuſe, and One Right. / One Obtuſe, and Two Acute.

C O N-

CONTENTS.

Part I. The KEY, *Of GEOMETRY.*

And therein of

D*Efinitions.*	*From* Page 1 *to* Page 7
Practical Problems	7 *to* 17
Theorems	18 *to* 27

Part II. The GATE, *Of TRIGONOMETRY.*

And therein of

1. D*EFINITIONS*	*From* Page 29 *to* Page 32
2. Right-Lines *applied to a* Circle, { Sines, Tangents, Secants, *Their* Construction }	32 *to* 38
3. *The* Affections *of* Right-lined, *or* Plain Triangles	38 *to* 43
4. *The* Mensuration (*or* Solution) *of* Plain Triangles	43 *to* 50
5. *SPHERICAL TRIANGLES,*	
{ Definitions, Theorems, }	50 *to* 56
Affections *of* Great Circles *of the* Sphere, *in order to the* Demonstrating *of* Spherical Triangles	56 *to* 60
6. *The* Solution *of* Right-Angled Spherical Triangles	60 *to* 68
7. Oblique-Angled Spherical Triangles, *and* Prenotions *concerning them*	68 *to* 70
8. *The* Solution *of them*	70 *to* 83
9. Oblique-Angled Spherical Triangles, *Resolved without regard had to a* Perpendicular	84 *to* 96

The CONTENTS.

10. Trigonometical Problems *Extraordinary* 96 *to* 105
11. Spherical Triangles Geometrically performed, *by* Projecting *of the* Sphere *in* Plano 106 *to* 114
12. Spherical Trigonometry, Instrumentally Performed *several ways*: *As, by the*
Steriographical, Orthographical } *Planispheres* 114 *to* 138

Part III. *ANCILLA: Vel,* Trigonometria Practica.

In Seven SECTIONS.

SECT. I. Of *GEOMETRY.*

And therein of

1. *ALTIMETRIA*; *Or the Taking of* Heights, (*as of* Towers, Trees, Steeples, *&c. Whether Accessible or Inaccessible*; *several Ways* *From* Page 141 *to* Page 147
2. *LONGIMETRIA*; *Or the Measuring of* Distances: *As of Remarkable* Places *upon the* Earth, Ships *upon the* Sea, *&c. Either Accessible, or Inaccessible* 147 *to* 152
3. *PLANOMETRIA*; *Or the Measuring of* Plains: *As of* Board, Glass, Pavements, Hangings, Wainscot, *&c.* 152 *to* 155
4. Geodæcia; *Or Measuring of* Land 155 *to* 158
5. *STERIOMETRIA*; *Or the Mensuration of* Solids *As* Timber, Stone, *&c. And of* Regulars: *As* Cubes, Globes, Prisms, Parallelipipedons, Pyramids, Cones, *&c. whether* Whole *or* Dissected 158 *to* 160

SECT. II. Of *COSMOGRAPHY.*

And therein of

1. *The* Material Sphere, *or* Globe, Celestial *and* Terrestrial: *And of such* Circles, Lines *and* Points, *as are described thereon, or Appendent unto it.* 161 *to* 167

2. COS-

The CONTENTS.

2. *COSMOGAPHICAL ELEMENTS, both* Geographical *and* Aſtronomical 167 *to* 175

SECT. III. Of *GEOGRAPHY.*

And therein of

The Uſe of the Terreſtrial Globe, *and thereupon to find the Situation of* Places *both in* Longitude *and* Latitude : *And to find their* Diſtances : *And alſo by* Trigonometrical Calculation 176 *to* 185

SECT. IV. Of *ASTRONOMY.*

And therein of

1. *ASTRONOMICAL TABLES* 185 *to* 197
2. Aſtronomical Problems, *relating to the* Sun *and* Fixed Stars *To Work them upon the* Globes; *And by* Trigonometrical Calculation *alſo* 197 *to* 256
3. Aſtronomical Problems, *relating to* Aſtrology 257 *to* 266

SECT. V. Of *SCIOGGRAPHIA*, or *DIALLING.*

And therein of

1. *DIALLING in General*; *with the* Situation *of all* Plains *on which* Dials *may be made* : *And to find out any ſuch* Situation. 267 *to* 272
2. *The Making of all ſorts of* Dials, *by finding out the places of* Subſtile, Stile *and* Hour-Diſtances, *by the* Globe : *And alſo, to find the ſame* Requiſites, *and* Hour-Diſtances, *by* Trigonometrical Calculation, *after a new Method* 272 *to* 296
3. *The Inſcription of the* Greater *and* Leſſer Circles *of the* Sphere *upon all ſorts of* Dial Plains

And ſuch are

- Parallels *of* Suns *Courſe throughout the* Zodiack,
- Parall. *of the* Days length, Suns Riſing, Setting, *&c.*
- Azimuth *or* Vertical Circles,
- Almicanters, *or* Circles *of the* Suns Altitude.
- Jewiſh, Italian, *and* Babyloniſh *Hours*

296 *to* 303

SECT.

The CONTENTS.

SECT. VI. Of *NAVIGATION*.

And therein of

1. *PLAIN SAILING*; *Or Sailing by the* Plain Sea-Chart: *And therein,* (1) *By laying down upon a* Blank Chart, *Places according to their* Longitudes *and* Latitudes: (2) *To find their* Distances, Difference *in* Longitude, Latitude, Rumb *and* Distance *upon the* Rumb. (3) *To Work a* Travers, *consisting of many* Courses. *With variety of* Problems, *useful in that Art.*— *All being performed* Geometrically *by Scale and Compasses*; *and by* Trigonometrical Calculation 304 *to* 324
2. *Sailing by* Mercators, *or the* True Sea-Chart 324 *to* 329
3. *Sailing by the* Middle Latitude 329 *to* 331
4. *The several Ways compared* 331 *to* 332

SECT. VII. Of *ASTRONOMY* Theorical.

And therein of

1. *The* Planetary System 333 *to* 335
2. *The* Theory *of the* Sun, *and other* Primary Planets 335 *to* 336
3. *Of the* Motions *of the* Planets *in their* Elliptical Orbs 336 *to* 339
4. *Finding the true Place of a* Planet *in its* Orb 339 *to* 340
5. *To Calculate the Place of a Planet Trigonometrically* 340 *to* 346
6. *Of the Magnitude of the Sun, Earth, Moon, and other Planets* 346 *to* 349
Geometrical Astronomy 349 *to* 366

CLA-

CLAVIS MATHEMATICÆ.

THE KEY TO THE Mathematical Sciences.

PART I.

OF Practical Geometry.

THIS First Part consists only of such DEFINITIONS, PROBLEMS and THEOREMS, GEOMETRICAL, which of Necessity ought to be understood and practiced, before farther Progress be made in any other Mathematical Science: And so I will begin It with these following

Geometrical Definitions.

I. *A* Point *is that which hath no Part.*

That is, it hath no Parts into which it may be divided; It being the least thing that by Mind and Understanding can be imagined or conceived; than which there can be nothing less: As the *Point* in the Margin, noted with the Letter A over it: It being neither *Quantity*, nor any Part of *Quantity*; but only the *Term* or *End* of *Quantity*.

A
.

II. *A* Line *is a Length without Breadth*; *as the Line* A B.

A————B

Unto *Quantity* there appertain Three *Dimensions*; viz. *Length*, *Breadth* and *Depth* (or *Thickness*;) of which, a *Line* is the first, and hath *Length* only, without *Breadth* or *Thickness*; as the *Line* C D,

E F

C|————|———|———|D

which may be divided into *Parts*; either *Equally*, in the Point E; or *Unequally*, in the Point F.

III. *The* Ends, *or* Limits, *of a* Line, *are* Points G————H.

For a *Line* hath its beginning from a *Point*, and likewise endeth in a *Point*: So the *Points* G and H are the *Ends* of the *Line* G H, and are no *Parts* of it.

IV. *A* Right Line *is that which lyeth* Equally *between its* Points.

Or, it is the *Shortest Distance* that can be drawn between *Point* and *Point*; so the *Right Line* G H, is the *Shortest Distance* between the *Points* G and H.

V. *Parallel (or Equidistant)* Right Lines *are such, which being drawn upon the same* Plain, *and infinitely produced, would never meet.*

And such are these Two *Lines*, A B and C D.

A————B
C————D

VI. *A* Plain Angle *is the Inclination (or Bowing) of Two* Right Lines, *one to the other, and the one touching the other; and not being directly joined together.*

Fig. I. So the Two *Lines* A B and B C incline one to the other, and touch each other in the Point B; in which Point (by reason of the *Inclination* of the said Two *Lines*) is made the *Angle* A B C; But if the Two *Lines* which incline one to the other, do (when they meet) make one *Streight Line*, then do they make no *Angle* at all: As the *Lines* D E and E F incline one to the other, and meet each other in the Point E, and yet they make no *Angle*.

And,

And here *Note*, That an *Angle* (generally) is noted with *Three Letters*, of which, the middlemost Letter represents the *Angular Point*; so, in this *Angle* A B C, the Letter B denotes the *Angular Point*. —And of *Angles* there are Three Kinds; *viz. Right*, *Acute* and *Obtuse*.

VII. *When a* Right Line *standing upon a* Right Line *maketh the* Angles *on either Side thereof* Equal, *then either of those* Angles *is a* Right Angle; *and the* Right Line *which standeth erected is called a* Perpendicular Line *to that* Line *upon which it standeth.*

So upon the *Right Line* C D, suppose there do stand another *Right Line* A B, in such sort, that it maketh the *Angles* A B C and A B D (on either side of the *Line* A B) equal; then are either of those *Angles*, A B C and A B D, *Right Angles*; and the *Line* A B, which standeth erected upon the *Line* C D, (without inclining on either side) is a *Perpendicular* to the *Line* C D. *Fig.* II.

VIII. *An* Obtuse Angle *is that which is* Greater *than a* Right Angle:

So the *Angle* C B E is an *Obtuse Angle*, it being greater than the *Right Angle* A B C, by the Quantity of the Angle A B E.

IX. *An* Acute Angle *is that which is* Less *than a* Right Angle.

So the Angle E B D is an *Acute Angle*, it being *Less* than the *Right Angle* A B D, by the Quantity of the Angle A B E.

X. *A* Limit *or* Term *is the End of any thing.*

Forasmuch as there is no *Quantity* (or *Magnitude*) of which *Geometry* treateth, but it hath *Bounds* or *Limits*: And as *Points* are the *Bounds* or *Limits* of *Lines*, so *Lines* are the *Bounds* or *Limits* of *Plains* or *Superficies*; and *Plains* (or *Superficies*) of *Solids* (or *Bodies*.)

XI. *A* Figure *is that which is contained under* One Term, *or* Limit; *or* Many.

So A is a Figure contained under one *Line* or *Limit*: B is a Figure under Three *Lines* or *Limits*: C under *Four*: D under *Five*, &c. which are their respective *Bounds* or *Limits*. *Fig.* III.

XII. *A* Circle *is a* Plain Figure *contained under* One Line, *which is called a* Circumference *or* Periferie; *unto which all* Right Lines *drawn from one certain* Point *within the* Figure *unto the* Circumference, *are equal one to the other.*

Fig. IV. So the Figure B C D contained under One crooked Line, is a *Circle*, whose *Circumference* or *Periferie* is B C D. In the middle whereof there is a *Point* A, from which all the *Right Lines*, A B, A C, A D, being drawn to the *Circumference* B C D, are *Equal*. And that *Point* A is called the *Centre* of the *Circle* B C D.

XIII. *The* Diameter *of a* Circle *is any* Right Line *drawn through the* Centre, *and ending at the* Circumference *on either Side, dividing the* Circle *into Two* Equal Parts.

Fig. V. So the Line E K F is a *Diameter*, because it passeth from the Point E of the *Circumference* on the one Side, to the Point F on the other Side; and passeth also by the Point K, which is the *Centre* of the *Circle*: And moreover, it divideth the *Circle* into Two equal *Parts*, *viz.* into the Part E G F above, and E H F below, the *Diameter*; which Two *Parts* are termed *Semicircles*.

XIV. *A* Section, Segment *or* Portion, *of a* Circle, *is a Figure contained under one* Right Line, *and a* Part *of the* Circumference; Greater *or* Lesser *than a* Semicircle.

So the *Right Line* L M divideth the *Circle* E G F M H L into Two unequal *Sections*; namely, into the *Section* L G M above, *Greater*, and the *Section* L H M below, *Lesser*, than a *Semicircle*.

XV. *The* Semidiameter *of a* Circle, *is half of the* Diameter *of that* Circle.

So K E or K F are *Semidiameters* of the *Circle* E G F H: And so is any *Right Line* drawn from K the *Center* to the *Circumference*; which Lines are frequently called the *Radius* of the *Circle*.

XVI. Right

XVI. Right Lined Figures, *are such* Figures *as are contained under* Right Lines; *of which none can consist of less than* Three; *and these are called* Trilaterals, *or* Triangles.

And *Triangles* are denominated partly from the Differences of their *Sides*, and partly from the Differences of their *Angles*: As for the *Differences* of their *Sides*, they may be, (1.) All *Equal*, and such a *Triangle* is called an *Equilateral Triangle*, as the Figure N. Or, (2.) *Two Sides* may be *Equal*, and the third unequal; and such a *Triangle* is called an *Isosceles Triangle*, as the the Figure O. Or, (3.) All the *Three Sides* may be unequal, and such a *Triangle* is called a *Schalenum Triangle*, as the Figure P. And these are the *Distinctions* in relation to their *Sides*. Fig. VI.

Now for the Distinction of *Triangles* in relation to their *Angles*, they are Three also: For, (1.) If a *Triangle* have *One Right Angle*, it will have Two *Acute* ones, as the Figure Q, where the *Angle* at A is a *Right Angle*, and the *Angles* at B and C are *Acute*, and such a *Triangle* is called *Orthogonium*. Or, (2.) If the *Triangle* have all the *Angles Acute*, as the Figure R; all whose *Angles* at D E and F are *Acute*; such a *Triangle* is called an *Oxogonium Triangle*. Or, (3.) If the *Triangle* have *One Obtuse Angle*, as the Figure S, whose *Angle* at I is *Obtuse*, and the other Two at G and H are *Acute*; such a *Triangle* is called *Ambligonium*. And these are the *Denominations* in relation to their *Angles*.

XVII. *Of* Four-sided Figures *(or* Quadrilaterals*) a* Quadrat, *or* Square, *is that whose* Sides *are* Equal, *and* Angles Right.

And as *Triangles* have their various *Denominations* from the Species of their *Sides* and *Angles*, so have *Quadrilateral Figures* also. For, (1.) If all the *Sides* be *Equal*, and all the *Angles Right Angles*, as the Figure O, such a *Figure* is called a *Quadrat* or *Square*. But, (2.) If of the Four Sides, Two be longer than the other, each to its Correspondent (or Opposite) and the *Angles* all *Right Angles*, as the Figure P; such a *Figure* is called a *Parallelogram*, or *Long Square*, (and sometimes) a *Rectangle*. But, (3.) If all the *Sides* be *Equal*, but the *Angles Unequal*, that is, *Two Acute*, and *Two Obtuse*, each to his Correspondent (as the Figure R) such *Figure* is called a *Rhombus* or *Diamond Form*. And farther, (4.) If such a *Figure* have *Two Sides Longer*, and *Two Sides Shorter*; and also *Two Angles Acute*, and *Two Obtuse*, Fig. VII.

each

each to it opposite corresponding, as the Figure Q such a *Figure* is called a *Rhomboyades*, or Diamond-like Figure. But, (5.) If *Quadrilateral Figures* have all their *Sides*, and all their *Angles*, *Unequal*, as the Figures S and T; such *Figures* are called *Trepezia*, or Table Forms.

PRA-

PRACTICAL PROBLEMS, GEOMETRICAL.

PROBLEM I.

To divide a Right Line A B, *into Two* Equal Parts A E *and* B E, *and at* Right Angles.

Practice. FIrst, Open your Compaſſes to any Diſtance greater than half the Length of the given Line A B.

2. With that Diſtance ſet one Foot in A, and with the other deſcribe the obſcure Arch *b c*,—and (with the ſame Diſtance) One Foot ſet in B, with the other deſcribe the obſcure Arch *d e*, croſſing the former Arch in the Points C and D.

3. Through the Points C and D, draw a Right Line C D, which will divide the given Line AB into Two *Equal Parts* in the Point E, and at Right Angles. The *Angle* A E C on one Side thereof, being equal to the *Angle* C E B on the other Side.

PROB. I.

Upon any Point *(as* O*) taken in the* Right Line Q R, *to erect a* Perpendicular O S.

Practice. FIrſt, Open the Compaſſes to any ſmall Diſtance; and ſetting one Foot in the given Point O, with the other Foot make Marks on either Side of O, as at T and V.

2. Open the Compaſſes to any Diſtance, greater than the former; and ſetting one Foot in T, with the other deſcribe the Arch *h h*. —Alſo, with the ſame Diſtance, ſet one Foot in V, and with the other deſcribe the Arch *g g*, croſſing the former Arch *h h* in the Point S. *Fig.* VIII.

3. Draw

3. Draw the Line O S, and it will be *Perpendicular* to the given Line Q R.

PROB. III.

From the End *of a* Line X Z, *to erect a* Perpendicular Z A.

Fig. X. *Practice.* FIrst, The Compasses being opened to any small Distance, set one Foot in Z, and with the other describe the Arch B C D, and upon it set the same Distance from B to C, and from C to D.

2. The Compasses still continuing at the same Distance, set one Foot in C, and with the other describe the Arch F D: Also set one Foot in D, and with the other describe the Arch C E, cutting the Arch D F in A.

3. From Z draw the Line Z A, which will be *Perpendicular* to the Line X Z.

PROB. IV.

Another Way to erect a Perpendicular *upon the* End *of a* Line.

Fig. XI. *Practice.* FIrst, With any small Distance of the Compasses set one Foot in H, and with the other describe the Arch I K, and set that same Distance from I to K.

2. Upon K (with the same Distance) describe the Arch I L M N.

3. Upon this Arch set the same Distance from I to L, from L to M, and from M to N.

4. A Line drawn from H, through N, shall be *Perpendicular* to the Line G H.

A Third Way.

Fig. XII. 1. THE Compasses opened to any small Distance, set one Foot in O, and pitch the other Foot down at Pleasure, as at Q; and one Foot resting upon Q, turn the other Foot about, till it cross the Line P O in S, and also describe the small Arch *b a*.

2. Lay a Ruler from S to Q, and it will cut the Arch *b a* in R.

3. A Line drawn from O, through R, shall be a *Perpendicular* to P O.

PROB.

PROB. V.

From a given Point *above, as* P, *to let a* Perpendicular *fall upon a given* Right Line *under it*, N O.

IN the Performance there are Two Varieties: For, (1.) The given Point may be scituate over (or about) the Middle; or over (or near) the End of the given Line

In the First Case.

Practice. LET N O be the *Right Line* given; and let P be the *Point* above, from whence the *Perpendicular* is to be let fall.

1. Open the Compasses to a Distance greater than is the nearest Distance between the Point above P, and the Line upon which the *Perpendicular* is to fall; and with that Distance, setting one Foot in P, with the other describe the Arch of a Circle, which will cut the given Line in Two Points, R and S.

2. Divide the Space between R and S (by the first Probl.) into Two equal Parts in the Point Q, then will a Right Line, drawn from P to Q, be *Perpendicular* to the given Line N O. *Fig.* XIII.

Note, To avoid the dividing of the Space between R and S into Two equal Parts, to find the Point Q, (if you have room either above or below the given Line) you may set one Foot of the Compasses in S, and opening the other to any convenient Distance, describe an Arch *y y*; and removing the Compasses to R, describe the Arch *z z*, crossing the former in A; and so, a Line drawn from A, through the given Point, shall be *Perpendicular* to the given Line N O.

In the Second Case.

Let V be the Point given, *Fig.* XIV.

Practice. FIrst, From any Part of the given Line N O, as from T, draw a Right Line to the given Point V, which divide into Two equal Parts in X.

2. Set one Foot of the Compasses in X; and with the Distance X T, describe the Semicircle V O T, cutting the given Line in O, so shall O be the Point in the Line N O; from which, if you draw a Line to V, it will be *Perpendicular* to N O.

C PROBL.

PROB. VI.

Unto a given Right Line P Q, *to draw another* Right Line S T, *which ſhall be* Parallel *to it; and at the* Diſtance *of the given* Line M.

Fig. XV. *Practice.* First, Take in your Compaſſes the Length of the given Line of Diſtance M.

2. Set one Foot in O, (near to one end of the give Line P Q,) and with the other deſcribe the Arch *a a*: Alſo, upon the Point L, (near the other end of the given Line) deſcribe the Arch *b b*; then,

3. By the Convexity (or Tops) of thoſe Two Arches, if you draw a *Right Line* S T, it will be *Parallel* to the given Line P Q, and at the *Diſtance* of the *Line* M.

PROB. VII.

To a given Line V W, *to draw another* Right Line Y A, *which ſhall paſs through a given Point* R, *and* Parallel *to* V W.

Fig. XVI. *Practice.* First, Set one Foot of the Compaſſes in the given Point R, and with the other deſcribe the Arch *c c*, ſo that it may only touch the given *Line* V W; and with the ſame Diſtance, ſet one Foot in any Point of the *Line* V W, as at X, and with the other deſcribe the Arch *d d*.

2. A Line drawn through the given Point R, by the Convexity of the Arch *d d*, as the Line Y R A, that Line ſhall be a *Parallel* to the *Line* V W.

PROB. VIII.

A ſecond Way to draw a Line Parallel *to a given* Line A B, *which ſhall paſs through a given Point* P.

Fig. XVII. *Practice.* First, With the Diſtance P A, upon the Point B, deſcribe the Arch *e e*.

2. With the Diſtance A B, upon P, deſcribe the Arch *f f*, croſſing the Arch *e e* in the Point C.

3. A *Right Line* drawn from the given Point P, through C, ſhall be *Parallel* to the given Line A B.

PROB.

PROB. IX.

To make an Angle D F E, *equal to a given* Angle A B C.

Practice. First, Upon the Angular Point B, at any Distance, describe an Arch *g g*. Then, *Fig.* XVIII.

2. Having drawn another Line, as F E, upon the end F (with the same Distance) describe the Arch *h h*.

3. Take the Distance *g g* in your Compasses, and set it from *h* to *h*: Then,

4. A Line drawn from F through *h*, as F *h* D, shall make the *Angle* D F E equal to the *Angle* A B C.

PROB. X.

To divide an Angle G H K *into Two* Equal Parts.

Practice. First, Upon the Angular Point H (with any Distance of the Compasses) describe an Arch cutting the Two Sides, containing the Angle, in the Points *b* and *s*. *Fig.* XIX.

2. The Compasses opened to the same (or any other) Distance, set one Foot in *b*, and with the other describe the Arch *k k*; and (with the same Distance) one Foot set in *s*, describe the Arch *i i*, crossing *k k* in the Point L.

3. Join H L, and so is the *Angle* G H K divided into Two *Equal Parts*, by the Line H L.

PROB. XI.

How to divide a given Right Line M N, *into any Number of* Equal Parts: *Suppose* Five.

Practice. First, From one end of the given Line, as N, draw the Line N O at Pleasure, making the Angle O N M. *Fig.* XX.

2. Upon the Point M (by Prob. IX.) make the Angle N M P equal to the Angle O N M; or (by Prob. VIII.) through the Point M, draw the Line M P Parallel to the Line O N.

3. With any small Distance of the Compasses, one Foot being set in N, run over *Four* of those Distances upon the Line N O, at the Points 1, 2, 3, 4.—Do the like upon the Line M P: Then,

4. If

4. If you draw the obscure Lines 1 4, 2 3, 3 2, and 4 1, they will cross the given Line M N in the Points *d*, *c*, *b*, *a*, dividing it into *Five* Equal Parts, as was required.

PROB. XII.

To make an Equilateral Triangle A B C, *whose* Sides *shall be Equal to a given* Line O.

Fig. XXI. *Practice.* MAke the Side B C equal to the given Line O, and with the same Distance of the Compasses, setting one Foot in B, with the other describe the Arch *m m*; and set also on C, describe the Arch *l l*, crossing the former Arches in A: Then join A B and A C, so have you constituted a *Triangle* A B C, whose Three Sides are severally equal to the given Line O.

PROB. XIII.

To make a Triangle G H K, *whose* Three Sides *shall be Equal to the* Three *given* Right Lines D, E, F.

Fig. XXII. *Practice.* First, Make the Line G H equal to the given Line D.

2. Take the given Line E in your Compasses, and setting one Foot in H, with the other describe the Arch *n n*.

D ————————
E ——————
F ——————

2. Take the Line F in your Compasses; and setting one Foot in G, with the other describe an Arch *o o*, crossing the former Arch in the Point K.

4. Join G K and H K; so shall you have constituted a *Triangle*; whose Three Sides G H, H K, and G K, are equal to the Three given Lines D E and F.

PROB. XIV.

To make a Geometrical Square B C D E, *whose Sides shall be equal to the given* Line A.

Fig. XXIII. *Practice.* First, Make the Line B C for one Side of the *Square*, equal to the given Line A, and on one End thereof, as

Fig.
XXIV.

Fig.
XXV.

th
vi

Fig. XXI. P

ti
fa
in
a
L

T

Fig. XXII. P

ye
F
fc

in
A

g
T

T

Fig. XXIII. P

is on C (by Prob. IV.) erect the Perpendicular C D, equal also to the Line A.

2. With the same Distance of A, set one Foot in B, and with the other describe the Arch *a a*, and on D, and describe the Arch *b b*, crossing *a a* in the Point E.

3. Join B E and D E, which will constitute the *Geometrical Square* B C D E.

PROB. XV.

To make a Parallelogram *(or* Long Square*)* G H I K, *whose* Length *and* Breadth *shall be equal to Two given* Lines E *and* F.

Practice. First, Make the Line G H equal to the given Line F, and on the end H, (by Prob. IV.) erect the Perpendicular H I, equal to the given Line E. *Fig.* XXIV.

2. With the Distance of the Line F, set one Foot in G, and with the other describe the Arch *c c*; and with the Distance of the Line F, with one Foot in I, describe the Arch *d d*, crossing *c c* in K.

3. Join I K and G K, and so have you formed the *Parallelogram* G H I K, whose *Sides* are equal to the given Lines E and F.

PROB. XVI.

To make a Rhombus, M N O P, *whose Four* Sides *shall be equal to a given* Line L.

Practice. First, Make M N, (for one Side of the *Rhombus*) equal to the given Line L, and with that Length, set one Foot of the Compasses in N, and with the other describe the *Arch* M O P Q. *Fig.* XXV.

2. Set the same Distance upon this *Arch*, from N to O, and from O to P.

3. Join M O, O P, and P N, so shall you have constituted a *Rhombus*, whose Four Sides are all equal to the given Line L.

PROB.

PROB. XVII.

To make a Rhomboyades C D E F, *whose* Sides *shall be equal to Two given* Lines A *and* B; *and the* Acute Angles *thereof* C *and* E, *equal to a given* Angle Z.

Fig. XXVI. *Practice.* FIrst, Make the Side of the *Rhomboyades* C D equal to the Line B.

2. On the end C, (by Prob. IX.) make an Angle F C D, equal to the given Angle Z; and the Side C F equal to the given Line A.

3. Upon the Point F, (with the Length of the given Line B) describe an Arch *h h*; and upon D, (with the Distance A) describe the Arch *g g*, crossing *h h* in the Point E.

4. Join F E and D E, and so have you constituted a *Rhomboyades*, whose Sides are equal to the Lines A and B, and its *Acute Angles* C and E, equal to the given *Angle* Z.

PROB. XVIII.

To make a Trapezia H I K L, *(or Figure of Four unequal Sides) which shall have one* Angle *at* I, *equal to a given* Angle G; *and the Four* Sides *equal to Four (possible)* Right Lines *given*, viz. *to the* Lines O, D, E, F.

Fig. XXVII. *Practice.* FIrst, Make the Line H I (for one of the Sides of the *Trapezia*) equal to one of the given Lines, as O.

2. Upon the Point I (by Prob. IX.) make an *Angle* H I K equal to the given *Angle* G, making the *Side* K I equal to the given Line D.

3. Take another of the given Lines in your Compasses (as the Line E) and setting one Foot upon H, with the other describe the Arch *k k*; also take the fourth given Line F, and setting one Foot of the Compasses in K, with the other describe the Arch *m m*, crossing *k k* in the Point L.

4. Join H L and K L; so shall you have constituted a *Trapezia*, whose Four *Sides* are equal to the Four given *Lines* O, D, E, F; and one of the *Angles*, *viz.* I, equal to the given Angle G.

PROB.

PROB. XIX.

To divide a Circle A F C G, *into any Number of* Equal Parts; *not exceeding* Ten.

Practice. First, Describe any Circle, and cross it with Two Diameters A C and F G, crossing each other at Right Angles (by Prob. I.) in the Centre E. Fig. XXVIII.

2. Make A B and A D, equal to E F, and join B D; so is B D the *Third Part* of the *Circle*.

3. Join A and F; so will A F be the *Fourth Part*.

4. Upon H, and Distance H F, describe the Arch F I, and join F I; which F I is the *Fifth Part*.

5. E F, E G, E A, E C, either of them, are the *Sixth Part*.

6. H D or H B are the *Seventh Part*.

7. Draw a Line from the Centre E through the Point M, extending it to K; join K A, which be the *Eight Part*.

8. Divide the Arch D A B into *Three Equal Parts* at S, and join S D, which will be the *Ninth Part*.

9. E I is the *Tenth Part*.

PROB. XX.

To any Three Points *given*, A, B, C, (*which are not in one Right Line,*) *to find a* Centre O, *upon which a* Circle *may be described, which shall pass through all the given Points* A, B, C.

Practice. First, Set one Foot of Compasses in one of the given Points, as in A, and extend the other Foot to another of the given Points, as to B, and on A, with the Distance A B, describe an Arch of a *Circle* G F D. Fig. XXIX.

2. The Compasses open still to the same Distance, set one Foot in B, and with the other Foot cross the former Arch in the Points D and E, and through them draw the Right Line D E.

3. Set one Foot of the Compasses in the third given Point C, (being still open to the former Distance) and with the other Foot cross the Arch first drawn in the Points F and G, through which Points draw the Right Line F G, which will cut the Line D E in the Point O, which is a *Centre*; on which,

4. If

4. If you set one Foot of the Compasses, and extend the other to any of the Three given Points, the Circle so described shall pass through all of them.

PROB. XXI.

Two Points X *and* Y, *within the* Circle ABD *being given, how to find the* Centre *of the* Arch *of a* Great Circle A X Y L, *which shall pass through those Two given* Points X *and* Y.

Fig. XXX.

Definition. A *Great Circle* of the *Sphere*, is such a *Circle* as divideth the *Sphere* or *Globe* into *Two Equal Parts*; and so the *Arch* of a *Great Circle* described upon a *Plain*, is such an *Arch* as divideth the *Periferie* or *Circumference* of the *Fundamental* or *Primitive Circle*, within which it is described, into *Two Equal Parts*:

Practice. LET G B H D be a *Primitive Circle* given, whose Centre is C, and let the Two *Points* within the same (through which the *Arch* of the *Great Circle* is to pass) be X and Y.

First, Through one of the given *Points*, as X, and the *Centre* C, draw a Right Line X C E, extending it infinitely towards F.

Secondly, Upon this Line, from the Centre C, erect the Perpendicular C B, and from B, through X, draw B X G; and from G, through C, draw the Diametre G C H.

Thirdly, Through the Points B and H, draw a Right Line, extending it till it cut the Line X C F in R, so have you found a *Third Point*, *viz.* R, through which the *Arch* of the *Great Circle* must pass: And now, having *Three Points*, X Y and R, you may through them (by the last Problem) draw the Arch A X Y L R, whose *Centre* will be at K; and this Arch doth divide the *Primitive Circle* into Two equal Parts in the Points A and L: And that it doth so is evident; for that the Right Line drawn from A to L, doth pass through the *Centre* C.

A Compendium.

It will often fall out that the third Point R will fall very remote from the *Centre* C of the *Primitive Circle*, notwithstanding (in all Cases) the Work may be performed without finding it at all. For,

Having found the Points E and G, as before, take the Distance E B, and set it upon the *Circle* from G to F, and from F let fall a Perpendicular to C E, as F O, extending it (if Need be) infinitely towards M; for in some Point of that Line will the Centre be. And to find that Point,

Divide the Line supposed between the Two given Points, X and Y, into Two Equal Parts at Right Angles, and that Line extended will cut the Line O M in the Centre, as here in K, as before.

PROB. XXII.

About a Triangle D E F, *to describe a* Circle.

Practice. First, (by *Prob.* I.) Divide any Two of the Sides of the *Triangle*, as D E and D F, each into Two equal Parts, at Right Angles in the Points G and H; through which Points the Lines *k k* and *m m* being drawn at Right Angles, will cross each other in the Point ⊙, which will be the *Centre* of the *Circle.* *Fig.* XXXI.

PROB. XXIII.

Within a Triangle G H K, *to* Inscribe *a* Circle.

Practice. Divide any Two of the *Angles* of the given *Triangle*, as the *Angles* at G and H into Two equal Parts, (by *Prob.* X.) by the Lines H O and G P, crossing each other in the *Point* ⊙, for that *Point* shall be the *Centre* upon which the *Greatest Circle* that the *Triangle* is capable to receive must be described. *Fig.* XXXII.

GEOMETRICAL THEOREMS.

I. *If a* Right Line *do fall upon* Two Parallel Right Lines, *it maketh the Alternate* Angles *Equal.*

Fig. XXXIII. SO the Right Line P Q, falling upon the Two Parallel Right Lines L M and N O, doth make the Alternate Angles Equal. As the Angle P R M, equal to the Angle N S Q; and P R L, equal to Q S O, *Elem.* L. 1. P. 29.

II. *If divers* Right Lines, *be cut by divers other* Right Lines, *which are* Parallel *one to the other, the* Segments *are* Proportional.

Fig. XXXI. Let the Two Lines R S and R T, be cut by the Four Parallel Lines V, X, Y, Z. I say then, the inner Segments R a and R b, as also a T and b S, are proportional one to the other: For if R *a* be one third Part of R T, R *b* shall be one Third of R S, *&c.* because the Right Line *a* X, cutteth off one third Part of the Space Æ V T Z, and therefore it cutteth off a third Part from every Line drawn within that Space.

III. *If Two* Right Lines *be multiplied into one another, there is made of them a Right Angled* Parallelogram.

Fig. XXXV. Let the Two Sides to be multiplied be A 7 and B 9. If of the Lines A and B, a rectangled *Parallelogram* be made, it will be such a Figure as C D E F: Or, if A 7 and B 9, be multiplied each by other, the Product will be 63, and so many little *Parallelograms* are there contained in the larger *Parallelogram* C D E F.

IV. If

nt
es
al

nd
ill
im
of
al-
wo
to

to
ed
all
im

I
wo
re
or,
ill

Fig. XXXVI.

of

led
im.
is
the
nal
an-
by
the

N

Fig. XXXIII.

S
As
eq

II.

Fig. XXXIV.

Li
as
be
ca
Æ
Li

III.

Fig. XXXV.

Li
fic
eac
lelo
C

V. *If Two* Rectangled Figures, *be made of One of the Sides of that Figure, and of any Two* Segments *of the other Side of the same* Figure; *those Two* Parallelograms *added together, shall be equal to that* Figure.

So if the *Parallelogram* G L I M be made of the Side B 9, and a Segment of the other Side; namely, 4, that *Parallelogram* shall produce 36, for 9 times 4 is 36. And if another *Parallelogram* shall be made of the whole Side B 9, and the other Segment of the other Side A 7, namely, 3, they will produce the other *Parallelogram* L H M K 27, for 3 times 9 is 27: And these Two *Parallelograms* 36 and 27 added together, do make 63 equal to the first.

V. *If Four* Right Lines *be proportional, (that is, as the first is to the second, so is the third to the fourth) the* Right-angled Figure *made of the Two* Means, *(or* Middle Terms) *shall be equal to the* Rectangled Figure, *made of the Two* Extream Terms.

Let there be Four Proportionals, as N 4, O 6, P 8, Q 12. I say, that the *Right angled Figure* R S T V, made of the Two *Means*, O 6, and P 8, shall be equal to the *Rectangled Figure* X Y Z Æ, made of the Two *Extreams*, N 4, and Q 12: For, as 6 times 8 is 48, so 4 times 12 is 48 also. And from hence will follow

Fig. XXXVI.

C O N S E C T A R Y I.

If Four Right Lines *(or Numbers) be proportional; if Three of them be given, the Fourth is also given.* For,

The *Rectangle Figure* made of the Two *Means*, being divided by one of the *Extreams*, the Quotient shall be the other *Extream*. ——As the *Rectangle Figure*, made of the Two *Means* 6 and 8, is 48, this divided by 4, the *Lesser Extream*, shall give you in the Quotient 12, for the *Greater Extream*.——Or, if the proportional Terms given had been (by Transposition) 4, 8, 6, 12, the *Rectangle* made of the *Means* 8 and 6, would be 48; which divided by 12, the *Greater Extream*, would give in the Quotient 4, the *Lesser Extream*.

CONSECTARY II.

Hence also it followeth, that equal Rectangled Figures, have their Sides reciprocally proportional: That is,

Fig. XXXVI. As the *Lesser Side* of the *First Figure*, is to the *Lesser Side of the Second Figure*; so is the *Greater Side* of the *Second*, to the *Greater Side* of the *First*, & Contra. As in the *Equi-rectangular Figures* R S T V, and X Y Z Æ, appeareth: For,

			So is	
	As R T : is to X Y : :		Y Æ : to R S	
	6 : : 4 : :		So is 12 : : 8	
Or,	As X Y : is to R T : :		R S : to X Z	
	4 6		8 12	

VI. *If Three* Lines *or* Numbers *be* proportional, *(that is, as the First is to the Second, so shall the Second be to a Fourth) the* Square *made of the* Means, *is Equal to the* Oblong *made of the* Extreams.

As let the proportional Numbers be 4 : 8 : 8, the Proportion will be

As 4 : is to : 8 : : So is 8 : to 16.

So the *Square* of the Two *Means* 8 is 64 : So also the *Oblong* made of the Two *Extreams* 4 in 16 is 64 also.

VII. *In a* Plain Triangle, *a* Line *drawn* Parallel *to the* Base, *cutteth the* Sides *thereof proportionally.*

Fig. XXXIV. As in the *Plain Triangle* R T S (in the Scheme of the Second hereof) if *c d* be Parallel to the *Base* T S, it cutteth off from the Side R S one third Part, and it cutteth from the Side R T one third Part also: And so they shall be proportional by the second hereof. For,

As R T : to R S : : So is R *c* : to R *d*. And
As. R *c* : to *c* T : : So is R *d* : to *d* S. Also,
As R *c* : to R *d* : : So is *c* T : to *d* S.

VIII. *If divers* Plain Triangles *be compared together, All* Equiangled Triangles *have the* Sides *about (or containing) the* Equal Angles *proportional: And the contrary,* Eucl. Lib. 6. P. 4.

For Illustration,

Fig. XXXVII. (1.) Let A B C and A D E be Two *Plain* and *Equiangled Triangles*: So that the Angles at B and D, at A and A, and also

also at C and E, be *Equal* one to another, each to its Correspondent: I say, the *Sides* about the *Equal Angles* are *Proportional*: For,

1. As A B : is to B C : : So is A D : to D E.
2. As A B : is to A C : : So is A D : to D E.
3. As A C : is to C B : : So is A E : to E D.

DEMONSTRATION.

(2.) Because the *Angles* B A C and D A E are equal, therefore if A B be applied to A D, then A C shall of Necessity fall in A E; and by such Application shall such a *Figure* be made.

In which *Figure*, because that A B and A C do meet together, and also the *Angles* at B and D, and at C and E, are *Equal*; therefore the other *Sides* B C and D E shall be *Parallel*, (*by the first hereof*). But in a plain *Triangle* a Right Line, *Parallel* to the *Base*, cutteth the *Sides* proportionally (*by the seventh hereof*); therefore in the *Triangle* A D E, the Right Line B C being *Parallel* to the *Base* D E, cutteth the *Sides* A D and A E proportionally; and therefore it follows, that

As A B : is to A D : : So is A C : to A E.

(3.) Again, By the Point B let the Right Line B F be drawn Parallel to the *Base* A E, and it shall cut the other Two *Sides* D A and D E proportionally in the Points B and F, (*by the last hereof*) and the Proportion will be

As A B : A D : : F E : D E; or,
A B : A D : : B C : D E.

For F E and B C are *Equal*, (*by the second hereof*). And since it is that,

As A B : A D : : A C : A E,

And so B C to D E, they shall also be,

As A C : A E : : B C : D E.

For those *Two Things*, which are *agreeable* to a *third Thing*, are *agreeable* one to another: Therefore it generally follows,

(1.) As

(1.) As A B : A D : : B C : D E
(2.) As A B : A D : : A C : A E
(3.) As A C : A E : : B C : D E

Or, by transposing the *Second* and *Third Terms* of the *Proportion*, thus;

(1.) As A B : B C : : A D : D E
(2.) As A B : A C : : A D : A E
(3.) As A B : B C : : A E : D E

And thus it is *demonstrated*, that all *Plain Equiangled Triangles* (as these, A B C and A D E are) have their *Sides* comprehending their *Equal Angles*, proportional.

IX. *If several* Plain Triangles, *(how many soever) be compounded, and be cut by* Parallel Right Lines, *the* Inter-Segments *are* Proportional. *As thus:*

Fig. XXXVIII. If the Two *Triangles* H F I and I F K be compounded, and be cut with the *Parallel Lines* G L M, and H I K, their *Inter-Segments* are proportional. For,

As G L : H I : : L M : I K. Or,
As G L : L M : : H I : I K

(by the second and seventh hereof) For the *Triangles* F G L and F H I are *Equiangled, (by the first hereof)* because G L and H I are *Parallel*: Therefore,

As F L : F I : : G L : H I

(by the seventh hereof) but *(by the same seventh hereof)*

As F L : F I : : L M : I K.

And those that are agreeable to a third, are also agreeable between themselves. Therefore,

As G L : H I : : L M : I K.

X. *If any one* Side *of a* Plain Triangle *be continued, the outward* Angle *(made by that Continuation) is equal to the Two inward and opposite* Angles *of the same* Triangle.

Fig. XXXIX. If, in the *Plain Triangle* N O P, the *Side* N P be continued to Q, the outward *Angle* O P Q shall be equal to the Two inward *Angles*, O N P and N O P: For,

If

If from the Point P, a Right Line P R, be drawn *Parallel* to N O, the outward Angle O P Q shall be compounded of the Angles O P R, and R P Q; but the Angles R P Q and R P O, are equal to the Two inward Angles O N P and P O N; that is to say, the Angle R P Q, to the Angle O N P, and the Angle O P R, to the Angle N O P, (*by the first hereof*) because of the the *Parallels* N O and P R; and therefore the *Angle* O P Q is equal to the Two inward and opposite *Angles*, O N P and N O P: *Which was to be demonstrated.*

XI. *The* Three Angles *of every* Plain (*or* Right Lined) Triangle, *are equal to Two* Right Angles.

As in the *Plain Triangle* N O P (in the Diagram before of the tenth) I say, the Three *Angles* N O P, O P N and O N P, are together equal to *Two Right Angles*. *Fig.* XXXIX

For the Angles (how many soever) *meeting in* One Point, *in one and the same* Right Line, *are equal to* Two Right Angles.

But the *Three Angles* N O P, O P N and O N P, are equal to the *Three Angles*, meeting in the Point P, upon the same *Right Line* N Q. For the *Angle* O P N is common to both, and the *Angles* R P Q and N O P (*by the last.*) Therefore, the *Three Angles* N O P, O P N and O N P, are *Equal* to *Two Right Angles*. And from hence will follow these

COROLLARIES.

1. That there can be but *One Right*, or *One Obtuse*, *Angle*, in any *Plain Triangle*.
2. And if *One Angle* be *Right*, or *Obtuse*, the other *Two* shall be *Acute*.
3. That the *Third Angle* of any *Plain Triangle* is the *Complement* of the other *Two* to *Two Right Angles*.
4. That if *Two Triangles* be *Equiangled* in any *Two* of their *Angles*, they are wholly *Equiangled*.

XII. *In every* Right-angled Plain Triangle, *The* Square *made of the* Side *which subtendeth* (*or is opposite to*) *the* Right Angle, *shall be* Equal *to both the* Squares, *which are made of the* Two Sides *which subtend the* Right Angle. Eucl. Lib. 1. Pr. 47.

As in the *Plain Right-angled Triangle* S T V, *Right-angled* at T; I say, the *Sides* S T and T V, including the *Right Angle* S T V, are *Fig.* XL.

are equal in *Power* to the *Hypotenuse* V S; that is, the *Squares* of the *Sides* S T and T V; namely, the *Squares* S W T Æ, and V T B C, added together, are equal to the *Square* of the *Hypotenuse* V S; to wit, the *Square* X Y S V.

For, if from the *Right Angle* at T, be let fall the *Perpendicular* T A Z, then out of the *Square* X Y V S is made *Two Rectangled Figures*, A S Y Z and A V X Z, the *Rectangle* (or *Oblong*) A S Y Z equal to the *Square* S W T Æ, and the *Rectangle* X Z V A, equal to the *Square* V T B C. ——And therefore the *Square* V X Y Z, compounded of those *Two Oblongs*, is equal to the *Two Squares* V T B C and S T Æ W.

But that the *Two Oblongs* A Z Y S and A V X Z are equal to the *Squares* S W T Æ and V T B C, shall thus be proved.

If Three Right Lines be Proportional, (as by the sixth hereof) *the Square of the* Means *is equal to the* Oblong *made of the* Two Extreams.

But the *Three Right Lines* S Y, (equal to S V) S T and S A, are *Proportional*; that is,

As S Y (= S V) : S T : : S T : S A:

Therefore the *Square* of S T is equal to the *Oblong* made of S Y (= S V) and S A; for the *Triangles* S T V and T A S are *Equiangled*, because of the common *Angle* at S, and the *Two Right Angles* at T and A: Therefore (*by the eighth hereof*)

As S V : S T : : S T : S A.

In the same manner it is also proved, that the *Oblong* V X Z A is equal to the *Square* V T B C: For the *Triangles* T S A and A T V are *Equiangled*, because of their common *Angle* at V, and the *Two Right Angles* at T and A. Therefore,

As S V : T V : : T V : A V.

And so (*by the sixth hereof*) the *Square of* T V is equal to the *Oblong* Z X A V. *Which was to be demonstrated.*

Hence

Hence it followeth: That,

If in a Right-angled Plain Triangle, *any* Two Sides *be given, the* Third *may be said to be given also.*

As if the *Two Sides* including the *Right Angle* T V 8, and T S 6 were given, their *Squares* 64 and 36 added together, make 100; the *Square Root* whereof is 10, for the *Third Side* (or *Hypotenuse*) S V —— On the contrary, if the *Hypotenuse* S V 10, and one of the containing *Sides* T S 6, be given; substract the *Square* of 6, *viz.* 36, from the *Square* of 10, *viz.* 100, the Remainder will be 64; the *Square Root* whereof is 8, for the other containing *Side* T V. *Fig.* XL.

XIII. *In every* Plain Triangle (*as well* Right *as* Oblique angled) *the* Sides *are in* Proportion *one to the other, as are the* Sines *of the* Angles *opposite to those* Sides, & contra.

DEMONSTRATION.

Let the *Triangle* A B C, be inscribed in a *Circle*, and from the Centre D, draw the *Radii* D F, D E, D G; bisecting as well the *Peripheries* as the Subtenses; and let there be also drawn the *Radius* D C. Now, because the *Angle* at the Centre E D C is equal to the *Angle* in the *Peripherie* A B C, and C D F, equal to C A B (*by the twentieth of the third of Eucl.*) Therefore shall the Halves of the *Sides* be as *Sines*; and what Proportion the *Side* C A hath to the *Side* C B, the same shall the *Sine* H C, have to the *Sine* C I: For what Proportion the *Whole* hath to the *Whole*, the same shall the *Half* have to the *Half*. *Which was to be demonstrated.* *Fig.* XLI.

And from hence follow these

CONSECTARIES.

I. *If the* Angles *of a* Triangle *be given, the* Reason *of the* Sides *is also given.*

II. *If* One Side *be given, besides the* Angles, *both the other* Sides *are also given.*

III. *If* Two Sides *of a* Triangle *be given, with an* Angle *opposite to one of them, the* Angle *opposite to the other of them is also* Given.

Theorems Extraordinary.

I. If in a Circle Two Right Lines be inſcribed, cutting each other, the Rectangles of the Segments of each Line, are equal: And the Angle at the Point of Interſection, is meaſured by the half Sum of its intercepted Arches.

II. If to a Circle Two Right Lines be adſcribed from a Point without, the Rectangles of each Line from the Point aſſigned, to the Convex and Concave are equal: And the Angle at the aſſigned Point is meaſured by the half Difference of its intercepted Arches.

III. If in a Circle Three Right Lines ſhall be inſcribed, one of them cutting the other Two: Then the Rectangles of the Segments of each Line, ſo cut, are directed proportional to the Rectangles of the reſpective Segments of the Cutter.

IV. If a Plain Triangle be inſcribed in a Circle, the Angles are one half of what their oppoſite Sides do ſubtend; And if it hath one Right Angle, then the longeſt Side of that Triangle ſhall be the Diameter of the Circle.

V. If in a Circle, any Plain Triangle be inſcribed, and a Perpendicular be let fall upon one of the Sides, from the oppoſite Angular Point. Then, as the Perpendicular, to one of the adjacent Sides; ſo is the other adjacent Side, to the Diameter of the circumſcribing Circle.

The End of the Firſt Part.

JANUA

JANUA MATHEMATICA.

THE

GATE

TO THE

Mathematical Sciences OPENED.

PART II.

OF

TRIGONOMETRY.

WHEREIN

The Doctrine of the Dimension of PLAIN and SPHERICAL TRIANGLES is Succinctly Handled, Geometrically Demonstrated,

And { Arithmetically, Geometrically, Instrumentally, } Performed.

By *William Leybourn*, Philomathemat.

LONDON:
Printed *Anno Domini*, MDCCIV.

JANUA MATHEMATICA.

SECTION I.

Of Plain (*or* Right Lined) Triangles.

CHAP. I.

DEFINITIONS.

OF *Triangles* there are Two Kinds; *viz. Plain*, (or *Right-lined*) and *Spherical*, (or *Circular*.) Either of which do consist of Six Parts; namely, of Three *Sides*, and as many *Angles*; but in this Place we shall only treat of the *Plain*.

I. *A Plain* (or Right Lined) *Triangle consisteth of Three Sides, and as many Angles:* And such are the Two *Figures* C B A and C D B; in which, in the first *Figure* A B, B C and C A, are the Three *Sides* of the *Triangle* C B A; and C A B, A C B and C B A, are the Three *Angles* of the same *Triangle* C B A. Fig. I.

¶ And Note here] *That an Angle* (in any Case) *is always noted with Three Letters; the middlemost whereof represents the Angular Point.* So in the *Triangle* A B C, if I would express the *Angle* at C, I would say, The *Angle* A C B; or, The *Angle* B C A.

Also in the second *Figure* D B C, the Lines D C, D B and C B, are the Three *Sides* of the *Triangle* D B C; and the *Angles* B D C, C B D and B C D, are the *Angles* of the same *Triangle* D B C.

II. *Any*

Fig. I. II. *Any Two Sides of a Triangle, are called the Sides of that Angle contained by* (or comprehended between) *them:* So the *Sides* B C and C A are the *Sides* containing the *Angle* B C A.

III. *Every Side of a Triangle, is the subtending Side of that Angle which is opposite unto it.* As in the *Triangle* A B C, B C is the *subtending Side* of the *Angle* C A B; the *Side* B A subtends the *Angle* B C A, and the *Side* C A subtends the *Angle* C B A.

¶ And here Note again,] *That the greatest Side, always, subtends the greatest Angle, the lesser Side the lesser Angle, and equal Sides subtend equal Angles.*

IV. *The Measure of an Angle is the Arch of a Circle described upon the Angular Point, and is intercepted between the Two Sides containing the Angle,* (increasing the Sides, if Need be). So in the *Triangle* A B C, the *Measure* of the *Angle* C B A, is *Arch* C F.

Fig. II. V. *Every Circle is divided into 360 Degrees, and every Degree into 60 Minutes,* (or rather into 100 or 1000 Parts, *&c.*) Which Degrees are so much the *Greater*, by how much the *Circle* is *Greater*; and those *Arches* which contain the same Number of Degrees in equal *Circles*, are *Equal:* But in unequal *Circles* they are termed *Like-Arches.* So the *Arches* C F and D E are *Equal Arches*, they being equal Parts of the same *Circle* D H F K. But the *Arches* C F and O P are *Like-Arches:* For, as C F is 40 Parts (or Degrees) of the *Greater Circle* D E F C, so is O P 40 Degrees of the *Lesser Circle* P O G.

VI. *A Quadrant* (or Quarter) *of a Circle, is an Arch of 90 Degrees.* As is the *Quadrant* (or Arch) H F.

VII. *The Complement of an Arch less than a Quadrant, is so much as that Arch wanteth of 90 Degrees.* So the *Complement* of the *Arch* C F 40 Degrees, is the *Arch* H C 50 Degrees.

VIII. *The Excess of an Arch Greater than a Quadrant, is so many Degrees as that Arch exceedeth 90 Degrees.* So the *Arch* D H C being 140 Degrees, is the *Arch* H C, 50 Degrees more than the *Quadrant* D H.

IX. *A Semicircle is an Arch of 180 Degrees.* As is the *Semicircle* D H F.

X. *The Complement of an Arch less than a Semicircle, to a Semicircle, is so much as that Arch wants of 180 Degrees.* So the Complement of the *Arch* D H C 140 Degrees, to the Semicircle D H E 180 Degrees, is the *Arch* C F 40 Degrees.

XI. *The*

XI. *The opposite Angles made by the crossing of Two Diameters in a Circle,* (or any Two other Right Lines crossing each other) *are equal.* So the *Angles* C B F and C B E, (made by the Intersection of the Two *Diameters* D F and C E in the *Center* B) are equal. Fig. II.

XII. *An Angle is either Right or Oblique.*

XIII. *A Right Angle is that whose Measure is a Quadrant or 90 Deg.* So the *Angles* H B D, and H B F, are *Right Angles*, their *Measures* being the *Quadrants* D H and H F.

XIV. *All Oblique Angles are either Acute or Obtuse.*

XV. *An Acute Angle, is that whose Measure is less than 90 Deg.* So the *Angles* H B C 50 Deg. and C B F 40, are Oblique *Acute Angles.*

XVI. *An Obtuse Angle, is that whose Measure is more than a Quadrant or 90 Deg.* So the *Angle* D B C (consisting of 90 and 50 Deg. *viz.* of 140 Deg.) is an Oblique *Obtuse Angle.*

XVII. *The Complements of Angles are the same, as are the Complements of Arches.*

XVIII. *All Angles concurring (or meeting) together upon One Right Line, all of them being taken together, are equal to a Semicircle, or 180 Deg.* So the *Angle* D B H 90 Deg. H B C 50 Deg. and C B F 40 Deg. (made by the concurring, or meeting, of the Three Lines D B, H B and C B, upon the *Diameter* D F, in the *Center* B) are all of them *Equal* to the *Semicircle* D H C F, or 180 Deg.

XIX. *A Triangle hath some of its Sides Equal, or else they be all Unequal.*

XX. *A Triangle of some Equal Sides is either* Equicrural *or* Equilateral.

XXI. *An* Equicrural Triangle *is that which hath only Two Equal Sides.* And such is the *Triangle* D B E; whose *Sides* B D and B E are Equal.

XXII. *An* Equicrural Triangle *is Equi-angled at the Base.* So in the *Equicrural Triangle* B D E, the *Angles* B D E and B E D, at the *Base* D E, are *Equal, viz.* each of them 70 Deg. for the *Angle* D B E being 40 Deg. that taken from 180 Deg. leaves 140 Deg. the half, whereof 70 Deg. is equal to the *Angle* B D E or D E B, and all the Three *Angles* equal to 180 Deg. or Two Right *Angles.*

This is *Demonstrated* in the XIth *Theorem* hereof.

XXIII. *An*

Fig. II. XXIII. *An* Equilateral Triangle, *is that whose Sides are all Equal, and whose Angles contain* (each of them) 60 *Deg.* So the *Triangle* E B K hath its *Sides* B E, B K and E K, all of them equal; and the *Angles* E B K, E K B and K E B equal also, and each of them equal to 60 Deg. and consequently all of them equal to 180 Deg.

XXIV. *A* Triangle *is either Right Angled or Oblique Angled.*

XXV. *A* Right-angled Triangle, *is that which hath one Right Angle.* And such is the *Triangle* C A B, *Right-angled* at A.

XXVI. *An* Oblique-angled Triangle *is that which hath all its Angles Oblique.* And such is the *Triangle* B C D.

XXVII. *An* Oblique-angled Triangle, *is either Acute-angled, or Obtuse-angled.*

XXVIII. *An* Oblique Acute-angled Triangle *is that which hath all its Three Angles Acute.* And such are the *Triangles* D B E, and E B K.

XXIX. *An* Oblique Obtuse-angled Triangle *is that which hath One Obtuse, and Two Acute Angles.* And such is the *Triangle* D B C, whose *Angle* D B C is *Obtuse*, and the *Angles* B D C and B C D *Acute.*

CHAP. II.

Of Right Lines, applied to a Circle.

FOrasmuch as the *Ratio* or *Proportion* of an *Arch Line* to a *Right Line*, is as yet unknown, yet it is absolutely necessary that *Right Lines* be applied to a *Circle*, for the *Calculation* of *Triangles*, wherein *Arch Lines* come in Competition: For the *Angles* of *Plain* (or *Right-lined*) *Triangles* are measured by *Arches* of *Circles*.

Now, the *Right Lines* applied (or relating) to a *Circle*, are *Chords*, *Sines*, *Tangents*, *Secants* and *Versed Sines.*

Fig. III. 1. A *Chord*, or *Subtense*, is a *Right Line*, joining the Extremities of an *Ark*, as A C is the *Chord* of the *Arks* A B C and A D C.

2. A *Right Sine*, which is singly called a *Sine*, is a *Right Line*, drawn from one end of an *Ark*, perpendicular to the Diameter drawn through to the other End: Or, it is half the *Chord* of twice the *Ark*; so A E is the *Right Sine* of the *Arks* A B and A D. The *Radius* (or *Sine* of 90 Deg.) is called the *Whole Sine*,

Sine, and is the greatest of all *Sines*: For the *Sine* of an *Ark* greater than a Quadrant, is less than the *Radius*; so F G is the whole *Sine* or *Radius*.

Fig. III.

3. A *Versed Sine* is the Segment of the *Radius* between the *Ark* and its *Right Sine*; so E B is the *Versed Sine* of the *Ark* A B, and of the *Ark* A G D.

4. The *Secant* of an *Ark*, is a *Right Line*, drawn from the Center through one end of an *Ark*, till it meet with the *Tangent*: That is, a *Right Line* touching the Circle at the nearest end of that Diameter which cuts the other end of the *Ark*. F M is the *Secant*, and B M the *Tangent*, of the *Ark* A B, or of A D.

5. The *Difference* of an *Ark* from a Quadrant, (or 90 Deg.) whether it be Greater or Less, is called the *Complement* of that *Ark*; so G A is the *Complement* of the *Arks* A B and A G D, and H A is the *Sine* of that *Complement*: G I the *Tangent* of that *Complement*: and F I the *Secant* of that *Complement*. All which (for Brevity) we write *Co-Sine*, *Co-Tangent*, *Co-Secant* of the *Ark*.

6. The Difference of an *Ark* from a *Semicircle* (or 180 Deg.) is called its *Supplement*; so the *Ark* A B is the *Supplement* of the *Ark* D G A, to a *Semicircle*.

7. That Part of the *Radius* which is between the Centre and its *Right Sine*, is equal to the *Co-Sine*. As F E is equal to H A, and F O is equal to the *Co-Sine* of the *Ark* D S.

8. If an *Ark* be Greater or Less than a Quadrant, the Sum or Difference, accordingly, of the *Radius* and *Co-Sine*, is equal to the *Versed Sine*: For F D and H A together, are equal to D E, the *Versed Sine* of the *Ark* D G A; and F B less by H A (or E F) is equal to E B, which is the *Versed Sine* of the *Ark* A B.

CHAP. III.

Some Short Problems, *to make* Canons *of* Sines, Tangents *and* Secants.

PROB. I.

The Sine *of an* Ark *being given, how to find out the* Sine *of the* Complement.

B C *being given, to find* A C.

Fig. IV. BEcause the Triangle A C B is a Rectangle (by the Definition of a *Sine)* and the Sides A C, B C, are of the same Power as the *Hypotenuse*; that is, the *Radius* A B: Therefore, if the *Square* of the *Sine* B C be substracted from the *Square* of the *Radius* A B, there remains the *Square* of A B, whose Side is the *Right Line* A C, the *Sine* sought for.

PROB. II.

The Sine *of an* Ark *being given, and likewise the* Sine *of the* Complement, *to find the* Sine *of half the* Ark.

R Q *and* A Q *being given, to find* B O *or* R O.

Fig. V. As A B : is to B O, so is : : B O : to B G; therefore B O will be the *Square* Side of the Plain from the *Radius* A B and B G, the half *Versed Sine*. For Q B, the *Versed Sine* of the *Ark* B R, is given; because A Q, the *Sine* of the Complement, and A B, the *Radius*, are given by the Supposition.

PROB. III.

The Sines *of Two* Arks, *and the* Sines *of the* Complements *being given, to find the* Sine *of the* Sum.

R Q, Q A, *and* S T, T A, *being given, to find* S P.

Fig. VI. As A R is to R Q, so is A T to T G, or C P.
As A R is to A Q, so is S T to S C.
B T and C P together make S P, the *Sine* of the Sum of Two *Arks*.

PROB.

Fig. VI.

PROB. IV.

The same being given, to find the Sine *of the Difference.*

R Q, Q A, *and* S P, P A, *are given, to find* S T.

As A Q is to Q R, so is A P to P O, from whence you may find O S. As A R is to A Q, so is O S to S T.

To these join the *Theorems.*

Theorem I. *The least* Sines *are, almost, in the same* Ratio *as their* Arks.

This *Theorem* will be prov'd to be true hereafter in the continued Bisections; but the least *Arks* are about one of the first Scruples, or less, and are almost in the same *Ratio* as their *Sines*, because they are, almost, contiguous amongst themselves, and almost of the same Quantity, as it appears, tho' not to every Search, yet to a very profound one.

Theorem II. *If the same* Line *be cut into unequal Parts in Number, the Number of the Parts of the first Section, is to the Number of the second, (reciprocally) as one Part of the second Section is to one Part of the first Section.*

Divide the same *Line*, first into Four, then into Three Parts; then it will be as 4 to 3, so $\frac{1}{3}$ Part to $\frac{1}{4}$ reciprocally; the Reason is, because 3 in $\frac{1}{3}$ makes 1, and 4 in $\frac{1}{4}$ makes 1; and because the Products are equal, the Multiplicands will be reciprocally proportionable.

The Construction *of the* Canon *of* Sines.

The *Sine* of the whole Quadrant is call'd the *Radius*, for it is the Semidiameter of a Circle. Set in the *Canon* a *Radius* of 100000 Parts, or 100000.00 for Necessity of Calculation; but for the better making of the *Canon* take a *Radius* 100000.00000 Parts; for by that Means, the Errors which oftentimes happen amongst the Right Hand Figures may safely be corrected without any Prejudice to the *Canon*.

Then

Fig. VI. Then bisect the Quadrant, and look for the *Sine* of the *Bisegment* by *Prob.* 2. and its *Co-sine* by *Prob.* 1. then bisect this *Bisegment* again, and look for the *Sine* of the second *Bisegment* by *Prob.* 2. and the *Co-sine* by *Prob.* 1. then bisect this second *Bisegment*, and look for its *Sine* or *Co-sine* by *Prob.* 2. and 1. and then bisect the third *Bisegment*, &c. and continue bisecting 13 times, till you find a *Sine* of $\frac{1}{8192}$ Part of the whole Quadrant, as 'tis here set in the *Table*. Now we come to the least *Arks*, where the Truth of the first *Theorem* is illustrated: For as the *Ark* of the Quadrant $\frac{1}{4096}$ is double to the *Ark* $\frac{1}{8192}$, so is its *Sine* almost to that *Sine*.

The Sine of the Quadrant.	100000.00000
The Sine of $\frac{1}{2}$ *of the Quad.*	70710 67811 +
The Sine of $\frac{1}{4}$ *of the Quad.*	38268.34323 +
The Sine of $\frac{1}{8}$ *Part of the Q.*	19509.03220 +
The Sine of $\frac{1}{16}$ *Part of the Q.*	9801.71403 +
$\frac{1}{32}$	
$\frac{1}{64}$	
$\frac{1}{128}$	
$\frac{1}{256}$	
$\frac{1}{512}$	
$\frac{1}{1024}$	
$\frac{1}{2048}$	
$\frac{1}{4096}$	
$\frac{1}{8192}$	

This

This leaſt *Sine* being thus found out, the *Sine* of one of the firſt Scruples, that is, of $\frac{1}{5400}$ part of the whole Quadrant, or of one hundredth part of a Degree; that is, of $\frac{1}{9000}$ part of the whole Quadrant is to be found. Therefore by the *ſecond Theorem*, as $\frac{5400}{9000}$ is to 8192, ſo the quantity of the firſt Part of this Diviſion is to the quantity of the firſt Part of that Diviſion, and by the firſt *Theorem*; ſo is the *Sine* of $\frac{1}{8192}$ Part, which you have in the Table at the *Sine* of $\left\{\begin{matrix}\frac{1}{60}\\ \frac{1}{100}\end{matrix}\right.$ part of Degree. Fig. VI.

Therefore the *Sine* of the firſt Minute, or of the firſt hundredth part thus form'd by *Prob.* 1. extract the *Sine* of the Complement, to wit, of the *Ark* 89 Deg. $\left\{\begin{matrix}\frac{59}{60}\\ \frac{99}{100}\end{matrix}\right.$ then by *Prob.* 3. find out the *Sine* of the ſecond Minute, and its *Co-ſine* by *Prob.* 1. and from thence you will find the *Sine* of the Sum of 2 Min. and 1 Min. that is 3 Min. by *Prob.* 3. and its *Co-ſine* by *Prob.* 1. and from the *Sine* and *Co-ſine* 2 Min. or from the *Sines* and *Co-ſines* 3 Min. and 1 Min. you may look for the 4th *Sine* by *Prob.* 3. and the *Sine* of the Complement by *Prob.* 1. Likewiſe from the *Sines* and *Co-ſines* of 2 Min. and 3 Min. or 4 Min. and 1 Min. you may find the 5th *Sine* and *Co-ſine* by *Prob.* 3. and 1. &c. even to $\left\{\begin{matrix}\frac{60}{60}\\ \frac{100}{100}\end{matrix}\right.$ or 1 Deg. and from the *Sine* of 1 Deg. you may by the ſame means find all the *Sines* of the 90 whole Deg. and from the *Sines* and *Co-ſines*, before found, of 60 ſingle Minutes, it will be eaſie by 3d *Prob.* taking the 4th *Prob.* too, when it will be uſeful, to extract the ſingle *Sines* of all the ſingle Minutes interſpers'd.

A Deduction of the Tangents *and* Secants *from the* Tables *of* Sines.

The *Tangents* are made thus:

As A C, the *Co-ſine*, is to C B the *Sine*; ſo is A E, the *Radius*, to E D the *Tangent*. Fig. VII.

But the *Secants* thus:

As A C, the *Co-ſine*, is to A B the *Radius*; ſo is A E, the *Radius*, to A D the *Secant*.

By this way the whole *Canons* of *Tangents* and *Secants* are extracted from the *Canon* of *Sines*.

I paſs

Fig. VII. I pass by all Rules of Calculation, for I don't undertake to make new Canons; forasmuch as some of the most excellent Artists, by their Study and Industry have sav'd me that Trouble, it sufficiently answers my Design, if the Reason of *Syntax*, whatever it be, be only understood, and the Truth of the Numbers put in the *Canon* which the Propositions above abundantly prove.

CHAP. IV.

Of the Affections *of Right-lined* Triangles, *in order to the* Calculating (*or* Resolving) *of them.*

A *Plain Triangle* is contained under Three Right Lines, and is either *Right-angled*, or *Oblique*.

2. In all *Plain Triangles*, Two *Angles* being given, the third is also given; and One *Angle* being given, the Sum of the other Two is given; because, *The Three Angles together are equal to Two Right Angles.* By the 32 Pro. Euclide. Lib. I. and by the 11th of Sect. I. hereof. Therefore,

In a Plain Right-angled Triangle, One of the Acute Angles is the Complement of the other to 90 Deg.

3. In the Resolution of *Plain Triangles*, the *Angles* only being given, the *Sides* cannot be found; but only the (*Reason*, or) Proportion of them: It is therefore necessary that one of the Sides be known.

4. In a *Right-angled Triangle*, Two Terms (besides the Right Angle) will serve to find the third; so that One of the Terms be a Side.

5. In *Oblique-angled Plain Triangles* there must be Three Things given to find a fourth.

6. In *Right-angled Plain Triangles* there are *Seven Cases*, and *Five* in *Oblique*: For the Solution of which, the Four following *AXIOMS* are sufficient.

AXIOME

AXIOME I.

In a Right-angled Plain Triangle: *The Rectangle made of Radius, and one of the Sides containing the Right Angle, is equal to the Rectangle, made of the other containing Side, and the Tangent of the Ang.e thereunto adjacent.*

DEMONSTRATION.

In the Right-angled Plain Triangle B E D, draw the Arch F E; *Fig.* VIII. then is B E Radius, and D the Tangent of the Angle at B: Make C A parallel to D E, then are the Triangles A B C and B D E like, becauſe of their Right Angles at A and E, and their common Angle at B. Therefore,

As B A : B E : : A C : E D.

And, B A in E D is equal to B E in A C.
That is, B A in *t* B is equal to Radius in A C.
Which which was to be demonſtrated.

AXIOME II.

In all Plain Triangles: *The Sides are proportional to the Sines of their oppoſite Angles.*

DEMONSTRATION.

In the Plain Oblique Triangle C B D, extend B C to F, making *Fig.* IX. B F equal to D C, and deſcribe the Arches F G and C H; then are the Perpendiculars F E and C A, the Sines of the Angles at D and B: and the Triangles B E F and B A C are like, becauſe of their Right Angles at E and A, and their common Angle at B. Therefore,

As B C : C A : : B F : F E.

That is,

As B C : *s* D : : D C (equal to B F) : *s* B.

Which was to be demonſtrated.

AXIOME

AXIOME III.

In all Plain Triangles: *As the half Sum of the Sides, is to their half Difference; so is the Tangent of the half Sum of their opposite Angles, to the Tangent of their half Difference.*

Otherwise,

In every Plain Triangle: *As the Sum of the Two Sides, is to their Difference; so is the Tangent of the half Sum of the Opposite Angles, to the Tangent of half their Difference.*

DEMONSTRATION.

Fig. X. In the Oblique angular Triangle A B C, let the known Sides be B A and B C; and the Angle A B C, comprehended by them: Where it is *Obtuse* in the *superior*, but *Acute* in the *inferior*, *Diagram.*

Continue the Side A B to H; so that H B may be equal to B C, and join C and H,—and make B I equal to A B: Also, from the Points B and I, draw the Right Lines B D and I G, parallel to the Side A C.

Then shall the exterior Angle C B H be equal to the Two interior and opposite Angles *(by the 32th of the 1st Euclid.)* For the Angle C B D is equal to the Angle A C B; and the Angle D B H, to the Angle C A B.

(Moreover, from the Point B, let fall a Perpendicular B E, which shall bisect C H at the Point E; then making B E the Radius, upon B, describe the Arch M E L. Therefore shall C E be the *Tangent* of half the *Sum* of the opposite *Angles*: And D E (to which F E is equal) the *Tangent* of half their *Difference.*

Now, because A C, B D, and I G, are *Parallel*, therefore shall C D, D G, F H, be *Equal*: As also, D F and G H: Therefore I say,

By Equality of Proportion:

As A H, the Sum of the Two Sides,
Is to I H, their Difference;
So is C E, the Tangent of half the opposite Angles,
To D E, the Tangent of half their Difference.

Otherwise

Fig. VIII.

Otherwise,

As the Sum of the Two Sides,
Is to the greater Side doubled;
So is the Tangent of half the Sum of the opposite Angles,
To the Sum of the Tangents of the half Sum, and half Difference of the Angles.

Otherwise,

As the Sum of the Two Sides,
Is to the lesser Side doubled;
So is the Tangent of half the Sum of the opposite Angles,
To the Difference of the Tangents of the half Sum, and the half Difference of the Angles.

AXIOME IV.

In all Plain Triangles: *As the Base, is to the Sum of the other Sides; so is the Difference of those Sides, to the Difference of the Segments of the Base.*

DEMONSTRATION.

Fig. XI.

In the Triangle B C D, let fall the Perpendicular C A; extend B C to F, and draw F G and D H.

Then is B F equal to the *Sum* of the Sides C D and CB; and H B equal to the *Difference* of those Sides; and G B is equal to the *Difference* between A B and A D, the Segments of D B, the Base: And the Triangles H D B and B G F are like; because of their equal Angles at D and F; the Arch H G being the double Measure of them both: And their common Angle at B.

Therefore,

As B D : B F : : H B : G B.

That is,

As D B : F B (the Sum of the Sides) : : H B (the *Difference* of the Sides C B and C D) : G B (the *Difference* of the Segments of the Base.)

ADVERTISEMENT.

A Table to Reduce Sexagenary Minutes, to Centesimal Parts; and the contrary.

Sexag. Minut.	*Cent. Parts.*	*Sexag. Minut.*	*Cent. Parts.*
1	.01.67	31	.51.67
2	.03.33	32	.53.33
3	.05.	33	.55.
4	06.67	34	.56.67
5	08.33	35	.58.33
6	.10.	36	.60.
7	.11.67	37	.61.67
8	.13.33	38	.63.33
6	.15.	39	.65.
10	.16.67	40	.66.67
11	.18.33	41	.68.33
12	.20.	42	.70.
13	.21.67	43	.71.67
14	.23.33	44	.73.33
15	.25.	45	.75.
16	.26.67	46	.76.67
17	.28.33	47	.78.33
18	.30.	48	.80.
19	.31.67	49	.81.67
20	.25 33	50	.83.33
21	.35.	51	.85.
22	.33.67	52	.86.67
23	.38.33	53	.88.33
24	.40.	54	.90.
25	.41.67	55	.91.67
26	.43.33	56	.93.33
27	.45.	57	.95.
28	.46.67	58	.96.67
29	.48.33	59	.98.33
30	.50.	60	1.00.00

WHereas in this *Part* of this *Book*, which treateth of Trigonometry, where the *Angles* of *Right-lined*, and the *Sides* and *Angles* of *Spherical Triangles* are meaſured by *Arches* of *Circles*; and thoſe *Arches* are uſually numbred (or accounted) by *Degrees*, *Minutes*, *Seconds*, &c. Of which one *Degree* contains 60 *Minutes*, One *Minute* 60 *Seconds*, &c. which is called the *Sexary Diviſion* of the *Degree*. Or otherwiſe, the *Degree* is ſuppoſed to be divided into 100 or 1000 *Parts*, which is called the *Centeſimal*, *Milleſimal*, &c. *Diviſion*, (and is the better of the Two in many Reſpects.)

Now, whereas in the following *Trigonometrical Calculations* of this *Book*, and alſo in the other *Parts* (which concern the *Doctrine of Triangles* applied to *Practice* in ſeveral *Parts* of the *Mathematicks*) I have, ſometimes, uſed the *Sexagenary*, and ſometimes the *Centeſimal*, Way of Dividing

Fig. XII.

Diagram M.

Fig. XII. *Case* I.

viding the *Degree:* (Not that I intended to make a *Difference*, (or rather a Confusion, by intermixing them) but because these several *Tractates* were not written at the same time, nor, at first, intended to be joined together, as here they are: But since it so is, I have here (in the beginning) inserted a short *Table*, by which *Sexagenary Minutes* are Reduced to *Centesimal* or *Millesimal Parts:* And also, *Centesimal Parts* to *Minutes* and *Seconds:* So that when the *Reader* meets with either of them, in any *Part* of these *Tractates*, he need not be at any Stand) (or Demurr about it) but readily know, to which Account it belongs, for the Manner of writing them will discover it: So 23 Deg. 30 Min. is written as here; but the same written *Centesimally*, thus, 23.5 Deg. Also 56 Deg. 43 Min. is written Sexagenarily, as here, but the same *Centesimally*, thus, 56.71 Deg. or *Millesimally*, thus, 56.716 Deg. And so of all others, as in the *Table*.

CHAP. V.

Of the Solution, Calculation, *or* Mensuration, *of* Right-angled Plain Triangles.

IN *Right-angled Plain Triangles*, I call those *Sides*, which comprehend the *Right Angle*, the *Legs*; and the *Side* subtending the *Right Angle*, I call the *Hypotenuse*; and the *Triangle* which I shall make use of in the *Resolving* the Seven CASES following shall be that *Diagram* noted with M, in *Fig.* XII. Whose *Sides* and *Angles* are,

Fig. XII.

Diagram M.

Side			
	A B	235.00	Parts and Centesms.
	B C	274.93	
	C A	142.72	

Angles			
	A	90.00	Deg. and Cent.
	B	31.27	
	C	58.73	

CASE I. *The* Legs A B *and* A C *given, to find the* Angle C.

Fig. XII. *Case* I.

PROPORTION.

As Log. A C : Radius : : Log. A B : *t* C. [By *Axi.* I.

OPE-

OPERATION.

As Log. C A, 142.72 Parts	2.1544848
Is to Rad. Tang. 45 Deg.	10.
So is Log. A B, 235.00 Parts	12.3710678
To Tangent C, 58.73 Deg.	10.2165830
Whose Compl. 31.27 Deg. is the Angle at B.	

Fig. XII. *Case* II. *CASE* II. *The* Angles C *and* B, *and the* Leg. A B *given*; *To find the* Leg. A C.

PROPORTION.

As Rad. : Leg. A B : : *t* B : Log. A C. [By *Axi.* I.

OPERATION.

As Radius, Tang. 45 Deg.	10.
To Log. A B, 235.00 Parts	2.3710678
So Tang. B, 31.27 Deg.	9.7883979
To Log. A C, 142.72 Parts	12.1544657

Fig. XII. *Case* III. *CASE* III. *The* Hypotenuse B C, *and the* Leg. A B *given*; *To find the Angle* C.

PROPORTION.

As Leg. B C : Rad. : : Leg. A B. *s* C. [By *Axi.* II.

OPERATION.

As Hypo. B C, 274.93	2.4392537
Is to Radius, Sine 90 Deg.	10.
So is Leg. A B, 235.00	12.3710678
To Sine C, 58.73 Deg.	3.9318141

Fig. XII. *Case* IV. *CASE* IV. *The* Hypotenuse B C, *and the* Angles B *and* C, *given*; *To find the* Leg. A B.

PROPORTION.

As Radius : Hyp. B C : : *s* C, : Leg. A B. [By *Axi.* II.

OPE-

OPERATION.

As Radius, Sine 90 Deg.	10.
To Hypotenuse B C, 274.93 Parts	2.4392537
So is Sine C, 58.73 Deg.	9.9318293
To the Leg. B A, 235.00 Parts	12.3710830

CASE V. *The* Angles B *and* C, *and the* Leg. A C, *given*; *To find the* Hypotenuse B C. Fig. XII. *Case* V.

PROPORTION.

As *s* B : Leg. A C : : Radius : Hypotenuse B C. [By *Axi.* II.

OPERATION.

As Sine B, 31.27 Deg.	9.7152273
Is to Leg. A C, 142.72 Parts	12.1544848
So Radius, Sine 90 Deg.	10.
To the Hypotenuse B C, 274.93 Parts	2.4392575

CASE VI. *The* Hypotenuse B C, *and the* Leg A C, *given*; *To find the other* Leg. A B. Fig. XII *Case* VI.

PROPORTIONS.

1.) As Leg. B C : Rad. : : Leg. A C : *s* B. [By *Case* I.

2.) As *t* B : Leg. A C : : Rad. : Leg. A B. [By *Case* I.

OPERATIONS.

1.) As Hypotenuse B C, 274.93 Parts	2.4392537
To the Radius, Sine 90 Deg.	10.
So is the Leg. A C, 142.72 Parts	12.1544848
To the Sine B, 31.27 Deg.	9.7152311
2.) As the Tangent of B, 31.27 Deg.	9.7833980
Is to the Leg. C A, 142.72 Parts	12.1544848
So is Radius, Tang. 45 Deg.	10.
To the Leg. A B, 235.00 Parts	2.3700868

CASE

Fig. XII. *Case* VII. CASE VII. *The two* Legs, B A *and* C A, *given*; *To find the* Hypotenuse B C.

PROPORTIONS.

1.) As Leg. A B : Rad. : : Leg. A C : *t* B. [By *Case* I.

2.) As *s* B : Leg. A C : : Rad. : Leg. B C. [By *Case* V.

OPERATIONS.

1.) As the Leg. A B, 235.00 Parts,	2.3710678
Is to Radius, Tang. 45 Deg.	10.
So is the Leg. A C, 142.72 Parts	12.1544848
To Tangent B, 31.27 Deg.	9.7834170
2.) As Sine B, 31.27 Deg.	9.7151857
Is to Leg. A C, 142.72 Parts	12.1544848
So is Rad. Sine 90 Deg.	10.
To the Hypotenuse B C, 274.93 Parts	2 4392991

The End of the Seven Cases of Right-angled Plain Triangles.

CHAP. VI.

Of the Solution (*or* Mensuration) *of* Oblique-angled Plain Triangles.

Fig. XII. *Diagram* S. THE *Triangle* which I shall make use of in the Solution of the Five CASES of *Oblique-angled Triangles*, shall be that noted with S; whose *Sides* and *Angles* are

Side	B D,	1270	Parts.
	C D,	865	
	B C,	632	

Angle	C,	115.18	Degrees and Cent.
	B,	38.05	
	D,	26.77	

CASE

ASE I. Two Sides, D C, *and* C B, *with the* Angle *at* D, *opposite to* C B, *given*; *To find the* Angle *at* B, *opposite to* C D.

Fig. XII. *Case* I.

PROPORTION.

As Log. CB : *s* D : : Log. C D, : : *s* B. [By *Axiom* II.

But here it must be known whether the *Angle* sought be *Acute* or *Obtuse*.

OPERATION.

As Log. C B, 632 Parts	2.800717
Is to the Sine of ∠ D, 26.77 Deg.	9.653608
So is the Logar. of C D, 865 Parts	2.937016
	12.590624
To the Sine of ∠ B, 38.05 Deg.	9.789907

ASE II. Two Sides, D C, *and* B C, *with the* Angle C, *comprehended by them, given*; *To find the other* Angles B *or* D.

Fig. XII. *Case*.II.

The given *Angle* C being above 90 Deg. *viz.* 115.18 Deg. substract it from 180 Deg. the Remainder is 64.82 Deg. Which must be made use of in the Calculation: And then, This the

PROPORTION.

As half the Sum of the given Sides D C and B C,
Is to half the Difference of those Sides;
So is the Tangent of half the Sum of the Angles at B and D,
To the Tangent of half the Difference of those Angles.
By *Axiom* III. Then,

To the half Sum of these Angles D and B, if you *Add* the half Difference thus found, you will have the *Greater Angle*; but being substracted, it will give you the *Lesser Angle*.

OPE-

OPERATION. *Parts.*

The Side { C D, is	——	865.0
The Side { C B, is	——	632.0
Their Sum		1497.0
The half Sum		748.5
Their Difference		233.0
The half Difference		116.5
The Sum of the Angles at B and D	——	64.82 Deg.
Their Difference		11.82 Deg.
The half Sum		32.41 Deg.
The half Difference		5.64 Deg.

Being thus prepared, say,

As the Log. of the half Sum of C B and C D, 748.5	2.8741918
Is to the Log. of half their Difference, 116.5	2.0663259
So is the Tang. of half the Sum of B and D, 32.41	9.8026808
	11.8690067
To the Tangent of 5.64 Deg.	8.9948149

This 5.64 Deg. is half the Difference between the Angle B and D,— Which added to the half Sum of those Angles 32.41 Deg. gives 38.05 Deg. for the *Greater Angle* at B. But being substracted therefrom, it leaves 26.77 Deg. for the *Lesser Angle* at D.

ig. XII. CASE III. *Two* Angles C *and* D, *with the* Side C B, *oppo-* se III. *site to* D, *given*; *To find either of the other* Angles.

PROPORTION.

As *s* D : Log. C B : : *s* C : Log. D B. [By *Axiom* II.

OPERATION.

As the Sine of the Angle at D, 26.77 Deg.	9.6536081
To the Logar of the Side C B, 632 Parts;	2.3007171
So is the Sine of the Angle C, 115.18 Deg. (or 64.82)	9.9566369
	12.7573540
To the Log. of the Side D B, 1270 Parts	3.1037549

CASE

CASE IV. *Two Sides,* D B *and* C B, *with the Angle* B, *comprehended by them, being given; To find the Side* C B. — Fig. XII. *Case* IV.

1.) Find the Angle C, by the Third *Axiome.*

2.) Find the Side C B, by the Second *Axiome.*

CASE V. *The Three Sides* D C, C B, *and* D B, *being given; To find the Angle at* D. — Fig. XII. *Case* V.

To the resolving of this Case, Two Operations are required. The First, To find the Segments of the Base D G and G B, by these

PROPORTIONS.

(1.) As the Log. of the Base D B,
Is to the Log. of the Sum of the Sides C D and C B:
So is the Log. of the Difference of those Sides D E,
To the Log. of D G. By *Axiome* 4.
And G D substracted from D B, shall be equal to G B,
and half G B, is equal to A G or A B. Then,

(1.) As Log. D C,
Is to Radius:
So is Log. D A,
To the Sine of the Angle D C A. By *Axiome* 2.
And A G and A B together, are equal to G B; and therefore, the Angle at B, may be found in the same manner.

OPERATION.

The Sum of the Sides C D and C B, is	1497
Their Difference is	.233
The Base B D is	1270

Being thus prepared, say

As the Log. of the Base D B, 1270	3.103804
Is to the Log. of the Sum of C D and C B, 1497	3.175222
So is the Log. of the Difference of C D and C B, 233	12.367356
	5.542578
To the Log. of the Segment D G 274.65	2.438774

H Sub-

Fig. VI. Substract D G 274.65, from the whole Base D B 1270, the Remainder will be G B 995.35; the half whereof, 497.675, is equal to A G or A B.

Then,

As the Logar. of D C, 865,	2.9370161
Is to the Radius, Sine 90 Deg.	10.
So is the Log. of A D, 772.32	12.8877973
To the Sine of D C A, 63.23 Deg.	9.9507812

Whose Complement to 90 Deg. is 26.77 Deg. for the Angle C D B.

In the same manner, may the Angle at B be found, by, first, finding the Angle A C B, which you will find to be 51.95 Deg. Whose Complement to 90 Deg. 38.05 Deg. is the Quantity of the Angle at B

The End of the Five Cases of Oblique-angled Plain Triangles.

JANUA

r-
i-

i-
s,
a-

of
fs
o,

he
s;
ta-

o-
it,

it Fig.XIII.
D,
ne
er
ny
ke
D

Fig. VI.

R
ec

A:

S

fu
D
ti

JANUA MATHEMATICA.

SECTION II.

Of Spherical Triangles.

CHAP. I.

Definitions *and* Theorems, *necessary to the right understanding of the* Doctrine *of the* Dimension *of* Spherical Triangles.

A *Spherical Triangle* (as well as a *Plain*, or *Right-lined Triangle*) consisteth of *Six Parts*; namely, of *Three Sides*, and as many *Angles*: The *Affections* whereof shall be demonstrated from these Twenty *Theorems* following.

I. *The Three* Sides *of every* Spherical Triangle *are the* Arches *of Three* Great Circles *of the* Sphere, *every one of them being less than a* Semicircle, *or consisting of fewer Degrees than* 180, *which is the* Measure *of a* Semicircle. *By the 9th of* Sect. II.

II. *A* Great Circle *of the* Sphere, *is such a Circle as divideth the whole* Sphere, *or* Globe, *into Two Equal Parts or* Hemispheres; *and so is in all Parts distant from the* Poles *thereof, by a* Quadrant *or* 90 *Deg. of the* Great Circle.

III. *If one* Great Circle *of the* Sphere, *do pass by the* Poles *of another* Great Circle, *those Two* Great Circles *do intersect, or cut, each other at* Right Angles: *And the contrary.*

Thus, Let A E C be a *Great Circle* of the *Sphere*; (and let it represent the *Horizon* of any Place) whose *Poles* let be B and D, (the *Zenith* and *Nadir* of the same Place, equi-distant from the *Horizon* A E C 90 Deg.) by which *Poles* B, and D, let another *Great Circle* pass; namely, B E D, one of the *Colures* (or any *Azimuth*, or *Vertical Circle*). Now, I say, that the *Great Circle* BED,

Fig. XIII.

Fig. XIII. B E D, cutteth the *Great Circle* A E C, at *Right Angles* in the Points E and F. For, if upon the *Pole* E or F another *Great Circle* A B C D be deſcribed, it is manifeſt, that A B, B C, C D, and D A, ſhall be the *Meaſures* of the *Angles* at E and F: *(by the 4th of Sect. II.)* But the *Arches* A B, B C, C D, and D A, are *Quadrants (by the 3d hereof)* and therefore, the *Angles* at E and F, are *Right Angles (by the 13th of Sect. II.) Which was to be demonſtrated.*

IV. *The* Meaſure *of a* Spherical Angle, *Is the Arch of a* Great Circle, *deſcribed upon the Angular Point*; *and intercepted between the Two* Sides, *they being continued out till they be* Quadrants: *(by the Firſt hereof.)*

Fig. XIV. So, In the *Spherical Triangle* A B C, the *Meaſure* of the *Spherical Angle* at A, is not the *Arch* B C, but the *Arch* E C, intercepted between the Two *Sides* A B and A C, continued till they be *Quadrants*; that is, to the Points E and D; becauſe the *Arch* B C is not deſcribed upon the angular Point A, but the *Arch* E D is; *(by the 1ſt hereof)* and therefore, the *Arch* B C, cannot be the *Meaſure* of the *Angle* B A C, *(by the 4th of Sect. II.)*

V. *The* Sides *of a* Spherical Triangle, *being continued till they meet together, do make Two* Semicircles, *and at their Interſection, do comprehend an* Angle, Equal *and* Oppoſite *to the* Firſt Angle.

Fig. XV. Thus, In the *Triangle* A B C, the *Sides* A B and A C, of the *Angle* B A C, being continued to D, do make the *Semicircles* A B D and A C D, which do comprehend the *Angle* B D C, equal to the *Angle* B A C, becauſe the ſame *Arch* G H (being diſtant from A and D 90 Deg.) meaſureth both thoſe *Angles*, *(by the 4th hereof.)*

VI. *Every* Spherical Triangle, *hath from every* Angle *thereof, another* Triangle *oppoſite thereunto, whoſe* Baſe *and* Angle *oppoſite to the* Baſe, *are the ſame*; *and the other Parts of that* Triangle, *are the* Complements *of the Parts of the other* Triangle.

Thus, the *Triangle* A B C, hath another *Triangle* B D C, oppoſite thereunto; whoſe *Baſe* B C, and *Angle* B D C, oppoſite thereunto, are the ſame, *(by the 5th hereof.)* And the *Sides* B D and C D, are the *Complements* of the *Sides* A B and A C, to a *Semicircle*:

circle: And the *Angles* D B C, and D C B, are the *Complements* of the *Angles* A B C, and A C B, to Two *Right Angles*, or 180 Deg. *(by the 18th of Sect. II.)* Fig. XV.

VII. *The* Sides *of a* Spherical Triangle *may be changed into* Angles, *and the contrary*; *the* Complement *of the* Greatest Side, *or* Greatest Angle, *to a* Semicircle, *being taken for the* Greatest Side, *or* Greatest Angle.

Thus *(in the Figure of Theorem IV. hereof)* the *Spherical Triangle* A B C, obtuse-angled at B: Let D E be the Measure of the *Angle* at A, and let F G be the Measure of the Acute *Angle* at B, (being the *Complement* of the Obtuse *Angle* at B, the *Greatest Angle* in the *Triangle*:) And let H I be the Measure of the *Angle* at C: Now, Fig. XIV.

K L	is equal to	D E	because	K D	and	L E	are Quadrants, and	their common Complement is	L D
L M		F G		L G		F M			L
K M		H I		K I		M H			K H.

Therefore, the *Sides* of the *Triangle* K L M are equal to the *Angles* of the *Triangle* A B C; taking for the *Greatest Angle* A B C, the *Complement* thereof F B G.

It may also be *demonstrated*, That the *Sides* of the *Triangle* A B C are equal to the *Angles* of the *Triangle* K L M: *(by the Converse of the former.)* For, the Side

A B	is equal to	O P	the Measure of the Angle	M I K	which is the Complement of the Obtuse Angle M K L. —
B C		F H		L M K	
A C		D I		D K I	

—For that	A D	and	C I	are Quadrants, and their common Complement is	C D
	A P		O B		A O
	B F		C H		C F.

Therefore, The *Sides* may be turned into *Angles*, and the *contrary*. *Which was to be demonstrated.*

VIII. *A Right-angled* Spherical Triangle *hath One* Right Angle, *or more than One.*

IX. Suppose the *Angles* at A and D, *viz.* B A C and B D C, to be *Right Angles*. Then, Fig. XV.

The Triangle	B A C	hath	One Right Angle, and Two Acute,
	B D C		Two Obtuse Angles, and One Right,
	C D E		One Obtuse, and Two Acute.

X. *A*

Fig. XV. X. *A* Right-angled Spherical Triangle, *hath*, *from the* Right Angle, *a* Right-angled Triangle *opposite thereunto*, *with* Two Obtuse Angles, *&* *contra*.

As in the *Triangles* B A C, and B D C.

XI. *The* Sides *of a* Right-angled Spherical Triangle, *with* Two Acute Angles, *are either of them less than* Quadrants.

As the Sides of the *Triangle* B A C, are all of them, less than *Quadrants*.

XII. *The Two Sides of a* Right-angled Spherical Triangle, *with* Two Obtuse Angles, *are* greater *than* Quadrants; *and the third* Side, *is* less *than a* Quadrant.

As in the *Triangle* B C D, *Obtuse-angled* at B and C; the *Sides* D B and D C are *greater* than *Quadrants*; and the *Side* B C *lesser*.

XIII. *A* Right-angled Spherical Triangle, *with* Two Acute Angles, *hath* (*from the* Acute Angle) *opposite to it*, *a* Right-angled Spherical Triangle, *with One* Acute, *and One* Obtuse, Angle.

As in the *Right-angled Spherical Triangle* E D F, having the *Angles* at E and F *Acute*, is opposite to the *Right-angled Triangle* C D E, with the *Acute Angle* E C D, and the *Obtuse* C E D.

XIV. *The* Sides *subtending the* Right Angles *of a* Spherical Triangle, *having divers* Right Angles, *are* Quadrants.

As in the *Triangle* A G H, If the Two *Great Circles* A G and A H, do cut the *Great Circle* G H at *Right Angles* in the Points G and H; then A is the *Pole* of the *Great Circle* G H (*by the third hereof*) and A G and A H are *Quadrants* (*by the second hereof*.) But if the *Angle* at A be also a *Right Angle*, then G H is also a *Quadrant*, (*by the* 13*th of Sect.* II. *and by the* 4*th hereof*.)

XV. *A* Spherical Triangle, *having divers* Right Angles, *hath either Two or Three* Right Angles; *and so of the* Sides, *it hath Two or Three of them* Quadrants.

As if the *Angle* at A be put for a *Right Angle*, the *Spherical Triangle* A G H shall have the *Three Right Angles* at A, G and H; and

and therefore the *Three Sides*, A G, A H and H G, ſhall be *Quadrants*. ---But, If you put the *Angle* at A for an *Acute Angle*, then the *Spherical Triangle* A G H ſhall have *Two Right Angles* at G and H; and thereupon the Two oppoſite *Sides*, A G and A H, *Quadrants*. *Fig. X*

XVI. *If the third* Angle *of a* Spherical Triangle, *having* Two Right Angles, *be* Acute; *the third* Side *is* leſs *than a* Quadrant; *but if* Obtuſe, *then it is* greater *than a* Quadrant.

As in the *Spherical Triangle* G H I, *Acute-angled* at G, the *third Side* H I is *leſs* than a *Quadrant*. ---But in the *Spherical Triangle* A G I, *Obtuſe-angled* at G, the *third Side* A I is *more* than a *Quadrant*.

XVII. *An* Oblique Spherical Triangle, *conſiſteth ſimply of* Acute *or* Obtuſe Angles, *or of both of them*.

XVIII. *A* Spherical Triangle, *with Two* Obtuſe Angles, *and One* Acute Angle; *is oppoſite to a* Spherical Triangle, *ſimply* Acute-angled, *& contra*.

As if the *Angles* at A and D, be ſuppoſed *Acute*; then the *Triangle* B C D, with *Two Obtuſe Angles* at B and C, and *One Acute Angle* at D, is oppoſite to the ſimple *Acute-angled Triangle* A B C.

XIX. *A* Spherical Triangle, *with Two* Acute Angles, *and One* Obtuſe; *is oppoſite to a* Spherical Triangle, *ſimply* Obtuſe-angled, *& contra*.

As if the *Angles* at A and D, be ſuppoſed *Obtuſe*; then the *Triangle* A B C, with *Two Acute Angles* at B and C, and *One Obtuſe Angle* at A, is oppoſite to the ſimply *Obtuſe-angled Triangle* B D C.

XX. *The* Three Angles *of every* Spherical Triangle, *are* greater *than Two* Right Angles.

This is evident in *Spherical Triangles*, having more *Right*, or *Obtuſe Angles*, than *One*: But in *Acute-angled Triangles*, it may be thus *Demonſtrated*.

D E-

DEMONSTRATION.

Fig. XVI. In the *Right-angled Spherical Triangle* A B C, *Right-angled* at C, and *Acute-angled* at A and B.

The *Measure* of the *Acute Angle*, { B A C / A B C or B D E } is the *Arch* { E F / H I } (*by the fourth hereof.*) But the *Arches* E F and E D together, are equal to a *Quadrant*; therefore the *Arches* F E and H I, added together, are more than a *Quadrant*; and consequently, the *Angles* answering to those *Arches*, namely, the *Angles* B A C and A B C together, are more than a *Quadrant*. But the *Angle* A C B is a *Right Angle*, by the *Proposition*: Therefore, in the *Spherical Triangle* A B C, consisting of *Two Acute Angles*, the *Three Angles* together are *greater* than *Two Right Angles*.

Fig. XVII. Again, In the *Acute-angled Triangle* K L M

The *Measure* of the *Acute Angle* { K L M / M K L / L M K } is the *Arch* { N O / V X / R Q }

But these *Three Arches* N O, V X, and R Q together, are more than *Two Quadrants*: For P Q, and P V, (being the *Complements* of the *Two Arches*, Q R and V X) added together, are *less* than the *Arch* N O, by the *Proposition*: Therefore the *Arch* N O, being the *Measure* of the *third Angle*, is more than the *Complements* of the other *Two Angles* added together; and consequently, the *third Angle* is *greater* than the *Complements* of the other *Two Angles*. And therefore, In *Acute-angled Spherical Triangles*, the *Three Angles* are *greater* than *Two Right Angles*. *Which was to be demonstrated.*

CHAP. II.

Such Affections of Great Circles *of the* Sphere, *as relate to the Solution of* Spherical Triangles.

1. A *Great Circle* of the *Sphere*, Is such a Circle, as divideth the whole Body of the *Globe* or *Sphere*, into Two Equal Parts.

2. A

2. A *Spherical Triangle* is that part of the *Superficies* of the Globe, as ſyes between the Arches of Three Great Circles of the Sphere interſecting one another. Fig. XVII.

3. A *Spheric Angle* is the ſame with the mutual Aperture or Inclination of the Plains of ſuch Two Great Circles which conſtitute the Angle.

4. When one Circle falls upon another Circle, or when the Arches of Two Great Circles interſect each other, the Sum of the Angles made thereby is equal to Two Right Angles: And the Vertical Angles made thereby are mutually Equal.

5. In all *Spherical Triangles*, the *Greater Angle* is always oppos'd to the *Greater Side*.

6. An *Iſoſceles Triangle*, hath its Two Angles at the Baſe mutually *Equal*; and, on the contrary, if a *Triangle* hath Two *Angles Equal*, it hath *Two Sides* alſo *Equal*.

7. Two *Triangles* mutually *Equilateral*, are alſo *Equiangular* one to the other.

8. If there be Two *Triangles*, and in each one Angle, and the Two Sides including it, reſpectively equal: Or, if One Side, and the Two Angles adjacent, be ſeverally equal, then are thoſe Two *Triangles* equal.

9. An *Arch* of a *Great Circle*, is the ſhorteſt Diſtance between Two Points on the Surface of a Globe: And ſo, any Two Sides of a *Spherical Triangle* taken together, are *Greater* than the *Third*.

10. All *Great Circles* cut each other into Two equal Parts; for their *common Section* is a Diameter of the *Sphere*, and conſequently the Two *Sections* of the Peripheries of Two *Great Circles* are at a Semicircle's Diſtance. Hence it follows, That

11. Every *Side* of a *Spherical Triangle*, is leſs than a *Semicircle*. So D B is leſs than the *Semicircle* D B C or D A C. Fig. XVIII.

12. The oppoſite *Angles* at the *Sections* of Two *Circles*, are *Equal*; as the Angle at D, is equal to that at C; for the ſame *Plains* conſtitute both *Angles*.

13. In any *Spherical Triangle*, if the Sum of the *Legs* containing an Angle be *Greater*, *Equal* to, or *Leſſer* than a *Semicircle*, the internal *Angle* at the *Baſe*, is (accordingly) *Greater*, *Equal* to, or *Leſſer* than the outward oppoſite; and conſequently, the Sum of the Two internal Angles at the Baſe, are *Greater*, *Equal* to, or *Leſſer* than Two Right Angles.

Fig. XVIII.

DEMONSTRATION.

If D B and B A together be *Greater*, *Equal* or *Lesser* than D C, then B A is *Greater*, *Equal* to, or *Lesser* than B C; and therefore the Angles at C and D are *Greater*, *Equal* or *Lesser* than the Angle B A C; and the Angles B D A and D A B, *Greater*, *Equal* to, or *Lesser* than the Angles B A C and D A B, equal to Two *Right Angles*.

COROLLARY.

In an *Isosceles Triangle*, if one of the *Equal Legs* be *Greater*, *Equal* to, or *Lesser* than a *Quadrant*, the Angle at the *Base* is *Greater*, *Equal* to, or *Lesser* than a *Right Angle*.

14. The *Sum* of the *Three Sides* of a *Triangle* is less than a *Whole Circle*, or 360 Deg. For B A is *Less* than B C and A C. Therefore D B, D A and B A together, are *Lesser* than D B C and D A C.

15. If from the Point of an Angle, as a *Pole*, you describe a *Great Circle*; or, if you describe a Circle at 90 Deg. on the angular Point, the Ark of that Circle so described, which is intercepted between the *Legs* of the *Angle*, is the *Measure* of that *Angle*.

Fig. XIX. 16. The *Poles* of the Sides of any *Triangle* G H D, constitute another *Triangle* *n x m*, which we may call *Supplemental* to the *Triangle* G H D, for the *Supplements* of the *Angles* and *Sides* of the *Triangle* *n x m* are equal to the *Sides* and *Angles* of the *Triangle* G H D.

DEMONSTRATION.

From the Points G, H, D, as *Poles*, describe Three Great Circles, *x* A Y, R T *m n*, *x* B *n* Z; then is Y *m* equal to a Quadrant, and equal to A *x*, because *m* is the Pole of H G Y, and *x* or E, the Pole of G A; therefore *m x*, equal to A Y, equal to the Supplement of C A, equal to the *Angle* H G D; and the Quadrant Z *n*, equal to B X; therefore *n x*, equal to B Z, equal to the Supplement of the *Angle* H D G, and *n* T, equal to a Quadrant, equal to *m* R; therefore *n m*, equal to T R, equal to the Supplement of the *Angle* D H G.

Now that the *Triangle* *n* E *m*, constituted between the Three next *Poles*, hath its Three *Sides* and *Angles*, equal to the *Angles* and *Sides* of the *Triangle* G H D, save that the Greatest Side *n m* is the Supplement of the Greatest *Angle* H, and the *Angle* E, of the Side G D.

17. Any

17. Any *Angle* of a *Triangle*, with the Difference of the other Two, is *Lesser* than Two *Right Angles:* For *x n* is *Lesser* than *x m* and *n m*, that is, Two *Right Angles*, wanting D, is *Lesser* than Two *Right Angles*, wanting G, and Two *Right Angles*, wanting H. Therefore, G and H wanting D, is less than Two *Right Angles*. *Fig.* XIX.

18. If Two *Triangles* are mutually *Equiangular*, they are also mutually *Equilateral*; for, because they are *Equiangular*, their *Supplemental Triangles* are *Equilateral* (*by the* 16*th*) and therefore *Equiangular* (*by the* 7*th*). And therefore the proposed *Triangles* are *Equilateral* (*by the* 16*th*.)

19. The *Three Angles* of every *Spherical Triangle*, are *Greater* than *Two Right Angles*, and *Lesser* than *Six Right Angles*. For, *n x* and *x m* and *n m* together, are *Lesser* than *Four Right Angles*, (*by the* 14*th*.) that is, *Six Right Angles*, wanting D, and G, and H, lesser than *Four Right Angles:* That is, *Two Right Angles* are lesser than D, and G, and H. Also, the *Sum* of the *Internal-angles* is less than the *Sum* of the *Internal* and *External Angles* taken together, for both of them make but *Six Right Angles*.

20. Of several *Arches* of *Great Circles* falling from the same Point of the *Spheres* Surface, on another Circle, the Greatest is that which passeth through the *Pole* of the *Circle*; and the next to this, is *Greater* than that which is farther off. For suppose P the *Pole* of the Circle C ϖ D, and ϖ the *Pole* of D P C; then is A D *Greater* than A B, A B *Greater* than A E, A E *Greater* than A C: And the *Ark* B ϖ C *Greater* than the *Ark* B P, and B P *Greater* than B D. *Fig.* XX.

21. A *Great Circle* passing through the *Poles* of another *Great Circle*, cuts it at *Right Angles*; And on the contrary, If it cut it at *Right Angles*, it passeth through its *Poles:* The *Angle* P B O is equal to a *Right Angle*, equal to P G D, equal to P D B, equal to ϖ A C.

22. In an *Oblique-angled Triangle*, if the *Angles* at the *Base* are *like*, or of the *same Kinds*; that is, both *Acute*, or both *Obtuse*, the *Perpendicular* falls *Within* the *Triangle*, and the *Quadrantal Arch* without. But if they be unlike, the *Perpendicular* falls *Without*; and the *Quadrantal Arch Within* the *Triangle*. For the *Triangle* A E F hath the *Angles* at E and F *Acute*, and the *Perpendicular* A C falls *Within*, and the *Quadrantal Arch* A ϖ *Without*. Also, the *Triangle* B A G hath the *Angles* at B and G, *Obtuse*, and the *Perpendicular* A D *Within*, and the *Quadrantal*

Fig. XX. *Arch* A ϖ *Without*. But the *Triangle* B A E hath the *Angles* at B and E of *different Kinds*; and the *Perpendicular* A C falls *Without*, and the *Quadrental Arch* A ϖ *Within*, the *Triangle*.

CHAP. III.

Of the Mensuration, or Solution, of Right-angled Spherical Triangles.

Fig. XXI. Diagram A.

I. IN a *Right-angled Spherical Triangle*, there are (besides the *Right Angle*) *Five* other *Parts*; whereof those *Three* which are more remote from the *Right Angle*, the Lord *Nepeir* changeth into their *Complements*: ——As in this Triangle A B C, Right-angled at A: For the *Three Remote Parts*, to wit, the *Angles* B and C, and the *Side* C B, he takes their *Complements*: These *Three Complements*, with the *Sides* C A and B A, do make *Five Parts*: Which, by an artificial Term, he calls

CIRCULAR.

Viz. The
- Side A B,
- Side A C,
- Complement of the Angle at B,
- Complement of the Angle at C,
- Complement of the Side B C.

But the *Right Angle* at A is set aside from being any of the *Circular Points*.

II. In the Solution of a *Right-angled Spherical Triangles*, there are always *Two* other *Parts*, or *Terms*, given (besides the *Right Angle*) to find out a *Fourth*.

III. These *Three Terms* (namely, the *Two* that are *Given*, and the *Third* which is *Required*) must be first looked upon according to their *Circular Parts*.

IV. Of which, *One* is named the *Middle* (or *Mean*) *Part*; the other *Two* are called the *Extream Parts*, borrowing their Appellation from the Scituation of the *Terms* themselves: For, of *Three Terms*, *One* must (of necessity) be in the *Middle*, and the other *Two* in the *Extreams*: Therefore the *Circular Part*

Fig. XXI.

60

Fig. XX

Fig. XXI
Diagram
A.

Part of the *Middle Term*, is called the *Middle Part*, and the *Circular Parts* of the *Extream Terms* are called the *Extream Parts*. *Fig.*XXI.

V. But the *Extream Parts* may be twofold, either *Conjunct*, or *Disjunct*: For those *Three Terms* (besides the *Right Angle*) do come in question, according as the *Two Extreams* are from either *Part* immediately *joined* to a third *Mean*, or are *dis-joined* from the same, by a *Side* or *Angle* interposed on both *Sides*: So are their *Circular Parts* named, *Extreams Conjunct*, or *Extreams Disjunct*.

VI. But of those *Three Terms* which may fall in question, we will subject all their Varieties in their *Circular Parts*, according as every one of them ought (in respect of each other) to be called, The *Middle Part*; and which, The *Extreams Conjunct* or *Disjunct*, as in this following *Synopsis* is fully demonstrated.

The Analysis or Synopsis.

	Middle Part.	*Extreams Conjunct.*	*Extreams Disjunct.*
If the	Side AB	Side AC and Comp. B	Com. BC and Com. C
	Side AC	Side AB and Comp. C	Com. BC and Com. B
	Compl. B	Side AB and Comp. BC	Side AC and Com. C
	Compl. C	Side AC and Comp. BC	Side AB and Com. B
	Compl. BC	Comp. B and Comp. C	Side AB & Side AC.

Which *Synopsis* is thus to be *Read* and *Understood*: *Example* of the *First Line*. If the *Side* A B be the *Middle Part*, then is the *Side* A C, and *Comp. of the Angle* B, the *Extreams Conjunct*: And the *Comp. of the Side* B C, and *Comp. of the Angle* C, the *Extreams Disjunct*. And so of all the rest.

And here it is to be noted, That the Sides A B *and* A C, *are supposed to be joined together, (as one entire Part,) because the* Right Angle *at* A, *is not reckoned amongst the* Circular Parts.

VII. Therefore, In the *Resolution* of a *Right-angled Spherical Triangle*, to know the *Mean*, and *Extream Parts*, you must observe, That

1. If *One* of the *Three Terms* (which, besides the *Right Angle*, come in question) doth stand *alone* by it *self*, severed from the

Fig. XXI. the other *Two* on both *Sides*; (as the *Side* B C, from the *Sides* C A and B A, by the *Angles* B and C interposed) that shall be the *Middle Term*; and so its *Circular Part* shall be called the *Middle Part*; and the other *Two Circular Parts* are the *Extreams Disjunct*. But,

2. If the *Three Terms* do immediately adhere together, the *Middle Term* doth easily shew the *Middle Part*, and the *Extream Terms*, the *Extream Parts Conjunct*.

These Things being all rightly understood, the whole *Trigonometry* of *Sphericals* will be absolved by this *One Proposition*; which therefore we will call *Catholick* or *Universal*.

Proposition Universal.

The Sine *of the* Middle Part *and the* Radius, *are* Reciprocally Proportional, *with the* Tangents *of the* Extream Parts Conjunct; *and with the* Co-sines *(or* Sines *Complements) of the* Extreams Disjunct. *That is*,

As the *Radius*,
To the *Tangent* of one of the *Extreams Conjunct*;
So is the *Tangent* of the other *Extream Conjunct*,
To the *Sine* of the *Middle Part*: & *contra*.

Then also,

As the *Radius*,
To the *Co-sine* of one of the *Extreams Disjunct*
So is the *Co-sine* of the other *Extream Disjunct*,
To the *Sine* of the *Middle Part*: & *contra*.

COROLLARY.

I. If the *Middle Part* be Sought, the *Radius* shall be in the *First Place* of the *Proportion*: But if one of the *Extream Parts* be Sought, then the other *Extream* shall be in the *First Place*. Of the *Second* and *Third Places*, it mattereth nothing how they be disposed.

II. If

II. If the *Extreams* (in any *Proportion*) be *Distinct* from the *Middle Part*, the *Proportion* will be performed by *Sines* only: But if the *Extreams* be *Conjunct* to the *Middle Part*, it must be performed by *Sines* and *Tangents* jointly. Fig. XXI.

The *Demonstration* of the *Universal Proposition* is obvious enough: For, where the *Extream Parts* are *Disjunct*, the *Proportions* differ nothing from the common ones: And in the *Extreams Conjunct*, where it is commonly said,

As *Radius*, to the *Tangent*,

We here say,

As the *Co-Tangent*, to the *Radius*.

And likewise *Inversly* and *Contrarily*; which is plainly the same thing: Because,

The Radius *is a* Mean Proportional, *between the* Tangent *of an* Arch, *and the* Tangent Complement *of the same* Arch.

Note, That when a *Complement* in any *Proportion* doth chance to concur with a *Complement* in the *Circular Parts*, you must then (always) take the *Sine* it self, or the *Tangent* it self; instead of the *Co-sine*, or *Co-Tangent*, in the *Circular Parts*: Because the *Co-sine* of the *Co-sine*, is the *Sine* it self, and the *Co-tangent* of the *Co-tangent*, is the *Tangent* it self.

As is the *Sixth Case* following; where C B and C A are given, and the *Angle* at B is required: Here C A is the *Middle Part*, and C B, and B, are the *Extreams Disjunct*: Wherefore (by the second Part of the foregoing *Corollary*) the *Proportion* will be performed by *Sines* only.——And (by this last) because the *Two Extreams*, CB, and B, fall upon *Complements* in the *Circular Parts*; therefore, instead of *Co-sine* B C, and *Co-sine* B, you must say *Sine* B C, and *Sine* B.

These Things premised, we will exemplifie in the Solution of *Right-angled Spherical Triangles* in all the Cases thereof.

CHAP.

CHAP. IV.

The Analogies, *or* Proportions, *for the* Solution *of the several* Cases *of* Right-angled Spherical Triangles, *by the* Universal Proposition.

Fig. XXI. Diagram B.

FOR the Performance hereof, I shall make use of this *Right-angled Spherical Triangle* A B C, *Right-angled* at A, the Quantities of whose *Sides* and *Angles* are adfixed to their respective *Circular Parts* in the Diagram noted with B, in Fig. XXI. And also in this *Table*, both in *Sexagenary Degrees* and *Minutes*; and in *Decimal Parts* also.

	D.	M.	*Cent.*
B C ——	66	30	66.50
C A ——	51	30	51.50
B A ——	50	10	50.16
∠ B ——	58	35	58.58
∠ C ——	56	52	56.85

And in every *Case* I shall distinguish the Two *Given Terms*, (besides the *Right Angle*, which is, always the *third*) by marking the *Sides* or *Angles Given*, by a short Stroak, (|), and the *Term Required*, I shall mark with (o.) All which are to be seen in Figure XXI.

The XVI *Cases* of *Right-angled Spherical Triangle*, Resolved,

The *Hypotenuse* B C, and the *Angle* at C, given; To find

Fig. XXI. Case I.

CASE I. The *Opposite Side* A B, the *Middle Part*.

As Radius 90 Deg.	10.0000000
To Sine C, 56 Deg. 52 Min.	9.9229834
So Sine B C, 66 Deg. 30 Min.	9.9623978
To Sine B A, 50 Deg. 10 Min.	19.8853312

CASE

CASE II. *The* Side Adjacent A C, Extream Conjunct. Fig. XXI. *Case* II.

As Co-tangent B C, 23 Deg. 30 Min. 9.6383019
To Radius, 90 Deg. 10.
So Co-sine C, 33 Deg. 8 Min. 19.7376611
To Tangent C A, 51 Deg. 30 Min. 10.0993592

CASE III. *The other* Angle B, Extream Conjunct. *Case* III.

As Co-Tangent C, 33 Deg. 8 Min. 9.8147277
To Radius, 90 Deg. 10.
So Co-sine B C, 23 Deg. 30 Min. 19.6006997
To Co-Tangent B, 31 Deg. 25 Min. 9.7859720

The Hypotenuse B C, *and* Side A C, *given, to find*

CASE IV. *The* Opposite Angle *at* B, Ex. Disj. *Case* IV.

As Sine B C, 66 Deg. 30 Min. 9.9623978
To Radius, 90 Deg. 10.
So Sine C A, 51 Deg. 30 Min. 19.8935444
To Sine B, 58 Deg. 35 Min. 9.9311466

CASE V. *The* Adjacent Angle C, Middle Part. *Case* V.

As Radius, 90 Deg. 10.
To Tangent C A, 51 Deg. 30 Min. 10.0993948
So Co-tangent B C, 23 Deg. 30 Min. 9.6383019
To Co-sine C, 33 Deg. 8 Min 19.7376967

CASE VI. *The* other Side A B, Extream Disjunct. *Case* VI.

As Co-Sine C A, 38 Deg. 30 Min. 9.7941496
To Radius, 90 Deg. 10.
So Co-sine B C, 23 Deg. 30 Min. 19.6006997
To Co-sine B A 39 Deg. 50 Min. 9.8065501

K *The*

Fig. XXI. *The* Side A C, *and the* Angle *opposite there to* B, *being Given*; *To find*

Case VII. *CASE* VII. *The other* Side B A, Middle Part.

As Radius, 90 Deg.	10.
To Tangent C A, 51 Deg. 30 Min.	10.0993948
So Co-Tangent B, 31 Deg. 25 Min.	9.7859004
To Sine B A, 50 Deg. 10 Min.	19.8852952

Case VIII. *CASE* VIII. *The* other Angle *at* C, Extream Disjunct.

As Co-sine C A, 38 Deg. 30 Min.	9.7941496
To Radius, 90 Deg.	10.
So Co-sine B, 31 Deg. 25 Min.	19.7170526
To Sine C, 56 Deg. 52 Min.	9.9229030

Case IX. *CASE* IX. *The* Hypotenuse B C, Extream Disjunct.

As Sine B, 58 Deg. 35 Min.	9.9311522
To Radius, 90 Deg.	10.
So Sine C A, 51 Deg. 30 Min.	19.8935444
To Sine B C, 66 Deg. 30 Min.	9.9623922

The Side C A, *and the* Angle C, *adjacent thereto, given*; *To find*

Case X. *CASE* X. *The* other Side A B, Extream Conjunct.

As Co-tangent C, 33 Deg. 8 Min.	9.8147277
To Radius, 90 Deg. 8 Min.	10.
So Sine C A, 51 Deg. 30 Min.	19.8935444
To Tangent B A, 50 Deg. 10 Min.	10.0788167

Case XI. *CASE* XI. *The* other Angle B: Middle Part.

As Radius, 90 Deg.	10.
To Sine C, 56 Deg. 52 Min.	9.9229334
So Co-sine C A, 38 Deg. 30 Min.	9.7941496
To Co-sine B, 31 Deg. 25 Min.	19.7170830

CASE

CASE XII. *The Hypotenuse* B C, Extream Conjunct. Fig. XXI. Case XII.

As Tangent C A, 51 Deg. 30 Min.	10.0993948
To Radius, 90 Deg.	10.
So Co-sine C, 33 Deg. 8 Min.	19.7376611
To Co-tangent B C, 23 Deg. 30 Min.	9.6382663

The Two Sides, A C and A B, *given*; *To find*

CASE XIII. Either Angle, *as* C; Extream Conjunct. Case XIII.

As Tangent B A, 50 Deg. 10 Min.	10.0787534
To Radius, 90 Deg.	10.
So Sine C A 51 Deg. 30 Min.	19.8935444
To Co-tangent C, 33 Deg. 8 Min.	9.8147910

CASE XIV. *The* Hypotenuse C B: Middle Part. Case XIV.

As Radius, 90 Deg.	10.
To Co-sine A C, 38 Deg. 30 Min.	9.7941496
So Co-sine B A, 39 Deg. 50 Min.	9.8065575
To Co-sine C B, 23 Deg. 30 Min.	19.6007071

The Two Angles B *and* C, *given*; *To find*

CASE XV. *Either of the* Sides, *as* A C: Extream Disjunct. Case XV.

As Sine C, 56 Deg. 52 Min.	9.9229334
To Radius, 90 Deg.	10.
So is Co-sine B, 31 Deg. 25 Min.	19.7170526
To Co-sine C A, 38 Deg. 30 Min.	9.7941192

CASE XVI. *The* Hypotenuse B C: Middle Part. Case XVI.

As the Radius, 90 Deg.	10.
To Co-tangent C, 33 Deg. 8 Min.	9.8147277
So Co-tangent B, 31 Deg. 25 Min.	9.7859004
To Co-sine B C, 23 Deg. 30 Min.	19.6006281

 Note,

Fig. XXI. *Note*, These are the *Proportions* answerable to the *Universal Proposition*, yet may (many of them) be varied; so that the *Radius* may be brought into the *first Place*, and that by the latter part of the foregoing *Corollary*: Which says,— *Radius is a Mean Proportional between the Tangent of an Arch, and the Tangent Complement of the same Arch.* —So that (in the XIIIth CASE) where it is said,

As the *Tangent* B A, Is to *Radius*:
So is *Sine* C A, To *Co-tangent* C.

It is all one, as if you should say,

As *Radius*, 90 Deg.	10.
To the *Co-tangent* B A, 39 Deg. 50 Min.	9.9222466
So the *Sine* of C A, 51 Deg. 30 Min.	9.8935444
To the *Co-tangent* of C, 33 Deg. 3 Min.	19.8157910

The like Course may be taken in the *Second*, *Third*, *Tenth* and *Twelfth* CASES.

And thus you have the whole *Doctrine* of the *Dimension* of *Right-angled Spherical Triangles*, performed by Help of this one *Catholick Proposition*.

CHAP. IV.

Some Prenotions concerning Oblique-angled, Spherical Triangles, *in order to the Solution of them.*

IN *Oblique-angled Spherical Triangles* there are XII *Cases*, Ten of which may be resolved by the *Universal Proposition*; but then the *Oblique Triangle* must be reduced into Two *Right-angled Triangles* by help of a *Perpendicular* let fall, sometimes *within*, sometimes *without*, the *Triangle*: And to know whether it fall *within* or *without*, the subsequent *Rules* are to be observed.

RULE I. *If the* Angles *at the Base of the* Triangle *be both of the same Affection; that is, both Acute or Obtuse; the* Perpendicular *let fall from the* Vertical Angle *shall fall within: But if of different Affections, without.*

As

B Fig.
ain XXII.
ith
Cit
C
Fit
B
is
he
pe- Fig.
iat XXII.
nt,
all.

at
on!
ne
be
E
de,
fe,
he
ah

fite
re-

de

B,
em

he Fig.
ied XXIII.
em
to
he
he

E

*Fig.*XXI.

a

R
C

I

th
g
u
it

R

As in the *Oblique angled Triangle* A B C, whose *Angles* at B and C are both *Acute*, the *Perpendicular* A D shall fall within the *Triangle*: For, if it fall not within, it must be the same with one of the Sides, or else it must fall without the *Triangle*: If it be the same with either of the Sides, then the *Angle* at B or C must be a *Right Angle*; which is contrary to the Proposition: If it fall without the *Triangle*, as suppose at E, then the *Angle* A E B shall be a *Right Angle*: But the *Angle* A B E is *Obtuse*, for it is the Complement of the *Acute Angle* A B C, and therefore the Side A E is greater than a Quadrant; And the *Angle* A C E being *Acute*, A E shall be also less than a Quadrant: But, that the same Side should be both *More* and *Less* than a Quadrant, is absurd: And therefore, in this *Case*, the *Perpendicular* shall fall within the *Triangle*.

Fig. XXII.

Fig. XXII.

But, In the *Triangle* A E B, *Obtuse-angled* at B, and *Acute* at E; the *Perpendicular* A D shall fall without the *Triangle* upon the Side E B, contrived: Or, if otherwise, it must be the same with one of the Sides, or fall within the *Triangle*: It cannot be the same with either of the Sides, for then the *Angle* at B or E should be a *Right Angle*: And it cannot fall within the *Triangle*, because then the *Angles* at B and E must either be both *Obtuse*, or both *Acute*, as hath been already proved. If therefore the *Angles* at the Base be of different *Affections*, the *Perpendicular* shall fall without; as was to be proved.

However this *Perpendicular* falleth, it must be always *opposite* to a *known Angle*; and for better Direction herein, take this *General Rule*.

RULE II. *Let your* Perpendicular *fall from the End of a Side given, and adjacent to an* Angle *given.*

As in this *Triangle* A B C, if there were given the Side A B, and the *Angle* at A; by the former and this *Rule*, the *Perpendicular* must fall from B upon the Side A C.

But if there were given the Side A C, and the *Angle* at A, the *Perpendicular* must fall from C, upon the Side A B, continued to E.——And to know whether the Side upon which the *Perpendicular* shall fall, must be continued or not, is no more than to ask whether the *Perpendicular* must fall *within* or *without* the *Triangle*. But, if the former Directions be not sufficient, the Calculation will determine it. For,

Fig. XXIII.

RULE

Fig. XXIII. RULE III. *If the Ark found at the first Operation (whether it be of a* Side *or an* Angle*) be more than the* Arch *given, the* Perpendicular *shall fall without; if less, within the* Triangle.

And this will plainly appear in the Solution of the following *Cases*.

General Rules to be observed in the Second Operation of the Solution of Oblique-angled Spherical Triangles, *when they are reduced into Two* Right-angled Triangles.

After the first Operation, whereby either the Segments of the Base, or the *Angle* at the *Cathetus*, or *Perpendicular*, is found; a diligent Care being had to the Addition or Substraction of them: The second Operation will be performed by one of the Four following *Rules*.

Fig. XXIV. RULE I. The Sines of the Complement of the Hypotenuses, to the Sines of the Complement of the Bases, are in direct Proportion. So,

cs BA : *cs* DA :: *cs* BC : *cs* DC.

RULE II. The Sines of the Bases, to the Tangents of the *Angles* at the Base, are in Reciprocal Proportion: So,

s BA : *s* DA : : *ct* B : *ct* D : : *t* B : *t* D.

RULE III. The Sines of the Complement of the *Angles* at the Base; to the Sines of the Complement of the *Angles* at the *Cathetus* (or Perpendicular) are in direct Proportion: So,

s BCA : *s* DCA : : *cs* B : *cs* B : *cs* D.

RULE IV. The Tangents of the Hypotenuses, to the Sines of the Complement of the *Angles* at the *Cathetus* (or Perpendicular) are in Reciprocal Proportion: So,

cs BCA : *cs* DCA : : *ct* BC : *ct* DC : : *t* DC : *t* BC.

CHAP.

CHAP. V.

The Solution of Oblique-angled Spherical Triangles: *By letting fall a Perpendicular, whereby the* Oblique Triangle *is Reduced into Two* Right-angled.

CASE I. *In the* Oblique-angled Spherical Triangle A B C, *there is given, the Two* Sides A B *and* B C, *with the* Angle C, *opposite to* B A; *to find the* Angle *at* B: If it be also known whether the enquired *Angle* be *Acute* or *Obtuse*.

Analogy.

As *s* BA : *s* BC : : *s* C : *s* A. Fig. XXV.

For by the Universal Proposition,

(1.) As Rad. : *s* A B : : *s* A : *s* B D.
(2.) As Rad. : *s* B C : : *s* C : *s* B D.

Therefore,

As *s* A B : *s* B C : : *s* C : *s* A.

CASE II. *Two* Angles, A *and* C, *and the* Side B C, *opposite to the* Angle A, *given*; *To find the* Side B A: If it be known whether the enquired *Side* be more or less than a Quadrant.

Analogy.

As *s* A : *s* C : : *s* B C : *s* B A.

CASE III. *In the* Oblique-angled Spherical Triangle A C D, *there is given, the* Two Sides A D *and* A C, *with the* Angle D A C, *contained by them: To find the third* Side D C.

In this *Case* the Perpendicular may fall from the Extremity of either *Side*, but opposite to the *Angle* given. Fig. XXVI.

Analogies.

(1.) As *ct.* A C : Rad. : : *cs.* D A C : *t.* A B.
And A D — A B = B D in *Triangle* I.
But A D + A B = B D in *Triangle* II.
(2.) As *cs.* A B : *cs.* A C : : *Ra.* : *cs.* B C.
(3.) As *Ra.* : *cs.* B C : : *cs.* B D : *cs.* D C.

There-

Therefore,

Fig. XXVI. As *c s.* A B : *c s.* A C : : *c s.* B D : *c s.* D C.

CASE IV. *Two* Sides A C *and* D C, *with the* Angle A D C, *opposite to* A C *given*; *To find the third* Side A D.

In this *Case*, Let the Perpendicular fall from the Concourse of the given *Sides*, upon the *Side* enquired, continued, if need be.

Analogies.

(1.) As *c t.* C D : Rad. : : *c s.* A D C : *t.* B D.
(2.) As *c s.* B D : *c s.* C D : : *Rad.* : *c s.* *C B.*
(3.) As *c s.* *C B* : *Rad.* : : *c s* A C : *c s* A B.

Therefore,

c s. C D : *c s.* B D : : *c s.* A C : *c s.* A B.
And, B D + A B = A D in the first } *Triangle.*
And, B D — A B = A D in the second }

CASE V. *Two* Sides C A *and* D A, *and their contained Angle* C D A *given*; *To find one of the other* Angles.

In this *Case* the Perpendicular may fall from the Extremity of either of the *Sides*, opposite to the *Angle* given.

Analogies.

(1.) As *c t.* A C : Rad. : : *c s.* D A C : *t.* A B.
And A D — A B = B D in the first } *Triangle.*
And A D + A B = B D in the second }
(2.) As *c t.* C A D : *s.* A B : *Rad.* : *t.* B C.
As *t.* B C. *Rad.* : : *s.* B D : *ct.* A D C.

Therefore,

As *s* A B : *ct.* C A D : : *s* B D : *ct.* A D C.

CASE VI. *Two* Angles, A D C *and* C A D, *and the Side* A C, *opposite to one of them being given*; *To find the* Side *between them*. If it be known whether the *Side* sought, or the *Side* opposite to the other given *Angle*, be *Acute* or *Obtuse*.

In this *Case*, Let the Perpendicular fall from the Extremity of the given *Side*, on the *Side* enquired, continued, if need be.

Analogies.

Analogies.

(1.) As *c t.* A C : Rad. : : *c s.* D A C : *t.* A B.
(2.) As *c t.* C A D : *s.* A B : : *Rad.* : *t.* B C.
(3.) As *Rad.* : *t.* B C : : *c t.* A D C : *s.* B D.

Fig. XXVI.

Therefore,

As *c t.* C A D : *s.* A B : : *c t.* A D C : *s.* B D.
And, A B + B D = A D in the 1st } *Triangle.*
But, D B -- A B = A D in the 2d }

C A S E VII. *Two* Angles D A C *and* A D C, *and the* Side A C, *opposite to one of them*, (viz. D) *being given*; *To find the other* Angle *at* C. If it be known, whether the *Angle* enquired, or the *Side* opposite to the other given *Angle*, be *Acute* or *Obtuse*.

Fig. XXVI. *Case* VII.

In this *Case*, Let the *Perpendicular* fall from the *Angle* enquired.

Analogies.

(1.) As *c t.* C A D : Rad. : : *c s.* A C : *c t.* A C B.
(2.) As *s.* A B C : *c s.* C A B : : *Rad.* : *c s.* B C.
(3.) As *c s.* B C : *Rad.* : : *c s.* B D C : *s.* B C D.

Therefore,

As *c s.* C A B : *s.* A C B : : *c s.* B D C : *s.* B C D.
And, A C B + B C D = A C D in the 1st } *Triangle.*
But, B C D — A C B = A C D in the 2d }

C A S E VIII. *Two* Angles, A C D *and* C A D, *with the* Side *between them*, A C, *being given*; *To find the third* Angle D.

Fig. XXVI. *Case* VIII.

In this *Case*, Let the *Perpendicular* fall from the Extremity of the given *Side*, and opposite to the *Angle* enquired.

Analogies.

(1.) As *c t.* C A B : Rad. : : *c s.* A C : *c t.* A C B.
And, A C D -- A C B = B C D in the 1st } *Triangle.*
But, A C D + A C B = B C D in the 2d }
(2.) As *s.* A C B : *c s.* C A B : : *Rad.* : *c s.* B C.
(3.) As *Rad.* : *c s.* B C : : *s.* B C D : *c s.* C D B.

Therefore,

As *s.* A C B : *cs.* C A B : : *s.* B C D : *cs.* C D B.

Fig. XXVI. *Case* IX. CASE IX. *Two* Sides, A C *and* C D, *with the* Angle A D C, *opposite to one of them,* (viz. A C) *being given*; *To find their contained* Angle A C D. If it be known, whether the enquired *Angle*, or the *Angle* oppofite to the other given *Side*, be *Acute* or *Obtuse*.

In this *Case*, Let the *Perpendicular* fall from the *Angle* enquired.

Analogies.

(1.) As *ct.* C D B : Rad. : : *cs.* C D : *ct.* B C D.
(2.) As *ct.* C D : *cs.* B C D : : *Rad.* : *t.* B C.
(3.) As *Rad.* : *t.* B C : : *ct.* A C : *cs.* A C B.

Therefore,

As *ct.* C D : *cs.* B C D : : *ct.* A C : *cs.* A C B.
And B C D + A C B = A C D in the 1st } *Triangle.*
But B C D − A C B = A C D in the 2d }

Fig. XXVI. *Case* X. CASE X. *Two* Angles, D A C *and* A C D, *with the* Side *between them*, A C, *being given*; *To find either of the other* Sides, D C *or* A D.

In this *Case*, Let the *Perpendicular* fall from the Concourfe of the *Side* given and fought, on the third *Side*, continued, if need be.

Analogies.

(1.) As *ct.* C A B : Rad. : : *cs.* A C : *ct.* A C B.
And, A C D — A C B = B C D in the 1st } *Triangle*
But, A C D + A C B = B C D in the 2d }
(2.) As *ct.* A C : *cs.* A C B : : *Rad.* : *t.* B C.
(3.) As *t.* B C : *Rad.* : : *cs.* B C D : *ct.* C D.

Therefore,

As *cs.* B C A : *ct.* A C : : *cs.* B C D : *ct.* C D.

CASE

CASE XI. *The Three* Sides *being given*; *To find an* Angle.

Fig. XXVI. *Case* XI.

For the resolving of this *Problem*, there must be some Preparation made; for that the *Universal Proposition* of the Lord *Nepier's* is not, of it self, sufficient for the Solution of this or the following *Case*. And therefore, the said Lord *Nepier's*, to bring these *Cases* within some Compass of this his *Universal Proposition*, he first finds the Difference of the Segments of that *Side*; which being made the Base of the *Triangle*, is divided into Two Parts, by letting fall of a *Perpendicular*; and by help of this

ANALOGY.

As the Tangent of half the Base,
Is to the Tangent of half the Sum of the other Two *Sides*;
So is the Tangent of half the Difference of those *Sides*,
To the Tangent of half the Difference of the Segments of the Base.

Thus then,

In the *Oblique-angled Spherical Triangle* A C D, there is given, the Two *Sides* (or *Legs*) A C and C D, together with the Base A D; to find the *Angle* C A D.

Analogies.

As the Tangent of half A D,
Is to the Tangent of half A C and A D;
So is the Tangent of half the Difference of A C and A D,
To the Tangent of half A E.

Fig. XXVII.

And half A D + half A E = A B in the 1st } *Triangle.*
But half A E − half A D = B E or B D in the 2d } *Triangle.*

Hence, to find the *Angle* at A.

As Rad. : *c t.* A C : : *t.* A B : *c s.* C A B.

CASE XII. *The Three* Angles *being given*; *To find a* Side.

Fig. XXVIII. *Case* XII

This Case is but the Converse of the last beforegoing, and is to be Solved after the same manner: If so be that we convert the *Angles* into *Sides*. Which, how to perform, is shewed in the 12th *Case* of the following *Tract* of the *Solution of Oblique-angled Spherical Triangles, without letting fall a Perpendicular.*

 And

Fig. XXVII. And for the *Proportions* or *Analogies* for the Resolving of this *Case* there are also Two Ways proposed, to which I referr you.

CHAP. VI.

The foregoing XII Cases, *of* Oblique-angled Spherical Triangles, *Geometrically Demonstrated, and Resolved, without letting fall a Perpendicular.*

FOR the Performance of what is here promised, these Six *Theorems* following, being demonstrated, will clear.

THEOREM I.

In any Spherical Triangle, whether Right or Oblique-angled.

The Sines of the Angles are Proportional, to the Sines of their opposite Sides: & contra.

This is demonstrated in *Sect.* 1. *Chap.* 4. of *Right-lined Triangles*, and is the same in *Spherical Triangles*, whether *Right* or *Oblique Angular.*

THEOREM II.

In all Oblique-angled Spherical Triangles, whose Three Sides together are less than a Semicircle, or 180 Degrees.

As the Sine of half the Sum of the Angles at the Base,
Is to the Sine of the half Difference of those Angles;
So is the Tangent of half the Base,
To the Tangent of half the Difference of the Sides.

And also,

As the Co-sine of half the Sum of the Angles at the Base,
Is to the Co-sine of half the Difference of those Angles;
So is the Tangent of half the Base,
To the Tangent of half the Sum of the Sides.

DE-

DEMONSTRATION.

I. It is already proved; That, Fig. XXVIII.

The Sine of the Sum of the Angles B *and* E,
Is to the Sine of the Difference of those Angles B *and* E;
As the Tangent of half the Base B E,
Is to the Tangent of half B C.

And multiplying the latter part of this Proportion, by the Tangent of half the Base B E: It is,

As the Sine of the Sum of the Angles B *and* E,
Is to the Sine of the Difference of those Angles B *and* E;
So is the Square of the Tangent of half B E,
To the Rectangle made of the Tangent of half B E, *and the Tangent of half* B C.

But, the Rectangle made of the Tangent of half B E, and the Tangent of half B C, is equal to the Rectangle made of the Tangent of the half Sum, and half Difference, of the Sides B A and A E: Therefore,

As the Sine of the Sum of the Angles B *and* E,
Is to the Sine of the Difference of the Angles B *and* E;
So is the Square of the Tangent of half B E,
To the Rectangle made of the Tangent of the half Sum, and half Difference of the Sides B A *and* A E.

And the former Part of this Proportion, being multiplied by the Sine of the Sum of the Angles B and E: It is,

As the Square of the Sine of the Sum of the Angles B *and* E,
Is to the Rectangles made of the Sines of the Sum, and Difference of those Angles;
So is the Square of the Tangent of half B E,
To the Rectangle made of the Tangents of the half Sum, and half Difference of the Sides B A *and* A E.

But, the Rectangle made of the Sines of the Sum and Difference of the Angles, is equal to the Rectangle made of the Sum and Difference of the Sines.

And therefore,

As the Square of the Sine of the Sum of the Angles B and E,

Is

Fig. XXVIII.

Is to the Rectangle made of the Sum and Difference of the Sines;
So is the Square of the Tangent of half B E,
To the Rectangle made of the Tangent of the half Sum, and half Difference of the Sides B A and A E.

II. *As the Difference of the Sines of* B *and* E,
Is to the Sum of the Sines of the Angles B *and* E;
So is the Tangent of the half Difference of the Sides,
To the Tangent of the half Sum of the Sides B A *and* A E.

And multiplying the former Part of this Proportion, by the Difference of the Sines of the Angles; and the latter Part thereof, by the Tangent of half the Difference of the Sides A B and A E: It will be,

As the Square of the Difference of the Sines of B and E,
Is to the Rectangle made of the Sum and Difference of the Sines of B *and* E;
So is the Square of the Tangent of the half Difference of B A and A E,
To the Rectangle made of the Tangents of the half Sum, and half Difference of the Sides.

But, by the First Section of this Demonstration it is proved, That,

As the Square of the Sine of the Sum of B and E,
Is to the Rectangle made of the Sum and Difference of the Sines;
So is the Square of the Tangent of half B E,
To the Rectangle made of the Tangents of the half Sum, and half Difference of the Sides.

Therefore,

As the Square of the Sine of the Sum of B and E,
Is to the Square of the Difference of the Sine of B and E;
So is the Square of the Tangent of half B E,
To the Square of the Tangent of half the Difference of the Sides.

And also,

As the Sine of the Sum of the Angles B *and* E,
Is to the Difference of their Sines;
So is the Tangent of half the Base B E,
To the Tangent of half the Difference of the Sides B A *and* A E.

But,

But,

As the Sine of the half Sum of B and E,
Is to the Sine of their half Difference;
So is the Sine of the Sum,
To the Difference of the Sines.

Fig. XXVIII.

And therefore,

As the Sine of the half Sum of the Angles B and E,
Is to the Sine of the half Difference of B and E;
So is the Tangent of half B E,
To the Tangent of half the Difference of the Sides B A and A E; which is the first Part of the *Proposition.*

III. Having already proved,—That the Sum of the Sines of the Angles B and E, is to the Difference of the Sines of those Angles; as the Tangent of the half Sum of the Sides, is to the Tangent of their half Difference. Therefore,

If you multiply the former Part of this Proportion, by the Sum of the Sines of the Angles of B and E; and the latter Part thereof by the Tangent of the half Sum of the Sides of A B and A E: Then it will be,

As the Square of the Sum of the Sines of B and E,
Is to the Rectangle made of the Sum and Difference of the Sines;
So is the Square of the Tangent of half the Sum of the Sides;
To the Rectangle made of the Tangent of the half Sum, and half Difference of the Sides.

But,

As the Square of the Sine of the Sum of B and E,
Is to the Rectangle made of the Sum and Difference of the Sines;
So is the Square of the Tangent of half B E,
To the Rectangle made of the Tangent of the half Sum, and half Difference of the Sides B A *and* A E.

Therefore,

As the Square of the Sine of the Sum of B and E,
Is to the Square of the Sum of the Sines of those Angles;
So is the Square of the Tangent of half B E,
To the Square of the Tangent of the half Sum of the Sides B A and A E.

And,

Fig. XXVIII.

And,

As the Sine of the Sum of the Angles B *and* E,
Is to the Sum of the Sines of the Angles B *and* E;
So is the Tangent of half the Base B E,
To the Tangent of the half Sum of the Sides A B and A E.

But,

As the Co-sine of the Sum of B and E,
Is to the Co-sine of the Difference of the Angles B and E;
So is the Sine of the Sum of the Angles B *and* E,
To the Sum of the Sines of the said Angles.

And therefore,

As the Co-sine of the Sum of the Angles B and E,
Is to the Co-sine of the Difference of the Angles B and E;
So is the Tangent of half the Base B E,
To the Tangent of the half Sum of the Sides A B and A E.
Which was to be Demonstrated.

THEOREM III.

In all Spherical Triangles:
As the Difference of the Versed Sines, of the Sum and Difference of any Two Sides (including an Angle,)
Is to the Diameter;
So is the Difference between the Versed Side of the Third Side, and the Versed Sine of the Difference of the other Two Sides,
To the Versed Sine of the Angle comprehended by the said Two Sides.

DEMONSTRATION.

Fig. XXIX.

Let the Sides of the Triangle Z S P be known, and let the Vertical Angle be S Z P: Then shall Z S, the one Side, be equal to Z R, and P R equal to their Sum; and P B the Versed Sine of P R, and P C, is the Difference of the Sides Z S and Z P, and the Versed Sine of P C, is P M.

Now then, M B is the Difference between B P, the Versed Sine of P R, the Sum of the Sides, and P M the Versed Sine of P C, the Difference of the Sides.

M H is the Difference between P H the Versed Sine of P S, and P M the Versed Sine of P C the Difference of the Sides: Q V is is the Diameter, and O V the Versed Sine of P Z S the Angle Sought: And the Right Lines N C, K L, and R G, being Parallel, by the Work, their Inter-segments M B and R C, and also M H and S C, are proportional.

Fig. XXIX.

And therefore,

M B : M H : : R C : S C;
R C : S C : : Q V : O V;

Therefore, M B : M H : : Q V : O V,
Or, M B : M H : : half Q V : half O V.

Which was to be Demonstrated.

THEOREM IV.

In all Spherical Triangles;
As the Rectangle of the Sines of the Sides containing the Angle enquired,
Is to the Square of the Radius;
So is the Difference between the Versed Sine of the Base, and the Versed Sine of the Difference of the other Two Sides;
To the Versed Sine of the Angle sought.

DEMONSTRATION.

Let the Sides of Triangle A E K be given, and let the Angle at A be enquired; and from O, let fall the Perpendicular O B.

Fig. XXX.

Now then, O E being the Difference of the Sides A E and A K, equal to A O; the Right Line O Q is the Right Sine thereof, and E Q the Versed Sine. In like manner, S M is the Right Sine, and E M the Versed Sine of E S; that is, of the Base E K; and M Q or O B, is the Difference of those Versed Sines.

O K, is the Versed Sine of the Angle O A K, in the Measure of the Parallel O F; and D X is the Versed Sine of the same Angle in the measure of a Great Circle, whose Diameter is H D. Now then, because of their like Arches, it shall be

As D C : D X : : O L : O K:

And because N E and O K are Parallel, as also E C and O B, the Angles B O K and C E N are equal: And the Triangles E C N and O K B like; and therefore the Sides N E, E C, O B, M Q and O K, are Proportional: And it will be

Fig. XXX.

As L O : D C : : O K : D X,
N E : E C : : O B : O K.

It will alſo be,

As L O * N E : D C * E C : : O K * O B : D X * O K :

Or rejecting the common Altitude O K, it will be

As L O * N E : D C * E C : : O B : D X,

The Verſed Sine of the Angle ſought. *Which was to be Demonſtrated.*

THEOREM V.

In all Spherical Triangles,

As the Rectangle of the Sines of the Sides containing the enquired Angle,
Is to the Square of the Radius;
So is the Rectangle of the Sines of the half Sum, and half Difference of the Baſe, and Difference of the Legs,
To the Rectangle made of the Radius, and half the Verſed Sine of the Angle enquired.

DEMONSTRATION.

Fig. XXXI.

It is already proved by the laſt *Theorem*,
That, As L O * N E : to the Square of the Radius:
So is O B : D X.

And therefore alſo,

L O * N E : Sq. of Rad. : : half O B : half D X.

And now in the following Diagram, O E G H, the Sum of E S and O E, is the double Meaſure of the Angle B S O; and the Arch O S is the Difference between the Baſe E S and E O, the Difference of the Sides A K and A E; And,

As R : half O S : : O H : half O B. And,
Half O S, and half O H, is equal to O B * Rad.

Therefore,

L O * N E : Sq. Rad. : : half O B * Rad. : half D X * Rad.

And

And also,

LO*NE : Sq. R. : : half O S * half O H * : half D X * R. Fig. XXXI.
Which was to be Demonstrated.

THEOREM VI.

In all Spherical Triangles,
As the Rectangle of the Sines of the Sides containing the enquired Angle,
Is to the Square of Radius;
So is the Rectangle of the Sines of the half Sum, and half Difference of the Base, and Difference of the Legs,
To the Square of the Sine of half the Angle enquired.

DEMONSTRATION.

It is already proved by the last *Theorem*, That
LO*NE : Sp. R. : . half O S * half O H : half D X * R. Fig. XXXI.

But the Rectangle made of Radius, and half the Versed Sine of an Arch, is equal to the Square of the Sine of half the Arch:
As in the foregoing Diagram; let the Arch given be D T, then is D X the Versed Sine of that Arch; and D F the Right Sine of half the Arch; and the Triangles D F R and D T X are like.

Therefore,

D R : D F : : half D T (= D F) : half D X:
And, D R * half D X, is equal to the Square of D F.

Therefore,

LO * N A : Rq. : : half O S * half O H : the Square of D F.
Which was to be Demonstrated.

CHAP. VII.

The Solution of the Twelve Cases of Oblique-angled Spherical Triangles, *without any Regard had to a Perpendicular let fall, whereby to reduce it into Two* Right-angled Triangles.

Fig. XXXII. AND the *Spherical Triangle* which I shall make use of is that noted with Z S P, whose *Sides* and *Angles*, both in Sexagenary Degrees and Minutes, and Decimal Parts also, are as in this Table is expressed.

	De.	*Mi.*	*Centes.*
S P —	42	04	42.07
Z S —	30	00	30.00
P Z —	24	04	24.06
∠ Z —	104	00	104.00
∠ S —	36	08	36.13
∠ P —	46	18	46.30

Fig. XXXII. Case I. CASE I. *Two Sides,* Z S *and* Z P, *with the Angle* P, *opposite to one of them, being given; To find the Angle* S, *opposite to the other.*

Analogy.

As the Sine of Z S, 30 Deg. *Co-Ar.*	0.3010299
Is to the Angle P, 46 Deg. 18 Min	9 8591186
So is the Sine of Z P, 24 Deg. 4 Min	9.6104465
To the Sine of S, 36 Deg. 8 Min.	x9.7705850

Fig. XXXII. Case II. CASE II. *Two Angles,* S *and* P, *with the Side* Z P, *opposite to one of them, being given; To find the Side* Z S, *opposite to the other*

Analogy.

As the Sine of the Angle S, 36 Deg. 8 Min. *Co-Ar.*	0.2293936
Is to the Sine of Z P, 24 Deg. 4 Min.	9.6104465
So is the Sine of the Angle P, 46 Deg. 18 Min.	9 8591180
To the Sine of Z S, 30 Deg.	x9.6989581

CASE

Fig.
XXXII.
Case III.

Fig. XXXII.

Fig. XXXII. *Case* I.

Fig. XXXII. *Case* II.

CASE III. *Two Sides,* ZP *and* ZS, *with the Angle* SZP, *contained between them, being given; To find the other Angles,* S *and* P.

Fig. XXXII. *Case* III.

		d.	′
Sides	ZS	30	00
	ZP	24	04
	Sum	54	04
	Difference	5	56
Half Sum		27	02
Half Difference		2	58
Half the Angle Z.		52	00

Analogy.

First Operation.

As the Sine of half the Sum of the Sides 27 d. 2′ *Co-Ar.*	2.3424577
Is to the Sine of half the Difference of the Sides 2.58	8.7139520
So is the Co tangent of half Z, 52 d.	9.8928098
To the Tangent of half the Difference of the ∠'s S and P, 5 d. 5 m.	18.9492195

Second Operation.

As Co-sine of half the Sum of the Sides 27 d. 2 m. *Co-Ar.*	0.0501479
To Co-sine of half their Difference 2 d. 58 m	9.9994175
So Co-tangent of half ∠ Z, 52 d.	9.8928098
To Tangent of half the Sum of the ∠'s S and P 41.31′.	19.9424752

	d.	m.
Which half Sum of the Angles S and P	41	13
Added to half Difference of the Angles	5	05
Gives the Quantity of the greater ∠ P	46	18
And substracted, gives the lesser ∠ S.	36	08

CASE

Fig. XXXII. *Case* IV.

CASE IV. *Two Angles*, S *and* P, *and their contained Side* S P, *being given*; *To find the other Two Sides*, Z S *and* Z P.

Angles { S	36	08
Angles { P	46	10
Sum	82	18
Difference	10	02
Half Sum	41	09
Half Difference	5	1

Analogy.

First Operation.

As Sine of half the Sum of the Angles S and P 41 d. 9 m. *Co-Ar.*	0.1817525
Is to the Sine of half their Difference 5 d. 1 m.	8.9402960
So is the Tangent of half the Side S P 21 d. 5 m.	9.5856859
To the Tangent of half the Diff. of Z S and Z P 2 d. 55 m.	x8.7077344

Second Operation.

As Co-sine of half the Sum of the Angles S and P 41 d. 9 m. *Co-Ar.*	0.1232111
To Co-sine of half their Difference 5 d. 1 m.	9.9983331
So is the Tang. of half the Side S P 21 d. 5 m.	9.5863624
To Tan. of half the Sum of Z S and Z P 27 d. 2 m.	x9.7076060

	d.	m.
To this half Sum of Z S and Z P	27	02
Add half their Difference	2	55
Their Sum is the Greater Side Z S	29	57
And subtracted gives the Lesser Z P	24	07

CASE

CASE V. *Two Sides* Z S, *and* Z P; *with the Angle* P, *opposite to one of them, being Given*; *To find the third Side* S P.

Fig. XXXII. *Case* V.

First Operation.

As the Sine of the Side Z S, 30 d. *Co-Ar.*	0.3010299
Is to the Sine of the Angle P, 46 d. 18 m.	9.8591185
So is the Sine of the Side Z P, 24 d. 4 m.	9.6104465
To the Sine of the Angle at S, 36 d. 8 m.	19.7705950

Then,

The Sides are				The Angles are	
Z S	30	00	P	46	18
Z P	24	04	S	36	08
Their Sum	54	04		82	26
Their Difference	5	56		10	10
Half Sum	27	02		41	13
Half Difference	2	58		5	05

Second Operation.

As the Sine of the half Difference of the Angles S and P, 5 deg. 5 min. *Co-Ar.*	1.0525439
Is to the Sine of half the Sum of those Angles 41 d. 13 m.	9.8188250
So is the Tangent of half the Difference of the Sides Z S and Z P, 2 deg. 58 min.	8.7139520
To the Tangent of half S P, 21 deg. 4 min.	19.5853209

Whose double is 42 deg. 9 min. for S P.

CASE VI. *Two Angles,* S *and* P, *with the Side* S Z, *opposite to one of them, being given*; *To find the third Angle* Z.

Fig. XXXI. *Case* VI.

First Operation.

As the Sine of the Angle at P, 46 d. 18 m. *Co-Ar.*	0.1408813
Is to the Sine of the Side Z S, 30 deg.	9.6989700
So is the Sine of the Angle at S, 36 deg. 8 min.	9.7700063
To the Sine of the Side Z P, 24 deg. 4 min.	19.6098576

Then,

Then,

Fig. XXXII.

The Sides				The Angles	
Z S	30	00	S	36	08
Z P	24	04	P	46	18
Their Sum	54	04		82	26
Their Difference	5	56		10	1[illegible]
Half Sum	27	02		41	1[illegible]
Half Difference	2	58		5	0

Second Operation.

As the Sine of half the Difference of the Sides Z S and Z P, 2 d. 58 m. *Co-Ar.* — 1.286047

Is to the Sine of half the Sum of the Sides, 41 d. 13 m. 9.81882[illegible]

So is the Tangent of half the Difference of the Angles S and P, 5 d. 5 m. — 8.94916[illegible]

To the Co-tangent of half the Angle Z, 52 deg. 10.05404[illegible]

Whose Double, 104 deg. is the Angle Z required.

Fig. XXXII. *Case* VII.

CASE VII. *Two Angles*, S *and* P, *and a Side* Z S, *opposite One of them, being given; To find the Side between them*, S[illegible]

First Operation.

As the Sine of the Angle at P, 46 deg. 18 min *Co-Ar.* 0.14095

Is to the Sine of the Side Z S, 30 deg. 9.6989[illegible]

So is the Sine of the Angle at S, 36 deg. 8 min. 9.7706[illegible]

To the Sine of the Side Z P, 24 deg. 4 min. 9.0105[illegible]

Then,

The Sides				The Angles
Z S	30	00	S	36
Z P	24	04	P	46
Their Sum	54	04		82
Their Difference	5	56		10
Half Sum	27	02		41
Half Difference	2	58		5

Second Operation.

the Sine of half the Difference of the Angles S and P, 5 deg. 5 min. *Co-Ar.* } 1.0525439
Is to the Sine of half their Sum, 41 deg. 13 min. 9.8188250
is the Tangent of half the Difference of the Sides Z S and Z P, 2 deg. 58 min. } 8.7145345

To the Tangent of half the enquired Side, 21 deg. 4 ½ min. } r9.5859034
[Th]e Double whereof is 42 deg. 9 min. for the Side S P.

[C]ASE VIII. *Two Sides,* Z S *and* Z P, *and the Angle* P, *opposite to the Side* Z S, *being given; To find the Angle* Z, *contained between the Two given Sides.*

Fig. XXXI. *Case* VIII.

First Operation.

[As] the Sine of Z S, 30 deg. *Co-Ar.* 0.3010299
Is to the Sine of the Angle P, 46 deg. 18 min. 9.8591186
is the Sine of Z P, 24 deg. 4 min. 9.6104465
To the Sine of the Angle at S, 36 deg. 8 min. r9.7705950

Then,

The Sides				The Angles	
Z S	30	00	S	36	08
Z P	24	04	P	46	18
Their Sum	54	04		82	26
Their Difference	5	56		10	10
Half Sum	27	02		41	13
Half Difference	2	58		5	05

Second Operation.

[T]he Sine of half the Difference of the Sides Z S and Z P, 2 deg. 58 min. *Co-Ar.* } 1.2860479
[I]s to the Sine of half their Sum, 27 deg. 2 min. 9.6575423
is the Tangent of half the Difference of the Angles S and P, 5 deg. 5 min. } 8.9491675
[T]o the Co-tangent of half Z, 52 deg. r9.8927577
[T]he Double whereof, 140 deg. is the Angle at Z required.

Fig. XXXI. Case IX.

CASE IX. *Two Sides Z S and Z P, with the Angle Z, comprehended between them; To find the third Side* S P.

You must first find the Two other Angles S and P (by the 3d *Case*) and then you may find the Side S P (by the 1st *Case* hereof.)

And so,

		d.	m.
The Angle { S } will be found to be	S	36	08
	P	46	18
And the Enquired Side S P		42	39

This *Problem* may be otherwise resolved at Two Operations, by a *Problem* next following the Twelve *Cases*.

Fig. XXXI. Case X.

CASE X. *Two Angles,* S *and* P, *with the Side between them,* S P *being given; To find the third Angle at* Z.

This Case must be resolved in the same manner as the foregoing, it being but the Converse thereof: By finding first the Two Sides, Z S and Z P, (by the 4th *Case*) and then the Angle Z (by the 1st *Case*.) So you shall find

		d.	m.
The Side to contain	Z S	30	00
	Z P	24	04
And the Enquired Angle Z		104	00

Fig. XXXI. Case XI.

CASE XI. *Three Sides,* Z S, Z P *and* S P, *being given; To find the Angle at* Z, *opposite to (the Base)* S P.

Note, I call that Side the Base (which ever it be) that is opposite to the Enquired Angle. Then,

First Operation.

As the Radius, or Sine of 90 deg.	10.
Is to the Sine of the Side Z S, 30 deg.	9.6989700
So is the Sine of the Side Z P, 24 deg. 4 min.	9.6104465
To a fourth Sine, *viz.* 11 deg. 46 min.	19.3094165

Then,

Fig. XXXI. *Case* XI.

Then,

		d.	m.
The Sides are	Z P	24	04
	Z S	30	00
	S P	42	09
Their Sum		96	13
Half Sum		48	06 $\frac{1}{2}$
The Base S P		42	09 Subst.
Difference		5	57 $\frac{1}{2}$

Second Operation.

As the Sine of the fourth Sine before found, 11 deg. 46 min. *Co-Ar.*	0.6905835
Is to the Sine of the half Sum, 48 deg. 6 $\frac{1}{2}$ min.	9.8718115
So is the Sine of Difference 5 deg. 57 $\frac{1}{2}$	9.0162186
To a seventh Sine.	19.5786136
Half this seventh Sine is	9.7893068

Which is the Sine of 38 deg. whose Complement 52 deg. is half the enquired Angle at Z, *viz.* 104 deg.

Another Way to resolve this Case.

Take half the Difference between the Two Sides that contain the required Angle, and add it to half the Base, and likewise substract it from the same, noting the Sum and Difference. Then,

The Arithmetical Complements of the Sines of the Sides comprehending the enquired Angle, added to the Sines of the Sum and Difference before found, half the Sum of them shall be the Sine of half the Angle required.

		d.	m.
The Side	Z S is	30	00
	Z P	24	04
Their Difference is		5	56
The half Difference is		2	58
Which added to half S P		21	04 $\frac{1}{2}$
The Sum is		24	02 $\frac{1}{2}$
And substracted, the Difference is		18	6

Then,

Then,

The Sine of the Side Z S, 30 deg. *Ar. Co.*	0.3013299
The Sine of the Side Z P, 24 deg. 4 min. *Ar. Co.*	0.3895535
The Sine of the Sum, 24 deg. 2 ½ min.	9.6190219
The Sine of the Difference, 18 deg. 6 ½ min.	9.4924864
The Sum	19.7933917
The half whereof is the Sine of 52 deg. 1 min. half of the Angle at Z.	9.8966958

CASE XII. *Three Angles* Z, P *and* S, *being given; To find any of the Sides.*

This is the Converse of the former, and is to be resolved after the same manner: But first, the Angles must be turned into Sides as followeth: For,

The Two Lesser Angles, S and P, are always equal unto Two Sides of another Triangle, comprehended by the Arches of Three Great Circles drawn from their Poles, and the Complement of the Greater Angle Z to a Semicircle (or 180 deg.) must be taken for the third Side. Therefore, in this Triangle thus converted, you shall by the preceding 11th Case find an Angle; that Angle so found, shall be one of the Three Sides enquired.

Fig. XXXIII. As in the Triangle A C D, the Poles of those Arches are H, R and Q; which connected, make the Triangle H R Q the Sides of the former Triangle, being equal to the Angles of the latter; taking for one of them the Complement of the Greater Angle to a Semicircle.

So A D is equal to the Angle at H; whose Measure is the Arch E N.

D C is equal to the Angle Q; whose Measure is the Arch M P.

And A C is equal to the Complement of the Angle H R Q; whose Measure is G L.

Therefore, if the Angles A, D and C, be given, the Sides Q R, Q H and R H, are likewise given.

If therefore we resolve the Triangle H R Q by the Directions of the 11th Case beforegoing, the Angle so found shall be the Side required.

CHAP.

CHAP. VIII.

A Problem *and* Theorem, *by Way of Supplement.*

In any Spherical Triangle, *Two Sides,* D B *and* B C, *with the Angle* B, *included between them being given; To find the other Angles,* B *and* C, *by Two Proportions.*

THIS Problem was invented by the Lord *Nepier*, and is celebrated, not less for its Subtilty, than Usefulness, in resolving this *Case*, without letting fall a Perpendicular, or any Ambiguity.

Fig. XXXIV.

ANALOGIES.

1. As the Sine of half the Sum of the Sides including the Angle given,
   Is to the Sine of half the Difference of those Sides;
   So is the Co-tangent of half the Vertical Angle,
   To the Tangent of half the Difference of the Two unknown Angles.

Then,

2. As the Co-sine of half the Sum of the Two given Sides,
   Is to the Co-Sine of half the Difference of those Sides,
   So is the Co-tangent of half the Vertical Angle,
   To the Tangent of half the Sum of the Angles.

   The half Difference before found, being *added* to this *half Sum* now found, gives the *Greater*, and *subtracted* therefrom, gives the *Lesser* of the Two unknown *Angles*.

DEMONSTRATION.

Suppose A, E, G, Poles of the Sides D B, D C, B C, therefore the Arch A E is equal to the Angle D; the Arch E G equal to the Angle C; the Arch A G equal to 180 deg. wanting the Angle B. Suppose the Ark E O, equal to the Ark E G, equal to the Ark E P. Then if the Points G A O E P be Stereographically projected, the Right Line A E will be equal to Tangent of half the Angle at D; and A G, equal to the Co-tangent of half B; and A O, equal to the Tangent of half the Difference of the Angles; and A P, equal to the Tangent of half the Sum of the Angles; and O G P will be a Semicircle, described from its Pole E, through G, whose

Fig. XXXIV. whose Center suppose n. Then take $B\lambda$, equal to BC, equal to BL; and draw the Diameter $Bm A\beta$, and DK, parallel to Bm, parallel to δH. Therefore, DB is equal to δB; and mK, equal to mH; and $F\lambda$, equal to DL. Draw $\lambda\omega$ perpendicular to $D\delta$, then the Triangle $\lambda\delta y$ is equiangular to the Triangle DLy; and DLK, equiangular to the Triangle $\lambda\delta\omega$; therefore, $Ly : \delta y :: DL : \delta\lambda$: And, as $DL : \delta\lambda :: LK : \delta\omega$; and therefore $L\delta$, parallel to $K\omega$. But the Points $\delta\lambda H\omega$ are in a Circle, whose Diameter is $\delta\lambda$: Therefore the Angle $\lambda\omega H$, is equal to (the Angle $\lambda\delta H$, equal to the Angle $L\delta D$, equal to) $K\omega y$. Therefore the Angle $K\omega H$, equal to $y\omega\lambda$, equal to a Right Angle: Therefore, $mH = m\omega = mK$. Also, $\lambda K : LK :: (\omega\lambda : \omega R ::)$ Tangent of the Angle $D\lambda\delta$ · the Tangent of the Angle $D\delta L$; that is,

As the Sine of the *Sum* of the Sides,
Is to the Sine of the *Difference* of the Sides;
So is the Tangent of half the *Sum* of the Sides,
To the Tangent of half the *Difference* of the Sides.

Therefore,

As the Sine of the *Sum* of the *Angles*,
Is to the *Sine* of the *Difference* of the *Angles*;
So is the *Tangent* of half the *Sum* of the *Angles*,
To the *Tangent* of half the *Difference* of the *Angles*.

But,

The *Sine* of the *Sum* of the Sides,
Is to the *Sine* of the *Difference* of the Sides,
(as the Sine of the Sum of the Angles : Sine of the *Difference* of the Angles;)
So is the *Tangent* of half the Sum of the Angles,
To the *Tangent* of half the *Difference* of the Angles.

That is, As $\lambda K : \lambda H :: AP : AO$.

Therefore,

$$\tfrac{1}{2}\lambda K + \tfrac{1}{2}\lambda H : \tfrac{1}{2}\lambda K - \tfrac{1}{2}\lambda H :: \tfrac{1}{2}AP + \tfrac{1}{2}AO : \tfrac{1}{2}AP - \tfrac{1}{2}AO.$$

That is, $\lambda m : m\omega :: An : nG$.

But

But also, the Angle $m \lambda \omega$, is equal to the Angle G A n; therefore (by 7 *Pr. 6 Lib. Eucl.*) the Triangles $\lambda\ m\ \omega$, and A n G, are equiangular; and therefore the Triangles λ H ω and A O G; and also the Triangles λ K ω; and A P G, are equiangular. Therefore,

Fig. XXXIV.

$\lambda \omega : \lambda H :: AG : GO$, and $\lambda \omega : \lambda K :: AG : AP$.

But, $D\lambda : \lambda\omega :: DL\ (LK =) : \lambda H$,

And $\delta\lambda : \lambda H :: AG : AO$, and $\lambda\omega : \lambda K :: AG : AP$.

But $D\lambda : \lambda\omega :: DL\ (LK =) : \lambda H$; and $\delta\lambda : \lambda\omega :: \delta L\ (LH) : \lambda K$.

Therefore,

$D\lambda : DL :: (\lambda\omega : \lambda H ::)\ AG : AO$.

And $\delta\lambda : \delta L :: (\lambda\omega : \lambda K ::)\ AG : AP$.

That is,

As the Sine of half the *Sum* of the Sides,
Is to the Sine of half the *Difference* of the Sides;
So is the Co-tangent of half the *Vertical Angle*,
To the Tangent of half the *Difference* of the *Angles*.

And,

As the Co-sine of half the Sum of the Sides,
Is to the Co-sine of half the *Difference* of the Sides;
So is the Co-tangent of half the *Vertical Angle*,
To the Tangent of half the *Sum* of the Angle.

COROLLARIES.

In any Spherical Triangle,

I. Tan. $\frac{1}{2}$ Base : Tan. $\frac{1}{2}$ Sum of the Sides :: Tan. $\frac{1}{2}$ Differ. of the Sides : Tan. $\frac{1}{2}$ Differ. of the Segments of the Base A G, made by a perpendicular Arch falling thereon from E.

$$AG : AO :: AP : Ad.$$

II. Tan. $\frac{1}{2}$ Base : Tan. $\frac{1}{2}$ Sum Sides :: Co-fine. $\frac{1}{2}$ Sum <'s : Co-fine $\frac{1}{2}$ Diff. <'s. $AG : AP :: \delta\lambda : \delta L$: For the Angle E A G = the Arch D B, and the Angle A G E = to the Arch B C.

III. Tan. $\frac{1}{2}$ Base : $t.\ \frac{1}{2}$ Sum Sides :: $s.\ \frac{1}{2}$ Sum of Ang. : $s.\ \frac{1}{2}$ Diff. <s.

$$AG : AO :: D\lambda : DL.$$

IV. If

Fig. XXXIV. IV. If the Angle E G A be supposed Right, then A P + A O = A G q. that is, In a Right-angled Spherical Triangle, E G A, the Tang. ½ Hypot. + ½ Per. + Tan. ½ Hypot. — ½ Per. is = the Square of the Tangent of half the Base.

CHAP. IX:

Some Problems *in Plain Triangles, which come not within the Limits of the Twelve foregoing Cases.*

PROBLEM I.

In the Oblique-angled Triangle C B D, *there is given the Angle at* C, (*116.20 deg.*) *the Side* D B, *opposite thereunto* (1270 *Foot*:) *And the Two Sides,* D C *and* C B, *in One Sum,* (1496 *Foot*:) *To find the other Angles* D *and* B, *and the Sides severally*

CONSTRUCTION.

Fig. XXXV. EXtend the Side of the Triangle D C, to A, making C A equal to C B; and draw the Line B A, so shall you have constituted a new Triangle A O B, in which you have given, (1.) The Side A D, equal to the Sum of the Sides D C and C B, (1497.) And (2.) the Side D B (1270.) And (3.) being you have the Angle D C B, (116 d. 12 m.) you have also the Angle B C A in the other Triangle, (63 80 d.) the Complement thereof to a Semicircle, or 180 deg.

Then in the Triangle A C B, having the Angle at (63.80) you have also the Sum of the other Two Angles, C A B and A B C, equal to the given Angle D C B, (116.20) the half whereof is the Angle C A B (58.10 d.) to which the Angle A B C is equal, because the Sides C A and C B subtending those Angles, are equal: And now in the Triangle D A B you have given, (1.) the Side D A, 1497. (2.) The Side D B, 1270. And (3.) the Angle, D A B (opposite to D B) (58.10 d.) by which you may find the whole Angle D B A (by *Case* I. of Oblique Triangles: For,

As Log. D B : s. C A B :: Log. D A : s. A B D.
1270 . 58.10 d. 1497. 95.53 d.

From

Fig.
XV.

Fig.
XVI.

Fig.
XXXI

Fig.
XXX

From which, the Angle C B A 58.10 d. being substracted, there will remain 37.43 d. for the Angle C B D, and then the Angle C D B will be 26.37 d. the Remainder of the other Two, C and B, to 180 d. *Fig.* XXXV.

And for the Sides, they may be found by the first Case of Oblique Triangles, thus:

As *s* C : Log. D B : : *s* B : Log. D C.
116.20 d. 1270 37.43 d. 865

As *s* B : Log. D C : : *s* D : Log. C B.
37.43 d. 865 26.37 d. 632

PROB. II.

In the Oblique-angled Triangle C D B *there is given the Angle at* C, *(*116.20 *d.) The Side* D B, *opposite thereunto (*1270 *Foot.) And the Difference of the other Two Sides* C D *and* C B, (233 *Foot.) To find the Two unknown Sides severally; and the Angles at* D *and* B.

CONSTRUCTION.

IN the Given Triangle, make C A equal to C B, and draw the Line B A; so is the Given Triangle C D B, reduced into Two Oblique-angled Triangles C A B and A D B; in which last is given the Two Sides, A D and D B, but never an Angle; and in the other, only the Angle at C, and never a Side. *Fig.* XXXVI.

But (by Construction) the Side C A being made equal to the Side C B, the Angles C A B and C B A are equal, and the Sum of them, equal to the Complement of the Angle at C, (116.20 d.) to 180 d. that is, to 63.80 d. the half whereof 31.90 d. is equal to the Angle C A B, and the Complement thereof to 180 d. is 148.10 d. equal to the Angle D A B.

And now in the Triangle D A B, there is given, the Sides A D and D B, and the Angle at A, to find the other Angles, by *Case* I. and II. of Oblique Triangles, Thus,

As Log. D B : *s* A : : Log. A D : *s*. A B D.
1270 148.10 d. 233 5.53 d.

Which added to the Angle C B A, before found, 31.90 d. the Sum will be 37.43 d. for the Angle C B D.

Fig. XXXVI.

Then,

As *s* DCB : Log. DB : : *s* CBD : Log. CD
116.20 d. 1270 37.43 d. 865

And,

As *s* DCB : Log. DB : : *s* CDB : Log. CB
116.20 d. 1270 26.37 d. 632

PROB. III.

In the Right-angled Plain Triangle ABC, *there is given, the Base* AB, *(28 Foot.) And the Hypotenuse* CB, *and Perpendicular* CA, *in one Sum,* (viz. *56 Foot) to find the other Sides and Angles severally.*

Fig. XXXVII.

IN the given Triangle ABC, extend the Perpendicular AC to D, making CD equal to CB; so shall you have constituted a new Triangle DAB; in which there is given, (1.) The Right Angle at A. (2.) The Base AB 28 Foot. (3.) The Side AD, equal to the Sum of the Sides AC and CB, by which you may find the Angles ADB, and ABD, *(by Case I. of R. Ang. Tri.)*

As Log. AB : Rad. : : Log. DA : *t* DBA
28 *t*. 45.00 d. 56 63.43 d.

Whose Complement 26.57 d. is the Angle ADB, to which the Angle CBD is equal, because the Sides CB and CD, subtending them, are equal by Construction. Then from the Angle DBA 63.43 d. before found, you substract the Angle DBC, 26.57 d. last found, the Remainder 36.86 d. is equal to the Angle CBA, and the Complement thereof, 53.14 d. to the Angle CAB.

And then,

As *s* ACB : Log. AB : : *s* CBA : Log. CA
53.14 d. 28 36.86 d. 21
And so : : Rad. : Log. CB.
90 d. 56

Other-

Otherwise, by this Theorem.

Fig. XXXVII.

If from the Square of the Sum of the Base and Perpendicular D A 56 (3136) *you substract the Square of the Base* A B 28, (784) *and divide the Remainder* (2352) *by double the Sum of the Hypotenuse and Perpendicular* 56, (viz. 112), *the Quotient* (21) *will give the Perpendicular: Which substracted from the Sum* (56), *leaves* (35) *for the Hypotenuse.*

PROB. IV.

The Three Sides of a Right-lined Triangle, *being given; To find the* Area.

RULE.

ADD the Three Sides of the given Triangle together, and take the half thereof: From which half Sum, subtract each Side of the Triangle severally, and note the several *Differences*: Then multiply the *Half Sum*, by any one of the *Differences*, and that *Product* multiply by another of the *Differences*, and that *Product* multiply by the third *Difference*: The *Square Root* of this third *Product*, shall be the *Area* of the given Triangle.

Example.

Fig. XXXVIII.

In the *Triangle* A B C, whose Three Sides are A B 20, A C 34, and B C 42: Their *Sum* is 96, the half whereof is 48: From which subtract the several Sides, 20, 34, and 42, and there will remain these *Differences*, 28, 14, 6.——Then, multiply 48 (the half Sum) by 28 (the first Difference) the Product will be 1344: Which multiplied by 14 (the second Difference (the Product will be 18816: And this Product multiplied by 6 (the third Difference) produceth 112896: The *Square Root* whereof is 336, for the *Area* of the Triangle A B C.

By Logarithms.

To the Logarithm of half the *Sum* of the *Sides*, add the *Logarithms* of the several *Differences* of the *Sides* from the half *Sum*: Half the *Sum* of those *Logarithms* shall be the Logarithm of the *Area* of the Triangle.

Example.

Fig. XXXVIII.

The half *Sum* of the Sides	48	whose Logarithm is	1.6812412
The Difference of A B	28		1.4471580
The Difference of A C	14		1.1461280
The Difference of B C	6		0.7781512
		Their Sum	5.0526784
		The Half Sum	2.5263392

Which is the Logarithm of 336, the *Area* of the *Triangle.*

PROB. V.

By the Three Sides *given; To find the Point in the longer* Side, *where a* Perpendicular *shall fall, and the Length of that* Perpendicular.

RULE.

TAKE the *Sum* and *Difference* of the Two *Sides* containing the *Angle* from whence the *Perpendicular* is to fall, and multiply them together, and divide the Product by the third *Side*, upon which the *Perpendicular* is to fall, the Quotient added to the third *Side*, or substracted from it, shall be the Double of the *Greater* or *Lesser Segment* on either *Side* of the *Perpendicular*.

Example.

Fig. XXXVIII. In the former *Triangle* A B C, the *Sum* of the *Sides* A B and A C, is 54, and their Difference is 14; which multiplied together, produce 756; which divided by the *Side* B C 42, the Quotient is 18; which added to the *Side* B C 42, gives 60; the half whereof 30, is the *Greater Segment* of the *Side* B C; or the Quotient 18, substracted from the *Side* BC, leaves 24; the half whereof 12, is the *Lesser Segment* B D, where the *Perpendicular* is to fall.

By Logarithms.

To the *Sum* of the *Logarithms* of the *Sum* and *Difference* of the *Sides* containing the *Angle* from whence the *Perpendicular* is to fall; substract the *Logarithm* of the *Side* upon which it is to fall: The Remainder shall be the *Logarithm* of a Number; which added to, or substracted from the *Side* on which the *Perpendicular* is to fall, shall be the Double of the *Greater* or *Lesser Segments* of the *Side* on which the *Perpendicular* is to fall.

Example.

Example.

Fig. XXXVIII.

The Sum of A B and A C is 54	1.7323937
The Difference of A B and A C is 14	1.1461281
Their Sum	2.8785218
The *Logarithm* of the third *Side* B C 42, substract	1.6232493
Remains	1.2552725

Which is the *Logarithm* of 18. Which 18 added to B C 42, makes 60; the half whereof 30, is the *Greater Segment* C D; or 18 substracted from B C 42, leaves 24; the half whereof 12, is the *Lesser Segment* B D.

For the Length of the *Perpendicular.*

RULE.

Multiply the *Sum* of A B and B D 32, by the *Difference* of A B and B D, 8; the Product will be 256, whose *Square Root* is 16; for the Length of the *Perpendicular* A D.

By Logarithms.

Half the *Sum* of the *Logarithms* of 32 and 8, the *Sum* and *Difference* of A B and B D is the *Logarithm* of the *Perpendicular*.

The *Logar.* of the Sum of A B and B D 32,	1.50515
The *Logar.* of the *Difference* of A B and A D 8.	0.90309
Their *Sum*	2.40824
The half	1.20412

Which is *Logarithm* of 16, the Length of the *Perpendicular* A D.

PROB. VI.

The Base B A *(or longest* Side) *of a* Plain Triangle B A D, *being* 47.8 P. *and a* Perpendicular D C, *let fall from the* Angle *opposite to that* Side 17.33 P. *being given; To find the* Area *of that* Triangle.

TO the *Logarithm* of half the given *Side*, add the *Logarithm* of the *Perpendicular*; the *Sum* of those *Logarithms* shall be the *Logarithm* of the *Area* of that *Triangle.*

Fig. XXXIX.

The

Fig. XXXIX.

The *Base* B A, is 47.8 —			
The half *Base*	23.9 —	*Log.*—1.3783978	
The *Perpendicular* D C	17.33—	*Log.*—1.2387986	
The *Logarithm* of	414.2	2.6171965	

Which is the *Area* of the *Triangle.*

After the same manner, if the *Base* of a *Triangle* were 42, and the *Perpendicular* 16, the *Area* will be found to be 336.

PROB. VII.

There is an Oblique-angled Plain Triangle A B C, *one of whose* Angles *at the Base* B C *is* 53.48 *d. And the* Sum *of the Two* Sides *opposite to the Base* B C, viz. A B *and* A C, *is* 129, *and the Base* C B, *is longer than the longer side* A C, *by* 16.8. *The respective* Sides *and* Angles *of the* Triangle *are required.*

CONSTRUCTION.

Fig. XL. DRaw a *Right Line* B C, at Liberty, for the Base; and upon one end thereof, at B, make an *Angle* A B C, to contain 53.84 d. Then the *Sum* of the Two *Sides* A B and A C being together 120, break it into any Two equal or unequal Parts, as into 43.2 and 76.8, then (by the *Proposition*) must the Base B C, be 93.6, which is greater than A C by 16.8.—So in the *Oblique-angled Triangle* A B C, you have given, (1.) The *Side* A B 43.2. (2.) The *Side* A C 76.8. (3) The *Angle* opposite thereto, A B C 53 d. 50 m. by which you may find the *Angle* A C B. For,

As A C 76.8 : Is to *s* A B C 53.84 d.
So is A B 43.2 : To *s* A C B 27.02 d.

PROB. VIII.

Of the Mensuration of the Area *of a* Spherical Triangle.

Lemma I. *The Lunary Superficies of the Hemisphericks, are as the* Angles *of the same Superficies.*

Fig. XLI. THE Proof of this *Lemma* (amongst many other Ways) may be this:

Let the Meridian *Semicircle* A E D be imagined to be equally moved over the Longitude of the *Equator* B E C, upon the *Poles* A and D. Therefore the *Angles* on the other side of A and D (to (to wit, F and G) ſhall be as the Times. The Superficies alſo, K and M, ſhall be as the Times; therefore F ſhall be to G, as the Superficies K, to the Superficies M. And G unto M, as the Superficies M, to the Superficies K, &c. howſoever they can be taken. For, Fig. XLI.

> *Thoſe Things that agree to a Third,*
> *Agree among themſelves.*

COROLLARY.

Therefore, by Compoſition,

As F ✠ G : G : : K ✠ M : M,
Or, As F ✠ G : F : : R ✠ M : K.

That is,

As the Two *Right Angles* are unto G,
So is half the *Spherical Superficies*, to M.
And, As Two *Right Angles*, are to F;
So is half the *Spherical Superficies*, to K.

Lemma II. *The Triangle* G, *is equal to the Triangle* H, *becauſe the Angles and Sides of one, are equal to the Angles and Sides of the other: To wit,* A = D, B = E, C = F. *Alſo,* L = O, M = P, N = Q. *Therefore they are Congruous and Equal.* Fig. XLII.

THEOREM.

The Exceſs of the Three Angles, over and above Two Right Angles, divided by 720, ſhews what the Area of the Triangle, is, in reſpect of the whole Spherick.

For by *Lemma* I. { 180 : A : : ½ Sph. : G ✠ R
180 : B : : ½ Sph. : G ✠ S
180 : C : : ½ Sph. : G ✠ T = H ✠ T.
(by the Second Lemma.)

Therefore, As 180 : A ✠ B ✠ C : : half the Spherick : to 3 G ✠ R ✠ S ✠ T, *(by 24 El. 5 Ev.)*

As 180 : A ✠ B ✠ C — 180 :
: : So half the Spherick : 3 G ✠ R ✠ S ✠ T—half the Sph.
But G ✠ R ✠ S ✠ T is equal to half the Spherick;

Therefore,

Fig. XLII. Therefore, 3 G + R + S + T — half the Sph. = 2 G.
So that, As 180 A + B + C — 180
: : So is half the Spherick : to 2 G.

And the Antecedent Terms being Quadropled, it shall be,

As 720. : A + B + C — 180 : : 2 Sphericks : 2 G.
And so the Spherick to G. Therefore,

$\frac{A + B + C - 180}{720}$ } shews what part the *Triangle* is of the whole Spherick.

These Things are likewise true in all *Spherical Polygons*, of what *Ordinate* Figure soever they be; or *In-ordinate*, so all the *Angles* be given. And the Reason is, because all *Polygons* may be resolved into *Triangles*. Therefore, this *Rule* shall hold in these *Multangles* also.

RULE.

Multiply 180 d. by the Number of the *Angles*; subduct the Product out of the Aggregate of all the *Angles* increased by 360 d. The Residue divided by 720 d. gives the *Area* of the *Polygon*.

Of the Completion of a Solid Body.

FRom the foregoing Mensuration of the *Area* of a *Spherical Triangle*, this Fruit ariseth.

If the Radius of the Sphere be 100000.00, the *Side* of an inscribed *Icosaedrum* shall be 105146.22, equal to the Subtense of 63 deg. 26 min. 10 sec. Therefore the *Plain Equilateral Triangle* F E O (in the *Icosaedrum*) answers to the *Equilateral Spherical Triangle* in the *Sphere*; whose Three *Spherical Triangles* are connected in the *Plain Angles*, in the same Points, F, E, O.

And the *Sides* of this *Spherical Triangle* are separately taken 63 deg. 26 min. 10 sec. to wit, because their Subtenses F E, E O, O F, in the *Plain Triangle*, are equal to one another.

Fig. XLIII. Let fall now the *Perpendicular* E P, the *Spherical Triangle* E P O shall be Rectangled; where, over and above the *Right Angle* at P, are given, E O and P O, equal to half E O; wherefore the *Vertical Angle* P E O shall be 36 deg. just, and the whole *Angle* at E, 72 deg. And the *Sum* of the Three equal *Angles*, E, F, O, shall be 216 deg. from whence taking Two *Right Angles*,

Angles, equal to 180 Deg. there remains 36 Deg. Therefore, the *Triangle* E F O is $\frac{36}{720}$, of the whole *Spherick*; that is $\frac{1}{20}$ Part: And this most truly, for 20 *Pylamides* F E O C, fill the Solid Place of the *Icosaedrum*. And so 20 *Spherical Bases*, (covered over with 20 *Triangular Plain Bases*,) compleat the whole *Spherick*.

Fig. XLIII.

F E O C is One of the 20 *Pyramids* in the *Icosaedrum*: The *Plain Triangle* B, is One of the *Hedra* or *Bases*: C is the *Center* of the *Body*, or *Sphere*, that circumscribes it.

4.77 Icosaedres 4.24 Dodecaedres 9.244 Octaedres 8.000 Cubes	Fill a *Solid Place*, as will appear out of Precedent Practice by *Triangles*, and *Spherical Polygons*.

That is to say, None of the Five *Regular Bodies* fill a *Solid Place*, the *Cube* only excepted.

Contrary to what *Potamon*, and from him *Ramus*, and all that have followed *Ramus*; to wit, *Snellius*, and others, have delivered.

SECTION II.

Spherical Trigonometry, *Geometrically Performed.*

Fig. XLIV.

AS the *Sides of Plain Triangles* are Three *Right Lines*, interſecting each other upon a *Plain*: And a *Right Line* is, The *Shorteſt Extenſion* between any Two *Points* upon a *Plain Superficies*: So the *Sides* of a *Spherical Triangle* are Three *Arches*, of Three *Great Circles* of the *Sphere*, interſecting each other upon the *Globe*. And an *Arch* of a *Great Circle*, paſſing through any Two *Points* upon a *Spherical Superficies*, is the neareſt *Diſtance* between thoſe Two *Points*.

In Purſuance of the Work in this *Chapter* intended, I ſuppoſe the Reader to be acquainted with the *Circles* of the *Sphere*; that is, to know their *Names* and *Situations* upon the *Globe*, to what *Uſe* each of them ſerveth; and alſo, how to project any of them upon a *Plain*, anſwerable to any *Poſition* of the *Globe*: For to ſuch as do not, this Section will be but of little Uſe; and therefore I would adviſe my *Reader*, before he enter upon this *Geometrical* Way of reſolving *Spherical Triangles*, to peruſe the Beginnings of the Second and Third *Sections* of the *Third Part* hereof, which treat of the *Circles* of the *Sphere*, and their ſeveral Poſitions and Affections; in which he may receive very much Satisfaction concerning thoſe Particulars. So that it ſhall ſuffice, in this Place, that I declare,

1. What a *Great Circle* is.
2. How to *Project* ſuch a *Circle* of the *Sphere* upon a *Plain*, ſuitable to any *Day* and *Time* of the *Day*, at any time of the *Year*, and in any *Latitude*.
3. To diſcover the *Triangle*, which is made by the Interſection of thoſe *Great Circles* ſo projected.
4. To find the *Poles* of thoſe *Great Circles*. And,
5. To *Meaſure* the *Sides* and *Angles* of the *Triangle* ſo laid down: Which to do, is uſually called, *The Doctrine of the Dimenſion of Triangles.*

I. *What*

107

Fig. IV.

Fig.
XLII

I. *What a* Great Circle *is*.

Fig. XLIV.

A *Great Circle* of the *Sphere* is ſuch a *Circle*, as divideth the whole *Sphere* into *Two Equal Parts* or *Hemiſpheres*: Of which, there are generally accounted Six, *viz.*

The { Meridian, Horizon, Æquinoctial } The { Æquinoctial, Solſtitial } Colure. Ecliptick.

Beſides theſe Six Principal *Great Circles* of the *Sphere* there are other *Circles*, which are *Great Circles* alſo. As are all *Azimuth* or *Vertical Circles*. Alſo all intermediate *Meridians* or *Home Circles*. And theſe are ſuch as we have moſt occaſion to make uſe of in this Place: And here note, That all *Great Circles* projected upon a *Plain*, *Steriographically* (or *Circular*,) are either, (1.) *Perfect Circles* (as the outward or *Primitive Circle* is. Or, (2.) *Semi-Great Circles*, projected within the *Primitive Circle*. Or, (3.) They are *Streight Lines*, paſſing through the Centre of the Primitive Circle, as afterwards will appear.

II. *How to* Project *the* Circles *of the* Sphere *upon a* Plain, *and ſwerable to a Prefixed* Time *and* Place.

PROBLEM.

LET it be required to *Project* ſuch *Circles* of the *Sphere* in *Plano*, upon the *Plain* of the *Meridian*, in the *Latitude North*, 40 Deg. Upon the 10th of *June*, at the time of the *Sun's Riſing* or *Setting*; and alſo at 10 in the *Morning*, or 2 in the *Afternoon*, the ſame Day: The Sun then having 23 Deg. 30 Min. of *North Declination*.

Firſt, With 60 Deg. of a *Scale* of *Chords*, upon the Point A, deſcribe the Primitive Circle Z H N O, repreſenting the *Meridian* of the Place.

Secondly, Draw the Right Line H A O for the Horizon of the Place; and at Right Angles thereto the Line Z A N, for the *Æquinoctial Colure*, Z being the *Zenith*, and N the *Nadir* Points.

Thirdly, Take 40 Deg. (the *Latitude* given) out of your Scale of *Chords*, and ſet them from O to P, from Z to Æ, from H to S, and from N to *æ*; and draw the Line P A S for the *Axis* of the World, and *Hour-Circle* of Six; and the Line Æ A *æ*, for the *Æquinoctial*.

Fourthly, Becauſe the *Sun's Declination* at the time given is 23 Deg. 30 Min. take 23 Deg. 30 Min. and ſet them from Æ to ♋, and from *æ* to ♋: And if you lay a Ruler from Æ or *æ*

P 2 to

Fig. XLIV. to ♋, it will cross the *Axis* of the World P A S, in the Point F; so have you Three Points, ♋ F ♋, through which you may draw the Circle ♋ F ♋; which will cut the *Horizon* H A O, in the Point ☉, the Place where the *Sun* will rise that Day.

Fifthly, This Point ☉ being found, you have Three other Points given, *viz.* P, ☉ and S; through which you may describe the Circle P ☉ S which is the *Hour* at which the *Sun Riseth*.

And thus have you *Projected*, so far of the *Problem*, as concerns the Time of the *Sun's Rising*. Now for his Place of being at Ten in the Morning, or Two in the Afternoon.

Sixthly, Ten in the Morning, or Two in the Afternoon, are (either of them) Two Hours, or 30 Deg. distant from the *Meridian*: Wherefore, take 30 Deg. from your Scale of *Chords*, and set them upon the *Meridian*, from Æ to G, then a Ruler laid from P to G, will cross the *Æquinoctial* Circle Æ A æ, in the Point B; so have you Three Points, S, B and P, by which you may *Project* the Circle S B P, for the *Hour Circle* of Ten in the Morning, or Two in the Afternoon: And the Circle S B P, will cut the *Tropick* of *Cancer* ♋ A ♋ (the Line which the Sun traces that Day,) in the Point C; in which Point the Sun will be at Ten in the Morning, or at Two in the Afternoon.

Seventhly, And now have you Three other Points, N, C and Z, through which you may draw the *Azimuth Circle* Z C N, for the *Azimuth*, that the Sun will be upon at Ten and Two a Clock.

III. *Concerning the* Spherical Triangles *that are made by the Intersections of these* Great Circles *thus projected.*

Many are the *Triangles*, that are made by the Intersections of these few *Circles* upon this *Projection*, but I shall exemplifie only in *Two* of them; namely, the *Triangle* P ☉ O, Right-angled at O, which is composed of Three *Arches* of *Great Circles* of the *Sphere*, *viz.* (1.) Of ☉ O, an *Arch* of the *Horizon*, comprehended between O, the *North* Point of the *Horizon*, and ☉, the Point in the *Horizon*, whereon the Sun *Riseth* that Day, or his *Amplitude* from the *North*. (2.) Of P ☉, an *Arch* of a *Meridian* or *Hour-Circle*, passing through P, the *North Pole* of the World; and ☉, the Place of the Sun's *Rising* that Day, equal to 66 Deg. 30 Min. the Complement

plement of the Sun's *Declination* that Day. And, (3.) Of P O, an *Arch* of the *Meridian* of the *Place*, comprehended between O, the *North* Point of the *Horizon*, and P, the *North Pole*, equal to the Latitude given, 40 Deg.

Fig. XLIV.

The other *Triangle* shall be the *Oblique-angled Triangle* Z C P, constituted by the Intersection of Three other *Great Circles* of the *Sphere*; namely, Of (1.) Z P, an *Arch* of the *Meridian* of the *Place*, comprehended between Z the *Zenith*, and P, the *North Pole*, equal to 50 Deg. the Complement of the *Latitude* of the Place. (2.) Of P C, an *Arch* of a *Meridian*, or *Hour-Circle* of Ten or Two a Clock, equal to 66 Deg. 30 Min. the Complement of the Sun's *Declination*; or, the *Sun's Distance from the elevated Pole* P. And (3.) Of Z C, an *Arch* of an *Azimuth* or *Vertical Circle*, passing through the *Hour-Circle* S C P, in the Point P, where the Sun is at Ten or Two a Clock that Day, and is the Complement of the *Sun's Altitude* at that time

IV. *To find the* Poles *of these* Great Circles, *they being thus Projected.*

The *Pole* of any *Great Circle* of the *Sphere*, is, always, 90 Deg. distant in all Places, from the same Circle, upon a Right Line, or Circle, which cuts it at *Right Angles*: So,

1. The *Pole* of the *General Meridian*, or *Primitive Circle* Z H N O, is A, the *Center* thereof, in all Places distant 90 Deg.
2. The *Poles* of the *Horizon* H O, are Z and N, the *Zenith* and *Nadir* Points.
3. The *Poles* of the *Æquinoctial* Æ æ, are P and S, the *Poles* of the World.
4. The *Poles* of the *Æquinoctial Colure*, or *Axis* of the World P S, are Æ and æ.
5. The *Poles* of the *Prime Vertical* Circle, or *Azimuth* of *East* and *West*, Z A N, are H and O, the *North* and *South* Points of the *Horizon*.

And thus you see, that all *Great Circles* of the *Sphere*, which, when *Projected*, become *Strait Lines*, their *Poles* fall all in the *Periferie* of the *Primitive Circle*: But for all other *Great Circles*, which consist of *Circular Arches*, as P ⊙ S, P B S, and Z D N, the *Poles* of them fall within the *Primitive Circle* Z H N O, upon those *Great Circles*, which cut them at *Right Angles*: So,

R

Fig. XLIV.	R / T / V	The *Pole* of the Great Circle.	P ☉ S / P B S / Z D N	will be upon the *Line*.	Æ A / A a / A O	at the Point	R. / T. / V.

Which *Points* or *Poles* may be thus found.

1. For the *Pole* of the *Hour-Circle* P ☉ Q S, lay a Ruler from P to Q, it will cut the *Primitive Circle* in *a*, set 90 Deg. of your *Chords*, from *a* to *b*; a Ruler laid from P to *b*, will cut the *Æquinoctial Circle* in R, so is R the *Pole* of the *Circle* P ☉ Q S.
2. For the *Pole* of the *Hour-Circle* P B S, lay a Ruler from P to B, it will cut the *Primitive Circle* in G, set 90 Deg. from G to *c*, a Ruler laid from P to *c*, will cut the *Æquinoctial* in T, so is T the *Pole* of the *Hour-Circle* of 10 and 2, *viz.* P B S.
3. For the *Pole* of the *Azimuth*, or *Vertical Circle* Z D N, lay a Ruler from Z to D, it will cut the *Primitive Circle* in *d*, set 90 Deg. from *d* to *e*, a Ruler laid from Z to *e*, will cut the *Horizon* H A O in V, so is V the *Pole* of the *Azimuth Circle* Z D N.

And in this manner may the *Pole* of any *Great Circle* be found. For the *Centers* of them, they always fall in the same Right Line, in which their *Poles* are, they being extended, if Need be: But no more of this in this Place: For I intend not here a *Treatise* of *Projection*, but of the *Measuring* of *Triangles*, to which I now proceed.

V. *How to* Measure *the* Sides *and* Angles *of* Spherical Triangles, *they being thus* Projected.

You must understand. That the Quantities of the *Sides* and *Angles* of *Spherical Triangles* are measured upon the *Periferie* of the *Primitive Circle*.

Then,

1. In the *Right angled Spherical Triangle* P ☉ O, *Right-angled* at O, there is given, (1.) The *Perpendicular* P O 40 Deg. equal to the *Latitude* given by the *Proposition*. (2.) The *Hypotenuse* P ☉, the Complement of the *Sun's Declination* 66 Deg. 30 Min. And let it be required to to find the *Base* ☉ O, whose Measure is ☉ O, the *Sun's Amplitude*.

Lay

Lay a Ruler to Z, the *Pole* of the *Horizon*, and to the *An-* *Fig.* XLIV.
lar Point ⊙, it will cut the *Primitive Circle* in the Point *f*, the Distance from *f* to O, measured upon the Scale of *Chords*, ill be found to contain 58 Deg. 38 Min. and that is the iantity of the *Side* ⊙ O; and is the *Amplitude* of the Sun's *sing* or *Setting* from the *North* Part of the *Horizon* at O; d its Complement, 31 Deg. 22 Min. is the Quantity of the *ch* A ⊙, the *Sun's Amplitude* from A, the *East* and *West* Points the *Horizon*.

The Canon for Calculation is,

As *c s* P O	: Radius	: : *c s* P ⊙	: *c s* ⊙ O.
50 d.	90 d.	: : 23 d. 30 m. :	31 d. 22 m.

2. By the same Things given to find the *Angle* at the *Perpendicular* ⊙ P O, whose *Measure* is the *Arch* of the *Æquinoctial* T *a*, or the *Hour* from *Midnight*.

Lay a Ruler to P, the *Pole* of the World, and T, it will cut *Primitive Circle* in *a*, and the Distance *a æ* measured upon the ile of *Chords*, will be found to be 68 Deg. 36 Min. And t is the Quantity of the *Angle* ⊙ P O: Which 68 Deg. 36 n. converted into *Time*, (by allowing 15 Deg. of the *Æqui-ial* for One *Hour*, and 4 Deg. for *One Minute* of *Time*) makes lours, and 34 Min. Which is the Hour from Midnight, or Time of the *Sun's Rising* in the Morning: Whose Complet to 12 Hours, is 7 Hours, 26 Min. the Time of the *Sun's ting*.

The Canon for Calculation is,

As Radius	: *t*. P O	: : *ct*. P ⊙	: *c s*. ⊙ P O.
90 d.	40 d.	23 d. 30 m.	21 d. 24 m.

. By the same Things Given, to find the *Angle* at the *Base* P ⊙ O; whose Measure is an *Arch* of a *Great Circle* intercepted between the *Hour-Circle* P ⊙ S, and the *Horizon* H ⊙ O.

orasmuch as the *Sides* ⊙ P, and ⊙ O, of the *Triangle* O, are less than *Quadrants*, we must measure the *Angle* O, by the *Angle* H ⊙ S, which is equal to it. To perform ch,

ay a Ruler to R, (the *Pole* of the *Hour-Circle* P Q S) and ⊙, *Angular Point*, and it will cut the *Primitive Circle* in *b*; set 90 Deg. from *b* to *a*: A Ruler laid from the *Pole* R, to

a,

Fig. LXIV. ☉, will cut the *Hour-Circle* P Q S in *k*: Then lay a Rulet from ☉ to *k*, it will cut the *Primitive Circle* in *l*, and the Diſtance from *l* to H, meaſured upon the Scale of *Chords*, will be found to contain 44 Deg. 30 Min. for the Quantity of the *Angle* H ☉ S, which is equal to the *Angle* P ☉ O required: And this is the *Angle* of the *Sun's Poſition*, (in reſpect of the *Pole* P, and *North* Part of the *Horizon* O,) at the Time of his *Riſing* or *Setting*.

The Canon for Calculation is

As *s*. P ☉	: Radius	:: *s*. P O	: *s* P ☉ O.
66 d. 30 m.	90 d.	:: 40 d.	: 44 d. 30 m.

And thus are all the *Sides* and *Angles* of the *Right-angled Spherical Triangle* P ☉ O meaſured: And the like may be done in any other *Right-angled Triangle*; of which, in this *Scheme* there are many.

VI. *And now for the* Oblique-angled Spherical Triangle Z C P.

In this *Triangle* there is given, (1.) The Side Z P, 50 Deg. equal to the Complement of the *Latitude* given. (2.) The *Side* C P, 66 Deg. 33 Min. equal to the Complement of the *Sun's Declination* given. And, (3.) The *Angle* Z P C, 30 Deg. the *Horary Diſtance* of the Sun from the *South* Part of the Meridian; namely, at Ten or Two of the Clock. And from theſe Things given, let it be required to find the third *Side* Z C, the Complement of the *Sun's Altitude* at the Time of the Queſtion, *viz.* at Ten in the Morning, or at Two in the Afternoon. Which, to perform,

Lay a Ruler to V, the *Pole* of the *Azimuth Circle* Z D N, and to the Angular Point C, it will cut the *Primitive Circle* in *m*, and the Diſtance from Z to *m*, by the *Scale* of *Chords*, will be found to 30 Deg. 7 Min. whoſe Complement D C, 59 Deg. 53 Min. is the *Sun's Altitude* at Ten or Two of the Clock, that Day.

The Canons for Calculation is

(1.) As *c t*. P Z	: Radius	:: *c s*. P	: *t* P A.
40 d.	: 90 d.	:: 60 d.	: 45 d. 56 m.

Which ſubſtracted from P C, 66 d. 30 m. there remains 20 Deg. 34 Min.

(2.) As *c s*. P A	: *s c*. Z P	:: *c s*. Rem.	: *c s*. Z C.
44 d. 4 m.	: 40 d.	:: 69 d. 26 m.	: 59 d. 53 m.

To

To find the Angle Z C P.

Fig. XLIV.

This is the *Angle* of the Sun's *Position* at the time of the *Question*, with Reference to the *Pole* and *Zenith*; and may be thus found.

In regard the Two Sides C Z and C P, comprehending the enquired *Angle* Z C P, are both of them less than *Quadrants*, we must make use of the alternate and opposite *Angle* S C N, which is equal to Z C P. And, whose *Measure* is the *Arch* of a *Great Circle* intercepted between the *Meridian* or *Hour-Circle* S C P, and the *Azimuth-Circle* Z C N; at 90 Deg. Distance from the *Angular Point* C: And to find the Quantity hereof, you must,

Lay a Ruler to V, the *Pole* of the Circle Z C D N, and the Point C, it will cut the *Primitive Circle* in *m*, set 90 Deg. from *m*, and it will reach to *n*: A Ruler laid from V to *n*, will cross the Circle Z D N in the Point X. —Again, Lay a Ruler from T, the *Pole* of the Circle P B S, to C, the Angular Point, and it will cut the *Primitive Circle* in *o*, set 90 Deg. from *o* to *p*: Then a Ruler laid from T to *p*, will cross the Circle P B S, in the Point Y. —*Lastly*, A Ruler laid from C, to X and Y, will cut the *Primitive Circle* in *r* and *s*, so the Distance between *r* and *s*, measured upon the Scale of *Chords*, will give 49 Deg. 46 Min. for the Quantity of the *Angle* Y C X, which is equal to the *Angle* Z C P enquired. And is the *Angle* of the *Sun's Position* at the time of the *Question*.

The Canon for Calculation.

As *s*. Z C	:	*s*. Z P C	::	*s*. Z P	:	*s*. Z C P.
30 d. 7 m.	:	30 d.	::	50 d.	:	49 d. 46 m.

To find the Vertical Angle C Z P.

This *Angle* is the *Sun's Azimuth* from the *North* Part of the *Meridian* Z O N, whose Measure is the Arch of the *Horizon* D A O; and to find the Quantity of it, lay a Ruler from Z, the *Pole* of the *Horizon*, to D, it will cut the *Primitive Circle* in *d*; so the Quantity of the Arch *d* N O, measured upon the Scale of *Chords*, will be found to be 113 Deg. 57 Min. And such is the Sun's *Azimuth* from the *North* Part of the *Meridian*: The Arch *d* H, 66 Deg. 3 Min. is the Sun's *Azimuth* from the *South*; and the Arch *d* N, 23 Deg. 57 Min. is the Sun's *Azimuth* from the *East* and *West*.

The Canon for Calculation

V. As *s.* Z C · *s.* Z P C :: *s.* C P · *s.* C Z P.
30 d. 7 m. 30 d. :: *66* d. 30 m. *66* d. 3 m.
(Or, 113 d. 57 m.

And thus are all the *Sides* and *Angles* of this *Oblique-angled Triangle* alſo Meaſured : And ſo may any other, being thus *Projected.*

And, to conclude, This *Projective* Way will give great *Light* to *Calculation*; for by the true delineating of your *Triangle* upon the *Projection*, you ſhall thereby diſcover whether your *Side* or *Angle* be *More* or *Leſs* than a *Quadrant*, and ſo be poſitive in your *Reſolution*; which otherwiſe you muſt have *Given*, or render a double *Solution:* As in this laſt *Angle* C Z P, whether it were *66* Deg. 3 Min. Or 113 Deg. 57 Min. The ſame *Sine* in the *Canon* anſwering to both.

JANUA

114

Fig.
XLIV.

JANUA MATHEMATICA.

SECTION IV.

Of Spherical Trigonometry, *Instrumentally Explained and Performed.*

THE *Explanatory Instrument* here described, is for the better Information of the Fancy, by Speculation; and it is deduced from the *Catholick*, or *Universal Proposition*, before treated of in *Part* II. *Sect.* II. *Chap.* III. of this Book. Notwithstanding, for the Convenience of the Reader, I shall here, again, insert it.

Proposition Universal.

The Sine *of the* Middle Part, *and the* Radius, *are* Reciprocally Proportional, *with the* Tangents *of the* Extream Parts Conjunct, *and with the* Co-sines *of the* Extreams Disjunct.

In every *Right-angled Spherical Triangle*, there are *Five Parts*, besides the *Right Angle*, and they are called *CIRCULAR PARTS*: Of which, those *Three* which lye most remote from the *Right Angle*, (as the *Hypotenuse*, the *Angle* at the *Perpendicular*, and the *Angle* at the *Base*) are noted by their *COMPLEMENTS*.

Of these *Five Circular Parts*, any *Two* of them (besides the *Right Angle*) being given, a *Third* may be found. And,

Of *Three Parts*, (Two given, and One required) *One* must (necessarily) be in the *Middle*, and must be called the *MIDDLE PART*.

Of the other *Two Extream Parts*, they must either *Join* to the *Middle Part*, or be *Separate from it*.

If they be	*Joined* to it,	then are they called *Extreams*	*Conjunct.*
	Separate from it,		*Disjunct.*

 If

If the *Extreams* be *Conjunct*, the *Proportion* must be performed by *Sines* and *Tangents* jointly.—But,

If the *Extreams* be *Disjunct*, the *Proportion* may be performed by *Sines* only.

And in both *Cases*,

If the *Part Sought* (whether *Side* or *Angle*) fall out to be the *Middle Part*, then the *Radius* must be the *First* (or leading) *Term* in the *Proportion*.—But,

If one of the *Extreams* (whether *Conjunct* or *Disjunct*) possess the *Middle Part*, then the other *Extream* must be the *First Term* in the *Proportion*.

Of the Instrument.

THis *Instrument* was the Coutrivance (many Years since) of the Right Worshipful Sir *Charles Scarborow*, M. D.: and divers of them have been made in *Silver* and *Brass* about the Bigness of a *Crown Piece*, with *Verses* in *Latin* about the Rimb, for the better bringing the *Rules* to be observed, in the Use of it, to Memory, which were to the same Effect with those which I have even now laid down in English.

The *Instruments* consists of Two *Parts*, the undermost is a round Piece, divided into *Five* Equal Parts, which represent the *Five Circular Parts* before spoken of, it hath Two Circular Margins, near the outer Edge thereof; in One of the Five Parts is engraven *Middle Part*, and under it the Word *Sine*. —On either Side of *Middle Part*, there is written or engraven *Extream Conjunct*; and under them, the Word *Tangent*. —In the Two other of the *Five* Divisions, which are opposite to *Middle Part*, are engraven *Extream Disjunct*, and under them, *Co-sine*. All which Words answer directly to the Words of the *Universal Proposition*: And this is all that is upon the Under Plate of the *Instrument*.

The Upper Plate of the *Instrument* is a Circle also, divided into Five Equal Parts as the former, and within that another Concentrick Circle, between which are Five Lines drawn from the Centre thereof. —Within the innermost of these Circles, there is described a *Right-angled Spherical Triangle*, the Five Circular Parts whereof do lye directly against the Five Lines, drawn from the Centre between the Two Circles: And upon those Three Lines which lye against the *Hypotenuse*, the *Angle* at the *Perpendicular*, and the *Angle* at the *Base*, is written,

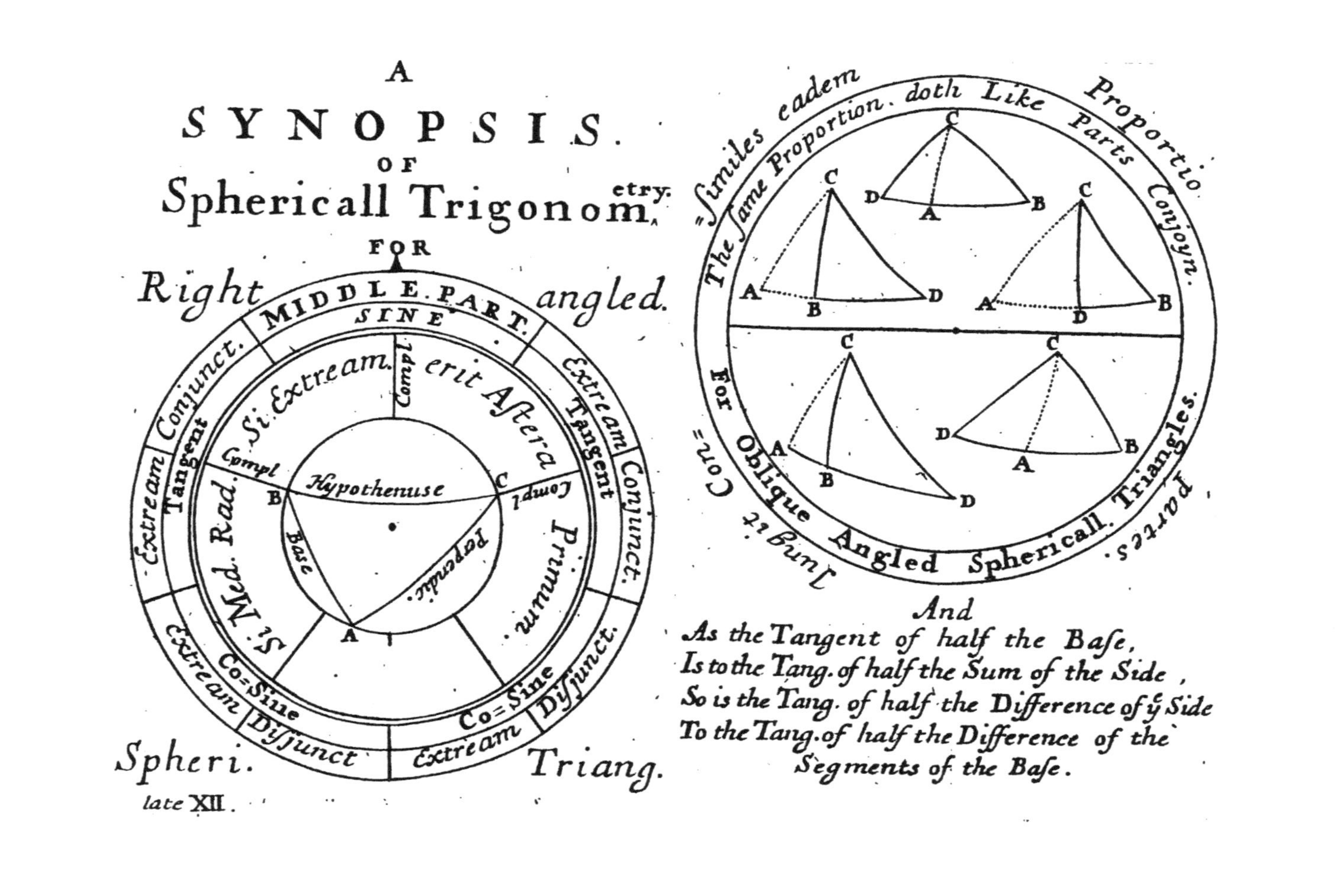
A
SYNOPSIS.
OF
Sphericall Trigonometry.
FOR
Right angled Spheri. Triang.
MIDDLE PART.
SINE
Si. Extream. erit Æstera
Si. Med. Rad.
Primum.
Compl
Hypothenuse
Base
Perpendic.
Extream Conjunct.
Tangent
Extream Disjunct
Co=Sine
Conjunct.
Disjunct.
A
B
C
D
Similes eadem Proportio Jungit Con= Partes.
The same Proportion. doth Like Parts Conjoyn.
For Oblique Angled Sphericall Triangles.
And
As the Tangent of half the Base,
Is to the Tang. of half the Sum of the Side,
So is the Tang. of half the Difference of ye Side
To the Tang. of half the Difference of the Segments of the Base.
late XII.

ten, or engraven, *Complement*; but upon the other Two Lines, which issue from the *Perpendicular*, and the Base, there is nothing written. —This Upper Plate is to move upon the Under Plate by a Rivet (or such like) through the Centers of both Plates: And so is your *Instrument* finished.

The Use of the Instrument.

THE Use of the Instrument is principally to give you by Inspection the *Order* of the *Terms* of the *Proportion* in any *Case*.

Example. In the *Triangle* A B C, suppose there were given the Perpendicular C A, and the *Angle* at the Base B, to find the *Angle* at the Perpendicular C. Fig. XLV.

Here it is evident, that the *Angle* at the Base B, is the Middle Part: Turn the Rundle about till you bring the *Angle* at the Base against Middle Part: Then shall you find that the Perpendicular C A, and the *Angle* at the Perpendicular C, will stand against *Extreams Disjunct*; which tells you, that C A and C, are *Extreams Disjunct*: And now (the Rundle thus resting) you see that One of the *Extreams*, as C, is sought; therefore, the other *Extream* C A must be the first *Term* in the Proportion, against which you find *Co-sine*. Wherefore say,

As the Co-sine of the Perpendicular A C,
Is to the Radius:
So is the Middle Part the Angle at the Base B, (against which stands *Sine* and *Complement*, that is *Co-sine* B.)
To the *Angle* at the Perpendicular C, against which stands *Co-sine* and *Complement*; that is *Sine*; for, *Co-sine Complement* is the *Sine* it self.

And so the Proportion in short is this:

As *c s*. C A : to Radius : : So *c s*. B : to *s*. C.

Another Example.

In this *Triangle* A B C let there be given the *Hypotenuse* C B, and the *Angle* at the Perpendicular C, to find the Base B A. Fig. XLVI.

It is evident that B A is the *Middle Part*. Bring B A against the *Middle Part*, then will C B and C stand against *Extreams Disjunct*: And (because the *Middle Part*) B A is sought, the *Radius* must be the *First Term* in the Proportion, and the Two *Extreams* the *Second* and *Third*, against both which stands *Comp. Cosine*, (which is *Sine*); and against B A the Middle Part

Part ſought, there ſtands *Sine* alſo: So that your Proportion will run thus:

As Radius,
To the Sine of the *Angle* the Perpendicular C:
So is the Sine of the *Hypotenuſe* C B,
To the Sine of the Baſe B A.

A Third Example.

Fig. XLVII. In this *Triangle* A B C, let there be given the Perpendicular C A, and the *Angle* at the Baſe B, to find the Baſe B A.

It is here evident that A B is the *Middle Part*: Turn the *Triangle* about till the Baſe A B lye againſt the *Middle Part*; then will the Perpendicular C A, and the *Angle* at the Baſe B, lye againſt the *Extreams Conjunct*; and ſeeing the *Middle Part* is ſought, the Radius therefore muſt be the *Firſt Term* in the Proportion: And becauſe the *Extreams* are *Disjunct*, the Proportion will be in *Sines* and *Tangents* jointly, as by the *Inſtrument* appears: And the Proportion will be

As the Radius,
Is to the *Tangent* of the Perpendicular C A:
So is the *Tangent* Complement of B,
To the Sine of the Baſe B A.

But if the ſame Things were given, and the *Angle* at C, the Perpendicular, had been required, then B had been the *Middle Part*, and C A and B, the *Extreams Disjunct*, and the Proportion in *Sines* only. As thus:

As the Co ſine of C A,
To Radius:
So is the Co-ſine of B,
To Co-ſine Comp. (that is, to *Sine*) of C.

And thus, by this *Inſtrument*, may the Proportions for the Solution of any *Right-angled Spherical Triangle* be readily ſet down, and the *Triangle* reſolved in any of the XVI. *Caſes*.

On the Back-ſide of this *Inſtrument*, there is another Contrivance of the forementioned Gentleman's, with *Latin Verſes* for bringing the *Rules* for the Solution of *Oblique-angled Spherical Triangles* to Memory.

I ſhall not here give you any Account of it more than the *Figure*, and the *Latin Verſes* (or *Rules*) in Engliſh: Referring you, for the farther underſtanding of it, to what is delivered in the IVth Chapter hereof, Page 68. *Of*

Of the Solution of Spherical Triangles, *by* PLANISPHERE.

A *Planisphere*, is a *Projection* of the *Sphere* (or *Globe*) in *Plano:* Of which, there are principally Two; One called *Stereographical*, projected by *Circles*; The other *Orthographical*, projected by *Ellipses*, and generally known by the Name of *Analemma*.

Fig. XLVIII.

To either of which *Planispheres*, there belongs proper *Indexes*, to move upon the Centres of the *Planispheres*. As to the *Stereographical*, an *Index* having Two *Legs* as a *Sector*; both which are to be divided, according to the *Tangents* of *half Arks*, as the Semidiametre of the *Planisphere* it self is divided, and must be numbred by 10, 20, 30, &c. to 90 Deg. both Ways, and on both *Legs*.

To the *Analemma*, or *Orthographical* Projection, there must be an *Index* of the whole Length of the Diametre thereof, to move about upon the Centre of the Projection, which must be divided as the Diametre of the *Planisphere* is; namely, as a *Scale* of *Natural Sines*, and must be numbred both Ways, from the *Centre*, by 10, 20, 30, &c. to 90 Deg. —Upon this *Index*, (by help of a *Groove* made through the former *Index*) another *Index* is to be made to move upon, and with, the former, and always keeping at *Right Angles* with it: And this *Index* is to be divided as a *Scale of Sines*, as one half of the other; and so numbred, by 10, 20, 30, &c. to 90, both Ways: And this second *Index* (in the Use of this *Planisphere*) I shall call the *Cursor*. And thus much for the Descriptions.

Concerning their Use, I shall only lay down such *General Rules* as are necessary for the counting of the Quantities of the *Sides* and *Angles* of *Spherical Triangles* upon the several *Planispheres* and their *Indexes*; in all *Cases*, both of *Right and Oblique Triangles*. Not insisting upon particular *Examples*, for that throughout all this Book there are such Variety. All which (by these few *General Rules* here delivered) may be wrought upon either of these Two Projections.

Fig. LXVIII.

The Solution of Spherical Triangles, by the Stereographical Projection.

I. Of Right-angled Spherical Triangles.

1. TO retain the Method before observed in the 16 *Cases* of *Right-angled Spherical Triangles*; I will here also follow the same Order, wherein I shall note the *Triangle* to be resolved by the Letters A B C, setting A at the *Right Angle*, and B and C at the other Two *Acute Angles*. So shall the *Base* be A B; The *Cathetus*, (or *Perpendicular*) C A; The *Hypotenuse* B C. —The *Angle* at the *Base* B; and the *Angle* at the *Cathetus* C.

2. There are therefore, besides the *Right Angle*, Five *Circular Parts*; namely, Three *Sides*, and Two *Angles*: Of which, Four come into the *Account* at once; Two of them are given, and the other Two found out.

3. *If the Four Parts, which at once come into the Account, be the Three Sides, and One Angle.* —Let that *Acute Angle* be evermore noted with the Letter B; and the *Triangle* may be resolved thus:

Set the *Angle* B at the Centre, reckoning it upon the *Limb* from the *Diameter*; and the *Base* B A upon the *Diameter* from the *Centre*; and the *Cathetus* C A upon the *Great Circles* from the *Diameter*, by help of the *Parallels*; and the *Hypotenuse* B C upon the *Index* from the *Centre*. —*Note*, That if any of those *Accounts*, fall not just upon some Line in the *Instrument*, either of the *Great Circles* or *Parallels*, the Excess is to be estimated in *Minutes* or *Parts* of a Degree.

Example. *The* Perpendicular C A, *and the* Angle *at the* Base B, *given*; *To find*, (1.) *The* Base B A. (2.) *The* Hypotenuse C B. (3.) *The* Angle *at the* Cathetus C.

1. Reckon on the *Limb*, from the *Diameter*, the Quantity of the given *Angle* B; and to the End thereof, set either of the *Legs* of the *Index*.

2. Upon the *Limb*, from the *Diameter*, reckon the *Cathetus* C A; and from the end of that Arch, estimate reasonably, a *Parallel Circle*, till it meet with the *Index*: For that Point of Intersection shall shew both the *Hypotenuse* B C, upon the *Index*: And the *Base* B A, upon the *Great Circle*, meeting also in that Point. —Then, to find the *Angle* at C: Take C A for the

the *Base*, and C A for the *Cathetus*; and to the end of that *Cathetus* apply the *Index*; so shall you have the *Angle* sought for C, upon the *Limb*. Fig. XLVIII.

3. Otherwise, you may set the *Angle* B at the *Pole*, reckoned upon the *Diameter* from the *Limb*; and B A upon the *Limb* from the *Pole*; and C A upon the *Index* from the *Limb*; and B C upon a *Great Circle* from the *Pole*.

4. But if the *Four Parts* which at once come into the Account, be the Two *Oblique Angles*, C and B, and the Two *Right Sides* C A and B A, then the *Triangle* may be resolved thus.

1. Reckon the *Lesser Angle* on the *Index* from the *Centre*, and the *Greater Angle* on a *Great Circle* from the *Pole*: So both of these, with the *Axis*, shall include a *Quadrantal Triangle*: And the *Greater Right Side* shall be on the *Limb* from the *Pole*: And the *Lesser Right Side* shall be upon the *Diameter* from the *Centre*.

Lastly, Two of the Four Parts being had, there will be no Difficulty in finding out of the *Hypotenuse*.

II. *Of* Oblique-angled Spherical Triangles.

1. In an *Oblique-angled Spherical Triangle* B C D, Five of the *Circular Parts* come into the Account at once; namely, *One Side* B D; and the *Two Oblique Angles* B and D, and the Two other *Sides* D C and B C: The third *Angle* C, opposite to B D, is not here enquired, but only the *Place* of the *Angular Point* C, both for D C and B C.

In the first of the Three *Schemes*, the inner *Angle* D is *Obtuse*, and B *Acute*: In the second the inner *Angle* D is *Acute*, and B *Obtuse*: In the third, both the inner *Angles* D and B are *Acute*. Fig. XLIX.

2. The *Side* B D, is evermore understood to be *Given*: And in every *Operation*, the first thing to be done, is to open the *Legs* of *Index* to the Wideness of B D, and thereto screw them fast: Also I call the Two *Legs* of the *Index*, the *Leg* or *Index* B, and the *Leg* or *Index* D, as they are noted with those Letters.

The *Angles* D and B, of every *Triangle* proposed, are reckoned upon the *Diameter* from the *Limb*, by the *Great Circles*; namely, the *Angle* B, and the *Conjunct* of the *Angle* D, from ♋; but the *Angle* D, and the *Conjunct* of the *Angle* B, from ♈: And the *Sides* D C and B C, are to be reckoned by the

Fig. XLIX. the *Parallels*; namely, the *Sides* themselves, from the *Pole*; or their Complements and Excesses, from the *Diameter*.

Note, That of a *Triangle*, the *Conjunct Angle*, or the *Outer Angle*, is all one.

4. *Two* Sides *of an* Oblique-angled Spherical Triangle, *whereof one is* B D, *with the* Angle *intercepted, being given: To find out, at one Work, the* Third Side; *and the other* Angle *at* B D.

There are these Two Varieties

Given,	D C and D B C and B	to find out	B C and B. D C and D.

5. The *Rule*. If the *Angles* D and B, be reckoned in the *Upper Semicircle* of the *Planisphere*, at the *Pole* N, and the *Angle* D be *Obtuse*, or B *Acute*; the Point C shall be taken in the former *Semicircle* towards ♋: But, if the *Angle* D be *Acute*, and B *Obtuse*, the *Point* C shall be taken in the *latter Semicircle* towards ♑; and contrariwise, for the lower *Pole* S. —Then, upon the *Diameter*, count the *Angle* given, D or B, either by it self, or by its Conjunct, according as was taught in *Sect.* 3. And from the end of that reckoning, imagine an Arch of a *Great Circle* aptly traced, until it meet with the *Parallel* of the Complement of the respective *Side* given. D C or B C, for that Point of *Concurrence* shall be the Point C, for the given *Side*, either D C or B C. To this Point C, thus found, apply the *Index* noted with the proper *Letter* of the *Angle* given, D or B; and mark the Distance of it from the *Centre*; for this Distance being exactly taken on the other *Index*, shall in the *Planisphere* give the true Place of the Point sought: Which being aptly estimated, will, upon the *Diameter*, shew the *Angle* sought; either B, or the Conjunct D; from ♋; or else, the Conjunct of B, or the *Angle* D, from ♑. And, it will also shew upon the *Limb*, the *Complement* of the *Side* sought.

By what hath been before shewed, (that the Point C, both of the *Sides* D C and B C, are evermore equally distant from the *Centre*,) there is opened a Way, whereby *Any Two of Five Circular Parts*; *together with the* Side D B, *being given, the other Two may be found at one Operation.*

Herein

Herein are Four Varieties.

Fig. XLXI.

Given,		to find out	
	D C and B C		D and B.
	D C and B		D and B C.
	D and B C		D C and B.
	D and B		D C and B C.

The RULE.

In the *Upper Semicircle* of the *Planisphere*, if the *Angle* D be *Obtuse*: Or if the *Side* D C be much *Lesser* than B C, the Point C shall be taken in the *former Quadrant* towards ♋: But, if D be *Acute*, or D C much *Greater* than B C; the Point C shall be taken in the *latter Quadrant* towards ♑. And, contrariwise, in the *Under Semicircle*. —And in both, if the *Angles* D and B be *Acute*, or if the *Sides* D C and B C, differ but little in length, the true Point sought, shall fall near the *Axis*, towards ♑, within a Space, equal to the *Index* opened at the Wideness of B D.

Set, therefore, the *Index* thus opened, so that the Two Points C, may fall within it. And among the Great Circles and Parallels, or both (according as Reason shall direct) reckon the Two Circular Parts given, tracing them proportionally, with a Pin in each Hand, until they shall both exactly meet with their proper *Legs* of the *Index*, at an equal Distance from the Centre, and as near to it as possibly may be: So have you upon the *Planisphere*, the true Places of both the Points C; and thereby, the Two Circular Parts sought, respectively, to each Point C; both of D C, and B C.

And note farther, That if any Point C, falleth, either among the Parallels, near the *Diameter*; or among the *Great Circles*, near the *Limb*; where they be almost equidistant, there may be some Uncertainty in finding the true Points exactly: The remedying whereof requireth the more Diligence in the *Computer*; but this may be remedied, by reducing the *Oblique Triangle* into Two *Right-angled*.

Other Inconveniencies may arise in the Use of this, as in all other *Instrumental Operations*; for the remedying whereof, *Usus optimus Magister*.

The

Fig. XLIX. *The Solution of* Spherical Triangles, *by the* Orthographical Projection, *or* Analemma.

I. *Of* Right-angled Spherical Triangles.

1. THE *Hypotenuse* is, always, represented upon the *Index*.

2. The *Right Angle*, at the Section of any *Meridian*, and the *Æquinoctial*.

3. One *Leg* in the *Meridian*, and the other in the *Æquinoctial*.

4. One *Angle* only entering the Question, is represented at the Centre, between the *Æquinoctial* and the *Index*, and is numbered in the *Limb*.

5. But the Two *Angles* entering the Question, you must turn the *Case* into an *Angle*, and its opposite *Leg*. For in the *Triangle* A B C, whose *Right Angle* is A; the other Two *Angles* A B C, and A C B, both entering the Question, you must lengthen the *Hypotenuse* to a full *Quadrant* to E; as also, the *Legs* adjacent to that Extremity of the *Hypotenuse*, from which it was lengthned. Thus in the *Scheme*, the *Leg* A C, adjacent to C, from which Extremity the *Hypotenuse* was lengthened: And so, in the *Triangle* C D E, *Right-angled* at E: The *Angle* D C E is equal the *Angle* B C A; and the *Leg* D E, equal to he *Complement* of the other *Angle* C B A. (Fig. L.)

II. *Of* Oblique-angled Spherical Triangles.

1. *Three* Sides *being given; To find an* Angle *opposite to any of them.*

Reckon the *Greatest Leg* from the Pole upon the *Limb*, and where it endeth, apply the *Index*; upon the *Index* reckon, from the *Limb*, the *Base*; (that is, the *Side* opposite to the *Angle* sought:) And to the Point of the *Index*, where this *Base* endeth; apply (or bring to) the *Cursor*: Then look where the *Cursor* cutteth the Parallel of the *Lesser Leg*, to be reckoned from the Pole, and observe what *Meridian* passeth by this Section of the *Cursor* and Parallel; for, where this *Meridian* cutteth the *Æquinoctial*, there you have the *Measure* of the *Angle* sought; to be accounted from thence to the *Limb*.

2. *Three* Angles *given, to find a* Side.

Turn the *Angles* into *Sides*, and deal with them, as with the *Sides*.

3. The

3. *The* Parts Given *and* Sought, *being altogether* Opposite. Fig. I.

Reckon the *Greater* of the Two first *Terms* upon the *Index*, and where this Number endeth upon the *Index*, apply that Point to the lesser of the said Two first *Terms*; which Parallel is to be reckoned from the *Æquinoctial*; then order the *Terms*, reckoning the *First* and *Third* both upon the *Index*; or both upon the *Parallel*; and so likewise do with the *Second* and *Fourth* Terms.

4. *Two* Sides, *with an* Angle *between them Given: To find the Third* Side.

Reckon the *Greater Side* given, from the *Pole* upon the *Limb*; and to the end of it, apply the *Index*. Reckon the other *Side Given* upon the *Parallels*, from the *Pole*: And the *Angle* given, upon the *Æquinoctial*, from the *Limb*; and the *Meridian* it comes to, pursue till it comes to the *Parallel* of the *Lesser Leg*: And to this *Section* of the said *Meridian* and *Parallel*, apply the *Cursor*, so it may justly lye on the *Index*, and mark what Point of the *Index* it pointeth out: For, this Point of the *Index*, counted from the *Limb*; is the *Third Side* required. And after the *Third Side* is found, you may find either of the Two unknown *Angles*, by the *Rate* for *Opposite Parts*.

5. *Two* Sides, *and an* Angle *opposite to the* Lesser *of them*, Given *To find the* Third Side.

Reckon the *Difference* of the given *Sides*, and the *Sum* of them (one and the same Way) from the *Pole*; and mark the Points where both of them do end. Count also, the *Angle* given, upon the *Æquinoctial* from the *Limb*, and mark what *Meridian* it cometh to. Then extending a strait Line (or applying a strait Ruler) between the Two Points marked in the *Limb*; it will cut the said *Meridian* in Two Places: So you are to observe the *Parallels*, where the strait Line (or Ruler) cutteth the said *Meridian*; for, these being reckoned from the *Pole*, will give you the *Quantity* of the *Third Side* required: For, this same *Third Side*, may be of a Twofold *Quantity*: The *Lesser* it is, when the *Angle* opposite to the *Greater* given *Side* is *Obtuse*, and the *Greater* it is, when the said *Angle* (opposite to the *Greater* given *Side*) is *Acute*.

6. Two

Fig. L. 6. *Two* Sides, *and an* Angle *opposite to the* Greater *of them Given*; *To find the third* Side.

Reckon the *Difference* of the given *Sides* from the *Pole*, one way; and the *Sum* of the same, the other way. Count also, the given *Angle* upon the *Æquinoctial*: And extend the strait Line (or lay a Ruler) cutting the *Meridian*, as in the last; for now it will cut but once; and so the *Third Side* will admit only of one single Answer.

The End of the Second Part.

POSTSCRIPT.

MAny are the Ways by which *Plain* and *Spherical Triangles* may be both *Geometrically* and *Instrumentally* performed: As by *Scales* of *Natural Sines*, *Tangents*, *Secants*, and *Equal Parts*, by *Protraction*: Also, by *Scales* of *Artificial Numbers*, *Sines*, *Tangents*, and *Versed Sines*, as Mr. *Gunter* long time since contrived them, to be used with Compasses: And I have now lately contrived an *Instrument*, which I call *TRISSOTETRAS*, which printed on a large Sheet of Paper, and pasted upon a Board, *all Triangles*, both *Plain* and *Spherical*, may be *Resolved* by Inspection, without *Pen*, *Compasses*, *opening Joints*, or other Moveable; save only the Extension of a *Thread*, or thin *Streight Ruler*, upon the Instrument; the Description and Use whereof I may hereafter publish by it self. But, the best and most absolute Way of *Resolving Triangles*; and to what Uses soever they be applied, is by the *Canons* of *Artificial Sines*, *Tangents* and *Logarithms*; both Decimal and Sexaginary: And such a Canon was intended to be joined to this Book, at this time; but must be referred till farther Opportunity, which may be shortly.

ANCILLA

ANCILLA MATHEMATICA.

VEL,

Trigonometria Practica.

PART III.

WHEREIN

The Doctrine of PLAIN and SPHERICAL TRIANGLES is applied to Practice: In

Geometry. } And therein of { Longimetria, Planometria, Geodæcia; Steriometria, } Or Measuring of { Heights, Distances, } &c. Plains, Land, Solids.

Cosmography. { Geographical } As to the { Distances of Places on Land or Sea. Astronomical } { Motion of the Sun, and Fixed Stars.

Sciographia: } or { Dialling, { By Calculation: After a new Method.

Navigation: } Or Sailing by { The Plain } Chart. Mercators } Middle Latitude.

By *William Leybourn*, Philomathemat.

LONDON:
Printed *Anno Domini*, M.DCC.IV.

ANCILLA MATHEMATICA.

VEL,

Trigonometria Practica.

SECTION I.

OF

GEOMETRY.

IN this *Section* of *GEOMETRY*, I shall treat only of such *Practical Parts* thereof, as the *Doctrine* of *Plain* (or *Right-lined*) *Triangles*, (both for their *Illustration* and *Demonstration*) becomes subservient: As,

I. In *ALTIMETRIA*: By which the *Height* of any *Object* (accessible or inaccessible) may be obtained; As of *Towers*, *Steeples*, *Trees*, &c.

II. In *LONGIMETRIA*: By which the *Distance* of one *Object* from any Place, or of many *Objects* one from another, (whether approachable, or in-approachable) may be known, their true Positions laid down, and a Map made of them.

III. In *PLANOMETRIA*: By which all Kinds of *Superficies*, (Regular or Irregular) as *Plains*, *Land*, &c. may be *Measured*.

CHAP. I.

Of ALTIMETRIA.

I. *Of an* Altitude *that is Accessible.*

Fig. I. SUppose A C to be a *Tower*, *Steeple*, or other upright *Object*, and you standing at B, were required to tell the Height thereof. —First, Measure the Distance from B, to the Foot of the *Object* A, which suppose to be 432.5 Feet. —Secondly, At B, (by a *Quadrant*, or other *Graduated Instrument*) look to the top of the *Object* at C, where we will suppose the Degrees found by the *Instrument* to be 32.25 deg. —By this *Observation*, the *Distance Measured*, and the *Object*, you have a *Right-angled Triangle* constituted; in which, there is *Given*, (1.) A B the *Distance Measured* 432.5 Foot. (2.) The *Angle* at B, observed by your Instrument, 32.25 deg. And, (3.) The *Right Angle* at A: To find the Leg C A, which will be the *Height* of the *Object*: (By *CASE* II. of *R*, *A*, *P*, *T*,) thus:

As Radius, Tangent 45 deg.
Is to the *Distance Measured* A B, 432.5 Feet;
So is the Tan. of the *Angle* observed by Instrument at B, 32.25 d.
To the Height of the *Object* C A, 272.89 Feet.

And thus having found the Height of the *Object* to be 272.89 F. you may find the Length of the *Visual Line* (which is the *Hypotenuse*) C B, (by *CASE* V. of *R*, *S*, *P*, *T*.) For,

As the Sine of the *Angle Observed*, B, 32.25 deg.
Is to the *Altitude* of the *Object*, 272.89 Foot;
So is the *Radius*, Sine 90 deg.
To the *Length* of the *Visual Line* C B, 511.39 Foot.

II. *Of an* Altitude *Un-accessible.*

Fig. II. Suppose D E to be an *Object*, as *Steeple*, *Tower*, or the like: And that you standing at G, were required to know the Height thereof; but (by reason of some broad Moat, or other Impediment, you cannot come to measure from G to E. In this Case, —First measure from G, towards E, as far as conveniently you can, suppose to F, 95.25 F. —Then making Observation at G, you

you find the Degrees of your *Quadrant* to be 52.50 d. and *Observing* at F, you find the Degrees to be 63.25 Now from this *Distance Measured*, and the Two *Observations* by the *Instrument* made, the *Altitude* D E may be attained unto by *Trigonometrical Calculation*, thus: Fig. II.

First, Upon a Sheet of Paper, draw a Line at Pleasure, as H K, upon which assume a Point for the Place of your first standing, as at G, and upon G, protract the *Angle* observed 52.50 d. drawing a Line through those Degrees at Liberty.

Secondly, By help of a Scale, set your *Measured Distance* 95.25 F. from G to F; and upon F, by a *Scale* of *Chords*, protract an *Angle* of 63.25 d. as you observed them to be; and through them draw another Line at Pleasure, which will cut the former Line drawn, in the Point D, which will represent the top of the *Object* to be measured; and a *Perpendicular* let fall from D, upon the Ground-line H K, will fall in the Point E, and so will the Line D E represent the *Object* it self.

Thirdly, By these Lines thus drawn, you will have constituted Two *Triangles*; one D E F, *Right-angled* at E; and the other D F G, *Obtuse-angled* at F; by the resolving of which, the height of the *Inaccessible Object* D E, will be found. For,

1. In the *Oblique-angled Triangle* D F G, you have given, the *Side* F G (which was the *Measured Distance*) 95.25 F. and the observed *Angle* D G F, 52.50 d. —And the *Angle* observed at F being 63.25 d. the Complement thereof to 180 d. *viz.* 116.75 d. is the Quantity of the *Obtuse Angle* D F G, so have you in the *Oblique-angled Triangle* D F G, Two *Angles* given; the Sum of which, *viz.* 169.25 d. taken from 180 d. there remains 10.75 d. for the *Angle* F D G: And now in the *Triangle* D F G you have given Two *Angles* F D G, and D G F, with the *Side* F G, opposite to F D G; whereby, you may find the *Side* D F, (by *Ax.* II.) For,

As the *Sine* of F D G, 10.75 d.
Is to the *Side* F G, 95.25 Foot.
So is the *Sine* of F G D, 52.50 d.
To the *Side* D F, 405.14.

2. In the *Right-angled Triangle* D E F, (having found the *Hypotenuse* as before) you have given, (1.) The *Observed Angle* at F, 63.25 d. (2.) The *Hypotenuse*, last found, D F 405.14 Foot; by which you may find D E, the *Height* of the *Object*, (by *Case* IV. of *R, A, P, T,*) thus:

As

Fig. II. As the Radius, *Sine* 90 d.
Is to the *Hypotenuse* D F, 405.14 F.
So is the *Sine* of D F E, (the *Angle observed* at F) 63.25 d.
To the Leg. D E, 367.77 Foot: Which is the Height of the Object.

And now (if you please) you may find the *Visual* Line G D, (by *Case* V. of *R, A, P, T,*) thus:

As the *Sine* of the *Angle* at G, 52.50 d.
Is to the height of the *Object* D E, 361.77 F.
So is Radius, *Sine* 90 d.
To the *Length* of the *Visual Line* D G 456.00 F.

And also, the Distance E F, (by *Case* II. of *R, A, P, T,*) thus:

As the *Sine* of the *Angle* observed at F. 63.25 d.
Is to the *Height* of the *Object*, D E, 361.77 F.
So is the *Co-sine* of the *Angle* at F, 63.75 d.
To the *Distance* E F, 182.35 Foot.

Unto which, if you add the *Distance* F G, 95.25 Foot, their *Sum* will be 277.6 Foot. And that is the whole *Distance* from G to E, the Foot of the *Object*.

III. *Of the* Altitude *of an* Object *standing upon a* Hill, *Un-accessible.*

Fig. III. Suppose M O to be such an *Object*; and you standing at L, were required to tell the *Height* thereof.

First, Upon Paper, or the like, draw a Right-line at Pleasure, as Q R; and therein, assume any Point, at Pleasure, for the Place of your standing, as L; where, with your Instrument directed to the top of the *Object*, you find the Degrees cut, to be 40.52 d. and directed to the bottom of the *Object* at O, the Degrees cut, to be 22.25 d. Wherefore upon L, protract an *Angle* of 40.52 d. and draw a Line L, at Pleasure: And also *b*, an *Angle* of 22.25 d. nd draw the Line L *c* at Pleasure.

Secondly, Go forwards, in a Right-line towards the *Object*, some considerable Distance, as to N, 212.5 Foot; and there, by your Instrument directed to the top of the *Object* at M, you find the Degrees cut, to be 61.82 d. through which draw a Line at Pleasure, as N *a*, crossing the Line L *b* in the Point M, which is the top of the *Object*: From whence, a *Perpendicular* let fall upon the Ground.

Ground-line Q R, as M P, that Line shall be equal to the *Altitude*, of the *Object*, and the Hill together. Fig. III.

Now, by the Intersections of these Four *Visual Lines*, L M, L O, N M, and N O, there are constituted Four *Right-lined Triangles*, viz. L M P, and N O P, both *Right-angled* at P: And L M N, and N M O, *Oblique-angled*: By the resolving of which, from the *Distance Measured*, L N, and the several *Angles observed*, at L and N, the required *Altitude* may be obtained. For,

1. In the *Oblique-angled Triangle* L M N, there is given, (1.) The *Angle* M L N, 40.52 d. (2.) The *Angle* L N M, 118.18 d. (it being the Complement of the *Angle* O N P 61.82 d. to 180 d.) (3.) The *Side Measured* L N, 212 5 Foot: And having the *Angles* at L and N, the Sum of them 158.70 d. taken from 180 d. there will remain 21.30 d. for the *Angle* L M N. From which Things given, the Two other *Sides*, L M and M N, may be found, by *Axiom* II. thus:

As the *Sine* of L M N, 21.30 d.
Is to the *Side* L N, 212.5 Foot;
So is the *Sine* of M L N, 40.52 d.
To the *Side* M N, 380.08 Foot

And so is the *Sine* of 61.82 d. (the Complement of M N L, to 180 d.)
To the *Side* L M. 515.66

In the *Right-angled Triangle* N M P, there is given, (1) The *Hypotenuse* M N, 380.08 Foot. (2.) The *Angle* M N P, 61.82 d. whereby you may find N P, (by *Case* IV. of *R, A, P, T*,) thus:

As the Radius, Sine 90 d.
Is to the *Hypotenuse* M N, 380.08 Foot;
So is the *Sine* of the *Angle* M N P, 61.82 d.
To the *Side* M P, 335.03 Foot:

Which is the *Height* of the *Object* and the *Hill* together.

And so is the *Co-sine* of M N P, *viz.* N M P, 28.18 d.
To to *Leg.* or *Side* N P, 179.49 Foot.
To which, if you add the measured Distance L N, 212.05, their Sum will be 391.54, for the whole length L P. Then,

3. In the *Triangle* L O P, (*Right-angled* at P) you have given, (1.) The *Side* L P, 391.54 Foot. (2.) The *Angle* O L P, 22.25 d. whereby

Fig. III. whereby you may find O P, (By *Case* II. of *R, A, P, T,*) thus:

As the Radius, Tang. 45 d.
  Is to the *Side* L P, 391.54 Foot,
So is the *Tangent* of the *Angle* O L P, 22.25 d.
  To the Height of the *Hill* O P, 160.18 Foot.

Which substracted from M P (the whole Height) 335.93, there remains 175.75 Foot, for the *Altitude* of the *Object* M O.

IV. *How the* Altitude *of the* Sun *may be taken, by the* Shadow *of a* Staff, *or other* Object, *of a known Length.*

Fig. IV. Upon A B, being a level Plain, let there be erected a streight *Staff*, or the like, of any Length, suppose 60.00 Inches, or 5 Foot) as C D; and the *Sun* shining, suppose it casts the *Shadow* thereof upon the plain Ground to E; which measured, suppose to contain 108 Inches, or 9 Foot.

Draw a Line at Pleasure, as A B; upon any Point thereof, as D, erect a *Perpendicular*, upon which set 60 Inches, the length of your Staff, from D to C; and the length of the Shadow thereof 108 Inches, from D to E; drawing the Line of *Umbrago* C E, and thus have you constituted a *Right-angled Plain Triangle* C D E, in which there is given, (1.) The Leg. C D, 60 Inches. (2) The Leg. D E 108 Inches, by which you may find the *Angle* C E D, (by *Case* I. of *R, A, S, T,*) thus:

As the Length of the *Shadow* C D 108.00 Inches,
  Is to the Radius, *Tang.* 45 d.
So is the Length of the *Staff* C D, 60.00 Inches,
  To the *Tangent* of the C E D, 29.05 d.
  And such is the *Sun's Altitude* at that time.

V. *How the* Height *of an* Accessible Object *may be obtained, by the* Length *of the* Shadow *of it.*

Fig. V. Suppose that the Sun should cast the Shadow of some upright *Object*, as F G, 84.5 Foot, from G to L. and at the same time, your Staff of 5 Foot, does cast its Shadow from L to K, 6.32 Foot: And from hence I would compute the *Altitude* of the *Object* F G.

The

The Two Lines of the *Object* F G, and *Staff* H L, together with the Two Lines of *Shadow*, G L and L K, being laid down, do constitute Two *Right-angled Plain Triangles*, viz. F G L, and H L K, *Equiangled*, and their *Sides Parallel*, and therefore *Proportional*; (by *Theorem* VIII. Lib. I.) And therefore, Fig. V.

As the Length of the *Shadow* of the *Staff* L K, 6.32 Foot,
Is to the Height of the *Staff* H L, 5.00 Foot;
So is the Length of the *Shadow* of the *Object* G L 84.5 Foot,
To the *Altitude* of the *Object* F G, 66.93 Foot.

CHAP. II.

Of LONGIMETRIA.

I. *How (standing upon an* Object *of a known Height) to find the* Distance *from thence, to some other remote* Object.

SUppose C A to be the Side of a *Fort* or *Bulwark* 22.5 Foot high, and being upon the *Platform* at C, you see a *Tree*, or other *Object* at B, whose *Distance* you would know, from the Foot of the Wall at A. Fig. VI.

The Lines A B and A C being drawn, and the Height of the Wall 22.5 Foot, set from A to C, where by your Instrument directed to B, you find the Degrees cut to be 71.25, which *Angle* lay down, so have you the *Right-angled Triangle* C A B, in which there is given, (1.) C A, the Height of the Wall 22.5 Foot. (2.) The *Angle* observed at C, 71.25 Deg. by which you may find the *Distance* A B, (by *Case* I. of *R, A, P, T.*) thus:

As Radius, Tangent 45 Deg.
To C A, the Height of the Wall 22.5 Foot;
So is the Tangent of A C B, the *Angle* observed, 71.25 Deg.
To the Distance A B, 66.28 Foot.

And if you would find the Length of the *Visual Line* C B, you may (by *Case* V. of *R, A, P, T,*) thus:

As the Sine of the *Angle* observed at C, 71.25 Deg.
Is to the Distance B A, 66.28 Foot;
So is the Radius, Sine 90 Deg.
To the *Visual Line* C B, 69.98 Foot.

 II. To

II. *To take a* Distance, *(Accessible or In-accessible) at Two Stations.*

Fig. VII. Being in a *Field* at E, there is a *Windmill* (or other *Object*) in another *Field* at F, (separated by a *River* or other Impediment) whose Distance is required.

In any other Part of the *Field*, remote from E, cause a *Mark* to be set up, as at D; and measure the *Distance* between E and D, which let be 115.00 Foot. —Then by your Instrument at E, find the Quantity of the *Angle* F E D, which suppose to be 106.50 Deg. And going from E to D, make Observation of the *Angle* F D E, which suppose to be 57.10 Deg.

By these Two *Angles*, and the *Measured Distance*, you have constituted an *Oblique Triangle* D E F, in which there is given, (1.) The *Measured Distance* E D, 115 00 Foot. (2.) The Two *Angles* at D and E, 106.50 Deg. and 57.10 Deg. And, (3.) The *Angle* at F, (the Complements of the other Two to 180 Deg.) 16.40 Deg. Whereby you may find the *Sides* E F and D F, (by *Axiom*. II.) thus:

As the Sine of the *Angle* at F, 16.40 Deg.
Is to the *Measured Distance* D E, 115.00 Foot;
So is the Sine of the *Angle* at F, 57.10 Deg.
To the *Distance* E F, 341.98 Foot.
And so is the *Angle* at E, 106.50 Deg. (or 73.50.)
To the Distance D F, 390.54 Deg.

III. *If there were* Three *(or more)* Ships *on the* Sea, *and you being upon the* Land, *desire to know how far those* Ships *are from you; and also, how far they are distant one from the other.*

Fig. VIII. LET the *Three Ships* be A, B, C; and you being upon the *Shoar* at M, are required to tell how far those *Ships* are from you, and also, how far from each other.

First, Being at M, make choice of some other Place upon the Shoar, at some considerable Distance, as at O 130.00 Fathom.

Secondly, Being at M, observe the Quantity of the *Angle* A M O, which we will suppose to contain 104.50 Deg. and then removing to O, and observing the *Angle* M O A, you find it to be 37.82 Deg. through which Degrees, Lines being drawn from M and O, will cross each other in A, which is the Place of the first *Ship*; which, with the Line of Distance M O, will form the *Oblique-*

Oblique-angled Triangle A M O: In which there is given, (1.) The Side M O 130.00 Fathom. (2.) The *Angle* A M O, 104.50 Deg. And, (3.) The *Angle* A O M, 37.82 Deg. And having Two *Angles* given, you have the third alſo given, *viz.* M A O, 37.68 Deg. whereby you may find the other Two Sides M A and O A (by *Axiom* II.) thus:

As the Sine of M A O, 37.68 Deg.
  Is to the Side M O, 130.00 Fathom:
So is the Sine of A O M, 37.82 Deg.
  To the Side M A, 130,41 Fathom, the Diſtance of the *Ship* at A, from M:
And ſo is the Sine of A M O, 104.50 (or 75.50 Deg.)
  To the Side O A, 205.91 Fathom, the Diſtance of the *Ship* at A, from O.

Again,

Obſerving from M and O, to B, you found the *Angle* B M O to contain 65.50 Deg. and M O B 68.00 Deg. wherefore, if upon M and O, you lay down thoſe *Angles*, you ſhall have another *Oblique-angled Triangle* M B O: In which you will have given as in the former, (1.) The *Angle* B M O, 65.50 Deg. (2.) B O M 68.00 Deg. And conſequently M B O, 46.50 Deg. together with the meaſured Diſtance M O 130 Fathom, by which you may find the other Two Sides M B and O B, by *Axiom* II. as in the former. For, *Fig.* VIII.

As the Sine of M B O, 46.50 Degrees,
  Is to M O, 130.00 Fathom:
So is to the Sine of B M O, 65.50 Degrees,
  To the Side B O, 163.08 Fathom:
And ſo is the Sine of the *Angle* B O M 68.00 Degrees,
  To the Side B M, 166.18 Fathom.

Laſtly,

Obſerving again from M and O, to C, you find the *Angle* C M O to be 11.75 Deg. and M O C 132.75 Deg. the which *Angles* laid down, you have a third *Oblique-angled Triangle* C M O, wherein there is given, as before, all the *Three Angles*, and *One Side*, M O, whereby you may find the other Two, M C and O C, (by *Axiom* II.) as in the former.

So will M C be 164.39 Fathom, and O C 45.62 Fathom.

Fig. VIII. And thus have you the *Distances* of all the *Ships*, from the Two places, M and O, on the *Shoar*.

Now for their *Distances* one from another:

And *First*, For the *Distance* A B.

In the *Oblique Triangle* A B M, there is given, the *Sides* A M 13041 Fathom, and B M 168.18 Fathom, and the *Angle* contained by them, A M B 39.00 Deg. whereby the third *Side* AB may be found, (By *Case* II. of *O, A, P, T,*) thus

As the Sum of the Sides, A M and B M, 296.59 Fathom,
Is to the Difference of those Sides, 35.77 Fathom:
So is the Tangent of half the *Angles* at A and B, 70.50 Deg.
To the Tangent of half the Difference of those *Angles*, 18.81 Deg.

Which added to 70.52 Deg. gives 89.31 Deg. for the greater *Angle* B A M; and substracted therefrom, leaves 51.69. Deg. for the lesser *Angle* A B M.

Then say, (By *Axiom* II.)

As the Sine of the *Angle* A B M, 51.69 Degrees,
Is to the Side M A, 130.41 Fathom:
So is the Sine of the *Angle* A M B, 39.00 Degrees,
To the Side A B, 104.59 F.

And in the same manner may the *Distance* from B to C. and from C to A, be found.

According to this Method may the *Distances* of many *Places* upon the *Land*, one from another be obtained, by making of *Observation*, by a *Theodolite*, *Semi Circle*, (or other *Graduated Instrument*) from Two Places, from whence all the other may be seen: An *Example* whereof I shall give, with the manner of making the *Observations*; and protracting of them, whereby the *Triangles* to be resolved for the Performance will be conspicuous: And for the *resolving* of them. (it being altogether the same with that foregoing) I shall leave to the Ingenuity of the Practitioner. Wherefore,

Let A B C D E be several *Places*, as *Churches* in a *Town* or *City*, or such like *Objects*.

Fig. IX. 1. Make Choice of Two such *Places*, from either of which, you may see all the *Places* whose *Distances* you require; which *Places* let be F and G, distant from each other 1000 Foot, more or less.

2. Set-

2. Set up your *Instrument* at F, and direct the *Sights* on the *Diameter* thereof, to the other Place at G. and there fix it: Which done, direct the *Sights* to the several *Places*, A, B, C, D and E, noting what *Degrees* of the *Instrument* are cut by the *Index* at every *Observation*. *Fig.* IX.

3. Then removing your *Instrument* to the second *Place* G, direct the *Sights* which are upon the *Diameter* thereof, back towards the *First Place* at F, where fix it: Then, turning the *Index* about, direct it to the several *Places*, A, B, C, D and F, as before; noting the *Degrees* cut by the *Index*. Which we will suppose to be such as are noted in this *Table*.

From F to	the Degrees cut were	From G to	the Degr. cut were
A	70.00	A	55.00
B	98.50	B	64 50
C	140.00	C	96 00
D	200.40	D	185.00
E	290.60	E	282.00

And the *Distance* from F to G 1000.

From these *Observations*, to make a *Plot* or *Map* of the Situation of the several *Places*.

1. Upon a Sheet of Paper, draw a *Right Line*, as F G, to contain 1000 Foot, of any *Scale*, which represents the Line of the *Distance* of the *Two Places*, where you *Observed* at F and G.

2. Place the *Center* of a *Protracter* upon F, and the *Diameter* thereof upon the *Line* F G; and there holding it fast, make *Marks* against the several *Degrees* that were cut by the *Index*, when the *Sights* were directed to the several *Objects* at A, B, C, *&c.* when you *Observed* at F, and through those *Points* draw *Right Lines*, at Pleasure; as F A, F B, F C, F D, F E.

3. Lay the *Center* of the *Protractor* to the Point G, and the *Diameter* thereof upon the Line F G, and there holding it fast, make *Marks* against those Degrees of the *Protractor*, as the *Index* did cut upon the *Instrument*, when you made *Observation* at G; and through those *Points*, and the *Point* G, draw *Lines*, as G A, G B, G C, G D, and G E, crossing the former *Lines* (drawn from F,) in the respective *Points*, A, B, C, D and E. Which *Points* will lye upon your *Paper*, in the same *Position* as the *Places* you took notice of, were situate on the Ground on which they stood: And being thus laid down, if you take with your Compasses

Fig. IX. passes the *Distance* between any *Two* of them, and measure it upon the same *Scale* you laid down the *Line* F G by, it will give you the *Distance* between those Two *Places*.

But, their *Distances* may be more exact and accurately attained unto by *Trigonometrical Calculation*.

For, in every *Triangle*, as in F G A, F G B, F G C, F G D, and F G E, there is given, (1.) The *Angle* A F G, observed at F. (2.) The *Angle* A G F, observed at G; and, (3.) The *Side* FG, (the *Stationary Distance*) included between them, to find the other *Sides*, A F and A G. The *Practice* whereof I commit to the Ingenuity of the *Practitioner*.

CHAP. III.

Of PLANOMETRIA.

OR,

Of the Mensuration *of* Plain Superficial Figures.

I. *Of the* Geometrical Square A B C D, *whose* Side A B *is* 27.32 *Foot*. By Logarithms.

Fig. X. As 1 F : to 27.32 F : : 27.32 F : to 746.38 F.

Logar. of 27.32	1.436481
Logar. of 27.32	1.436481
Logar. of 746.38	2.872962

The *Superficial Content* of the *Square* in *Feet*.

II. *Of the* Parallelogram (*or* Long Square) E F G H, *whose* Length E F *is* 27.25 *Yards*, *and* Breadth E G, 6.29 *Yards*.

Fig. XI. As 1 : 27.25 : : 6.29 : 17.03

Logar. of 27.25	1.4353665
Logar. of 6.25	0.7958800
Logar. of 17.03	2.2312565

III. Of

153

XII.

XIII.

XIV.

p51

Fig. IX

Fig. X

Fig. X

III. *Of the* Triangle K L M, *whose* Longest Side L M *is* 267.26 Pole *or* Perches; *and its* Perpendicular L N 160.25 Perches.

As 1 P : ½ L M 133.63 P : : L N 160.25 P : 21414 P. *Fig.* XII.

Logar. 160 25	2.204798
Logar. 133.63	2.125904
Logar. of 21414	4.330702

IV. *Of the* Trapezia (*or Figure of Four unequal* Sides) O P Q R *The Diagonal, whereof* O Q *is* 27.32 Perches, *the* Perpendicular P S 9.53 P. *and the* Perpendicular R T, 21.06 P.

As 1 : ½ O P 13.66 : : P S + R T 30.59 : 417.85 *Fig.* XIII.

Logar. of half O Q 13.66	1.135451
Logar. of P S and R T, 30.59	1.485579
Logar. of 417.85 *Perches*	2.621030

Which is the *Area*, or *Content* in *Perches*.

IV. *Of an* Irregular Plot, *consisting of many unequal* Sides *and Angles*: *As the Figure* A B C D E F G.

Before such *Irregular Figures* can be measured, they must be reduced in *Triangles* or *Trapezia's*, by drawing of *Lines* from *Angle* to *Angle* at the best Advantage, as in this, by the *Lines* F C and F C, which Two *Lines* divides the whole *Plot* into the Two *Trapezia's* A B C F; F C D E, and One *Triangle* F E G. And then the *Bases* and *Perpendiculars* being such as are expressed in the *Figure*, they may be *measured*, as is shewed in the Two foregoing *Sections* hereof; and according to the following *Operations*. *Fig.* XIV.

I. *For the* Trapezia A B C F.

Logarithm of	half A C, 12.16	1.084933
	B H and H F, 8.13	0.910089
	the *Area* of A B C F, 98.86	1.995023

II. *For the* Trapezia F C D E.

Logarithm of	half F D, 12.96	1.112605
	C L and E M, 23.11	1.363800
	the *Area* of F C D E 299.51	2 476405

III. *For*

III. *For the* Triangle F E G.

Logarithm of	half E F, 9.50	0.977723
	G N, 5.82	0.764923
	the *Area* of F E G, 55 29	1.742646

The *Area* of	A B C F	is	98.86
	F C D E		299.51
	F E G		55.29
The *Area* of A B C D E F G			453.66

V. *Of a* Circle A B C D, *whose* Diameter B C *let be* 14.00.

Fig. XV. The Proportion of the *Diameter* of any *Circle*, to the *Circumference* thereof, is (in the least Terms) as 7 to 22: But in greater Numbers, as 113 to 355; and of these I shall make use.

I. *By the* Diameter, *to find the* Circumference.

As 113 : 355 : : 14 Diam. : 43.98 *Cir.*

Logar. of 113	2.053078
Logar. of 355	2.550228
Logar. of 14.00 the Diameter	1.146128
	3.696356
The Logar. of 43.98 the *Circumference*	1.643278

II. *By the* Circumference, *to find the* Diameter.

As 355 : 113 : : 43.98 *Circum.* : 14.00 *Diam.*

Logar. of 355	2.550228
Logar. of 113	2.053078
Logar. of 43.98 *Circ.*	1.643278
	3.696356
Logar. of 14.00, the *Diam.*	1.146128

III. By

III. *By the* Diameter, *to find the* Area. Fig. XVI.

As 113 in 4 (*viz.* 452.)	2.655138
Is to 355	2.550228
So is the *Diam.* 14 in 14 (*viz.* 196)	2.292256
	4.842484
To the Logar. of 159.94, the *Area*	2.187346

IV. *By the* Circumference, *to find the* Area.

As 355 ✢ 4 (*viz.* 113)	3.152288
Is to 113	2.053078
So is 43.98 ✢ 43.98, *Area* Squa.	1.643255
The same again	1.643255
	5.339588
To the Logar. of the *Area* 153.49	2.187300

V. *By the* Diameter, *to find the* Area.

As 10 : 8.862 : : 14 Di. : 12.41 : : 12.41 : 154.00

Log. 8.862	0.947532
Log. 14	1.146128
Logar. 12.41	1.093660
The double of it	2.187320 Log. of 154 the *Area.*

CHAP. IV.

Of GEODÆCIA: *Or,* Land Measuring.

I. *How* (*by any* Graduated Instrument) *to take the true* Plot *of any large* Piece *of* Ground, *as* Common Field, Park, Wood, *&c.*

IN going round about a *Field* to Survey it, there are Two ways; for in going round about it, you must either go on the *inside* or on the *outside*; and sometimes you may be constrained to go sometimes *within*, and sometimes *without*.

Let A B C D E F, be such a Field to be *surveyed* and *plotted.* Fig. XVII.

1. Begin at any *Angle* thereof, as at A, and there setting up your *Instrument*, lay the *Index* and *Sights* upon the *Diameter* thereof; and turning it about, direct the *Sights* to B, and there

Fig. XVII. screw it fast; and turn the *Index* about, till through the *Sig*[hts] you see the *Angle* at F; and there note what Degrees of the [In]*strument* were cut by the *Index*, which we will suppose to [be] 300 Deg. for the Quantity of the *Exterior Angle* F A B, [with]out the Field: Or, 60 Deg. for the Quantity of the *Interior* [An]*gle* within the Field: Then measure the Side A F, which [sup]pose to be 41.00 *Pole* or *Perches*, and the Side A B 30.00 *Per*[ches]. These *Distances*, with the Quantity of the *Angle* before obse[rved] 300, or 60 Deg. set down in a *Book* or *Paper*, as you see in [the] following *Table*.

2. Remove your *Instrument* to B, and laying the *Index* [on] the *Diameter*, turn the *Instrument* about, till through the *S*[ights] you see the Place where the *Instrument* last stood at A, and screw it fast; and turn the *Index* about, till, through the *Sights* [you] see the *Angle* C, and note the Degrees which the *Index* cut[s], which are here 145 Deg. for the *Exterior*, or 215 Deg. f[or the] *Interior Angle* A B C; and the *Distance* B C measured is [...] *Perches*: Both which, note down as before.

3. Remove your *Instrument* to C, and placing the *Index* [on the] *Diameter*, look back to B, and there screw it fast; and turn the [*Index*] about, till you see the *Angle* at D; where the *Index* [cuts] 270.25 Deg. for the *Exterior*, or 89.75 Deg. for the *Interi*[or An]*gle* B C D, and the measured Distance C D is 23.40 Deg.

4. Remove the *Instrument* to D, the *Index* on the *Diamet*[er], look back to C, and then screw it fast; then turn the *Inde*[x] till by the *Sights* you see the *Angle* at E; the *Index* cutti[ng] [...] Deg. for the *Exterior*, or 97 Deg. for the *Interior Angle* [...] the *Measured Distance* D E being 28.00 *Perches*: Whi[ch set] down as the other.

5. Place your *Instrument* at E, the *Index* on the D[iameter], look back to D, and then screw it fast, and turn the *Ind*[ex] till through the *Sights* you see the *Angle* at F, the *Ind*[ex] cutting 220 Deg. for the *Exterior*, or 140 Deg. for the [Interior] *Angle* D E F; the *Measured Distance* E F being 42.80 [...]

6. Place the *Instrument* at F, and the *Index* lying on [the Dia]*meter*, look back to E, and then screw it; and turn the [Index a]bout, till through the *Sights* you see the *Angle* at A, (o[r] where you first placed your *Instrument*) where the *In*[dex cuts] 210 Deg. for the *Exterior*, or 150 Deg. for the *Interi*[or Angle] E F A, the *Measured Distance* F A being 41 00 *Perches*, [...] All which being noted down according to these *Observ*[ed and] *Measured*, will stand as in this *Table*.

Fig. XVII.

The Angle at	Exterior.	Inferior.	The Distance from	Perches.
A	300.00	60.00	A to B	30.00
B	145.00	215.00	B to C	28.40
C	270.25	89.75	C to D	23.40
D	263.00	97.00	D to E	28.00
E	220.00	140.00	E to F	42.80
F	210.00	150.00	F to A	41.00

These *Observations* of *Sides* and *Angles*, as they were taken in the *Field*, being noted down in a *Book* or *Paper*, as in this *Table*; a *Plot* of the *Field* may be drawn upon *Paper* or *Parchment*, by the following *Directions*.

To Protract *the former* Work.

1. Upon a Sheet of Paper or Parchment draw a Line A B, to contain 30.00 Perches, of any *Scale*: And upon the end thereof A, place the Centre of a *Protracter*, laying the Diameter of it upon the Line A B: —Then, the *Exterior Angle* at A, being 300.00 Deg make a Mark against 300.00 Deg. of your *Protractor*; and through that Point, and the Point A, draw a Right Line, downwards, as A F, to contain 41.00 Perches, of the same *Scale*.

2. Apply the Centre of the *Protractor* to the Point B, and the Diameter upon the Line A B; then against 145.00 Deg. (the *Angle* at B) make the Mark, and through that Mark, and B, draw a Right Line B C, to contain 28.40 Perches.

3. Lay the Centre of the *Protractor* upon C, and the Diameter upon B C, and against 270.25 Deg. make a Mark; through which, and the Point C, draw the Line C D, to contain 23.40 *Perches*.

4. Apply the Centre of the *Protractor* to the Point D, and its Diameter to the Line C D; making a Mark against 263.00 Deg. through which, and the Point D, draw the Line D E, to contain 28.00 *Perches*.

5. Lay the Center of the *Protractor* upon the Point E, and its Diameter upon D E, and against 220.00 Deg. make a Mark; through which, and the Point E, draw a Right Line; which (if you have committed no former Error) will cut the Line, first drawn, in the Point F: And, by this Means, you will have upon your *Paper*, the exact *Figure* of your *Piece of Ground*: Which you may cast up, and find the Quantity thereof in *Perches*, (by the Directions of the Third Chapter hereof) which being divided by 160, will give the Content in *Acres* and odd *Perches*.

Fig. XVII. In this *Example*, we have wrought by the *Exterior*, (or *Outward*) *Angles:* But if you would *Protract* the same by the *Interior* (or *Inward*) *Angles*; it is done, by substracting the *Exterior Angle* from 360 Deg. and the Remainder will be the *Interior Angle:* So, the *Angle* at A, being 300 Deg. that substracted from 360 Deg. leaves 60 Deg. for the *Interior Angle* at A. And so of all the rest, as they are set down in the *Table*.

CHAP. V.

Of STEREOMETRIA: OR, Mensuration of Solids.

Fig. XVIII.

I. *Of the* Cube ABCDEF, *whose* Side *is* 5.20 *Foot*.

As 1 : 5.20 : : 5.20 27.04 : : 27.04 : 140.61 *Foot*.

Logar. 5.20	0.716003
Logar. 5.20	0.716003
Logar. 27.04	1.432006
Logar. 5.20	0.716003
Log. of 140.61	2.148009

Equal to the *Solidity* of the *Cube* in *Feet*.

Fig. XIX. This might have been done more easily, for the *Logar.* of the *Side* 5.20, which is 0.716003, multiplied by 3, would have produced 2.148009, the *Logar.* of 140.61, the *Solidity*.

II. *Of a* Long Cube *or* Parallelipipedon, GHIKL, *whose* Side *of the* Square *at the* End, KI, *is* 1.27 *Foot*, *and of its* Length GH, 5.32 *Foot*.

As 1 : 1.27 : : 1.27 : 1.61 : : 5.32 : 8.58 Foot.

Log. 1.27	0.103804
Log. 1.27	0.103804
Log. 5.32	0.725911
Log. 8.58	0.933519

The *Solid Content*.

III. Of

III. *Of an* Oblong Parallelipipedon M N O P Q R, *whose* Breadth *at the* End N O *is* 3.40 *Foot, and* Depth M N 6.50 *Foot, and the* Length *thereof* M P, 16.25 *Foot.* Fig. XX

As 1 : 3.40 : : 6.50 : 22.10 : : 16.25 : 159.12 Foot.

Logar. 3.40	0.531479	
Logar. 6.50	0.812913	
Logar. 16.25	1.210853	
Logar. 359.12	2.555245	The Solidity.

IV. *Of a* Pyramis S T V X N, *whose* Side *of its* Base S T, *&c. is* 1.12 *Foot, and* Altitude N X 12.90 *Foot.* Fig. XXI.

As 1 : 1.12 : : 1.12 : 1.25 : $\frac{1}{3}$ N X : 4.30 : 5.39 the Sol. Con.

Logar. 1.12	0.049218	
Logar. 1.12	0.049218	
Logar. 4.30	0.633468	
Logar. 5.39	0.731904	the *Solid Content.*

V. *Of a* Cone A B C D, *whose* Diameter B C *is* 6.12, *and* Altitude A D 12.90 *Foot.* Fig. XXI.

As 7 in 4 (28) : 22
So 6.12 in 6.12 (37.45) : 29.30
The *Area* of the *Circle* of the *Base.*

As 1 : 29.30 : : So 4.30 : 126.00
The *Solidity.*

Logar. 28	1.447058	
Logar. 22	1.340422	
Logar. 37.45	1.573452	
	2.913874	
Logar. 29.30	1.466816	the *Area.*
Logar. 4.30	0.633468	
Logar. 126.00	2.100284	the *Solidity.*

VI. *Of the* Frustum *of a* Pyramis *or* Cone. Fig. XXIII. XXIV.

Let A B C H G F be the *Frustum* of a *Square Pyramid*, the *Side* of the *Square* at the *Lesser* end A B, is 1.60 Foot; and therefore

fore the *Area* of that *Square* A B C D is 2.56 Foot: —The *Side* of the *Square* at the *Greater* End F G, is 2.30 Foot; and therefore the *Area* thereof is 5.29 Foot, and the *Altitude* D E 39.3 Foot.

Then,

The Logarithm of A B C D 2.56 is		0.408240
The Logarithm of E F G H 5.29 is		0.723455
Their Sum		1.131695
Their half Sum		0.565847
Which is the Logarithm of	3.68	
The Sum of the Three Products	11.53	
The Logar. of the Sum	11.53	1.061829
The Log. of ⅓ of the Altitude	13.31	1.117271
The Log. of	151.04	2.179100

Which is the *Solidity* of the *Frustum* A B C D E F G H.

In the same manner the *Frustum* of a *Cone* of the same *Altitude*, and the *Area's* of the *Circles*, at both *Ends*, the same with the *Area's* of the *Squares*; the *Solidity* of such a *Frustum Cone*, will be found to be the same.

Fig. XXV.

VII. *Of a* Globe *or* Bullet, *whose* Diameter *is* 1.75 *Foot.*

The Logar. of the Diameter 1.75	0.243038
Multiplied by 3	0.729114
The Logar. of 355	2.550228
Their Sum	3.279342
Logar. of [678 always] Subtract	2.831229
Rests the Logar. of 2.806 Foot	0.448113

Which is the *Solidity* of the Bullet.

ANCILLA

Fig.
XXV.

ANCILLA MATHEMATICA.
VEL,
Trigonometria Practica.

SECTION. II.

OF

COSMOGRAPHY.

COSMOGRAPHY, in Greek Κοσμογραφία, is derived *à* Κοσμ@, i. e. *Mundus*, & γραφὴ; i. e. *Scriptum*, the Description: So that *COSMOGRAPHY* is a Description of the World; consider'd under the Names of the Heavens, as the *Cœlestial*; and of the Earth and Water, as the *Terrestrial* Parts thereof. A true aud perfect Represention of both which, may be (and usually is) described upon Two Round Bodies, called *GLOBES*: And therefore we shall first treat of them.

CHAP. I.

Of a material Sphere *or* Globe; *and of such* Circles, Lines *and* Points, *as are described upon it, and that are appendant to it. And how to rectifie the* Globe, *fitting it for the resolving of such Problems as are, to be performed by it.*

I. *Of the Circles of the Sphere.*

A *Globe* or *Sphere* is an Artificial Representation of the *Starry Heaven*, or of the *Earth* and *Water*, under the Form or Figure

Fig. XXV.

Fig. XXV. gure of *Roundness*, which they are supposed to have: Shewing in a just *Proportion* every particular *Constellation* in the *Heavens*: And such a *Globe* is called [A *Cœlestial Globe*] and every particular *Region* or *Country* on the *Earth*, and *Island* in the *Sea*: And such a *Globe* is called [A *Terrestrial Globe*.]

Upon the *Convex Superficies* of either of these *Globes* (besides the *Constellations* of the *Stars*, upon the *Cœlestial*; and the *Countries*, *Kingdoms*, *Sea-Coasts*, *Islands*, &c. upon the *Terrestrial*) there are described several *Circles*; some *Great Circles*, which divide the whole Body of the *Globe* into Two *Equal Parts*; and some *Lesser Circles*, which divide the Body of the *Globe* into Two *Unequal Parts*.

The *Great Circles* described upon the Body of either *Globe* are,

The *Æquinoctial* } { The *Æquinoctial* } *Colure*.
The *Eclyptick* } { The *Solstitial* }

The *Lesser Circles* are,

The Two *Tropicks*; and, The Two *Polar Circles*.

I. *Of the* ÆQUINOCTIAL.

The *Æquinoctial* is a *Great Circle*, dividing the Body of the *Globe* into Two *Equal Parts*, passing through Æ and æ; which *Points* are, each of them, equidistant from P and S, the Two *Poles* of the *World*; and so the *Line* or *Axis* P A S passing through the Body of the *Globe*, cutteth this Circle always at *Right Angles*.

II. *Of the* ECLIPTICK *and* ZODIACK.

The *Zodiack* is a *Great Circle*, crossing both the *Horizon* and *Æquinoctial* in A, the *East* and *West* Points of the *Horizon* at *Oblique Angles*, viz. the *Horizon* at an *Angle* equal to the *Sun's* greatest *Meridian Altitude* in any *Latitude*, but the *Æquinoctial*, always, at an *Angle*, equal to (23.5 Deg.) the *Sun's greatest Declination*. This *Circle* is supposed to have Breadth; for, through the middle of it, there passeth a *Line*, noted in the Figure with the Twelve *Signs* of the *Zodiack*, ♈ ♉ ♊ ♋ ♌ ♍ on the upper Part, and with ♎ ♏ ♐ ♑ ♒ ♓ on the lower Part; which *Line* is called the *Ecliptick*, or *Via Solis*, for that the *Sun* always keepeth in this *Line*.

III. *Of*

III. *Of the Two* COLURES.

These are Two *Great Circles*, which pass through the *Poles* of the *World*, where they cut each other at *Right Angles:* One of these cutteth the *Horizon* in A, and the *Æquinoctial* also in the Point ♈ ♎, and is therefore called the *Æquinoctial Colure.* The other *Colure* cutteth the *Horizon* in H and O, the *North* and *South* Points thereof, and also the *Ecliptick* in the *Points* ♋ and ♑, and is therefore called the *Solsticial Colure.* In the Figure the *Æquinoctial Colure* is noted with P A S (and is the same with the *Sixth Hour-Circle.*) the *Solsticial Colure*, with H P O S, and is the same with the *Meridian.*

Fig. XXV.

IV. *Of the* TROPICK *of* CANCER.

This is a *Lesser Circle* of the *Sphere*, described upon the Superficies of the *Globe*, parallel to the *Æquinoctial*, on the *North Side* thereof, at 23.5 Deg. distant therefrom, equal to the *Sun's greatest Declination.* It is called the *Northern Tropick*, for that it lyeth towards the *North Pole*, and is represented in the *Figure* by the *Small Circle*, ♋ *6* ♋.

V. *Of the* TROPICK *of* CAPRICORN.

This is a *Lesser Circle* also, described upon the *Globe*, parallel to the *Æquinoctial* on the *South* Side thereof, at the Distance of 23.5 Deg. towards the *South Pole*, and is called the *Southern Tropick:* It is noted in the *Figure* with ♑ *6* ♑.

VI. *Of the Two* POLAR CIRCLES.

These are *Two Lesser Circles* described about the *Poles* of the *World*, parallel to the *Æquinoctial* and *Tropicks*, and so far distant from either *Pole*, (*viz.* 23.5 Deg.) as the *Tropicks* were from the *Æquinoctial:* That described about the *North Pole*, noted with B *6* C, is called the *Artick Circle*; and the other, noted with G *6* E, described about the *South Pole*, is called the *Antartick Circle.*

Besides these *Circles*, which are described upon the Convex Superficies of either *Globe*, there are others, which are framed without the Body of the *Globe*; and those are, principally, Two,

 of

Fig. XXV. of the *Meridian*, and the *Horizon*; the one of *Brass*, the other of *Wood*. —*Circles*, indeed they are not so properly called; for, in the rigorous Sense, no *Line* is supposed to have any *Breadth*, but both these have *Breadth* allowed them; that such Things might be written, or engraven upon them, as might render them more useful in all Positions of the *Globe* And therefore, (they being of a *Circular Form*) notwithstanding the Impropriety of Speech, we will have it so; and we must call them, The *Meridian* and *Horizontal Circles*. The *Meridian* in the Figure, is noted with the Letters Z Æ N æ, and the *Horizon* with H A O.

Unto this *Brazen Meridian*, there belong Two other Appendants, *viz.* An *Hour-Circle*, with its *Index*; and a *Quadrant* of *Altitude*.

Through the Body of either *Globe*, there runs a strong *Wyre*; the Ends whereof are so fixed in the *Brazen Meridian*, that the *Body* of the *Globe* may turn about together with the *Wyre*. —This *Wyre* is called the *Axis* of the *World*, and the Ends thereof, the *Poles* of the *World*, one the *Artick* or *North Pole*, noted with P; the other, the *Antartick*, or *South Pole*, noted with S.

Unto this *Brass Meridian* also, there belongs another Appendant, called *A Quadrant of Altitude*; which is a thin Plate, divided into 90 equal Parts or Degrees, and fitted with a *Nut* and a *Screw*, to move to any Degree upon the *Meridian*; but generally in the Vertical Point Z, the Zenith of any Place.

II. *Of the several* Positions, *that a* Globe *or* Sphere, *may be posited in its* Horizon.

THere are but *Three Positions*, in which a *Globe* may be seated in its *Horizon*; viz. (1.) *Direct*. (2.) *Parallel*. (3.) *Oblique*. Of which, the Two first are *Particular*, the third more *General*.

I. *Of* Direct Position

The *Globe* may be so placed in the *Frame*, that both the *Poles* thereof may rest upon (or lye directly in) the *North* and *South* Points of the *Horizon*, neither *Pole* having any Elevation: The *Zenith* Point being in the *Æquinoctial Circle*, and the *Axis* of the World in the Plain of the *Horizon*. And this is called *Direct Position*.

II. Of

II. *Of* Parallel Position.

Fig. XXV.

The *Globe* may be so placed in the Frame, that one of the *Poles* shall be in the *Zenith*, and the other in the *Nadir Points*; the *Poles* having 90 Deg. of *Elevation* above the *Horizon*: And in this Situation, the *Æquinoctial Circle* will be in the *Horizon*. This *Position* is called *Parallel Position*.

III. *Of* Oblique Position.

The third *Position* of the *Globe* is more *General*; for it hath relation to all *People* living between the *Æquinoctial* and either *Pole*: And according as the *Poles* are *Elevated* or *Depressed*, accordingly are the *People* said to be *situate*: That is, if the *Globe* be placed in the *Horizontal Frame*; so that the *Pole* be elevated above the *Horizon* 10 Deg. then is the *Globe* elevated, or fixed, to resolve such *Questions Geographical* or *Astronomical*, as relate to such *Places* and *People*, who have the *Pole* elevated (or who live in the *Latitude* of) 10 Deg. And this *Position* of the *Globe* is called *Oblique*, because the *Axis* of the *World*, the *Æquinoctial*, and all the *Parallels* of the *Sun's Declination*, are cut by the *Horizon* at *Oblique Angles*; whereas in the Two former *Positions*, they cut one the other *Right Angles*.

III. *How to Rectifie the* Globes, *fitting them for* Use *in any* Latitude *or* Part *of the* World.

Being provided of a *Pair of Globes*, the *Meridian*, *Horizon*, and *Hour-Circle*, truly turned and divided; also the *Balls* truly hung or poized upon the *Axis*, and the *Meridian* and *Horizon* (in all Positions) cutting each other at *Right Angles*; the *Papers* truly joined in their pasting upon the *Bodies*, &c. All which are to be performed by the *Workman*, (yet the *Buyer* ought also to have Inspection thereunto) you may proceed to *Rectifie* them for *Use* in this manner.

1. Put the *Brass Meridian* into the Two *Notches* that are cut in the *North* and *South* Parts of the *Horizon*; the *Graduated*, or *Divided*, Part thereof, towards the *East Part* of the *Horizon*; and the plain, or undivided Side thereof, towards the *West*, and let the *Meridian* also rest in the *Notch* which is in the Foot, or bottom, of the *Frame* of the *Horizon*.

Fig. XXV. 2. Place the *Brass Hour-Circle*, or *Wheel*, about the Pole; so that the *Hour-Lines* of 12 and 12 do lye directly over the *East*, or *Graduated*, Side of the *Meridian*; and that the Point of the *Axis* of the *Globe* do pass directly through the *Centre* of the *Hour-Wheel*; then shall the Two *Twelves* represent the Two Hours of 12 and 12: That towards the South Part of the *Meridian* 12 at Noon, and the other towards the North Part, 12 at Midnight: And the Two Sixes shall represent the Two Hours of Six a Clock: That towards the *East*, 6 in the *Morning*, and the other 6 at *Night*. Then put the *Index* (or *Pointer*) upon the end of the *Axis*; so that as the *Globe* being turned, which way soever, the Pointer may move with it; and so is your *Hour-Circle* rectified.

3. Elevate the Pole of your *Globe* (whether *North* or *South*) according to the *Latitude* of the Place of that what part of the World you are in: As suppose *London*, which hath 51 Deg. 30 Min. of *North Latitude*: The *Meridian* being in the Notches of the *Horizon*, and also in the Notch at the bottom of the Frame, as is before directed. Move the *Meridian* upwards or downwards in the Notches, till you find 51 Deg. 30 Min. of the *Meridian*, justly to touch the upper part of the *Horizon*, on the *North* part thereof: And so is your *Globe* rectified to the *Latitude* of 51 Deg. 30 Min.

4. The next thing to be rectified is the *Quadrant* of *Altitude*; which must be done, by having respect to the *Latitude* also: Wherefore, the *Latitude* being 51 Deg. 30 Min. Count 51 Deg. 30 Min. *North*, upon the *South* part of the *Meridian*, from the *Equinoctial Circle*, towards the *North* (or elevated) *Pole*, and there put on the *Nut*, which is at the end of the *Quadrant*; so that the edge of the *Divisions* of the *Quadrant*, may be directly under the Degrees of the *Latitude*, viz. 51 Deg. 30 Min. and there screw the *Nut* fast. And thus is your *Globe* Rectified for the Solution of all such Questions Cosmographical, as are to be wrought thereby in that *Latitude* of 51 Deg. 30 Min.

CHAP.

CHAP. II.

Cosmographical ELEMENTS,

Necessary to be known.

I. THere are Two kind of *Motions* in the Heavens; the first is called the *Common Motion* of the *fixed Stars* and *Planets* together; by which they go all about in 24 Hours from *East* to *West*. The second Motion is called the proper *Motion*; by which they go about, every one in his own *Time* or *Period*, from *West* to *East*.

Fig. XXV.

II. These Two *Motions* are the Original of Two *Circles*, the *Æquinoctial* and the *Ecliptick*; for the *Diurnal Motion* is done about the Pole of the *Æquinoctial* either in the *Æquinoctial* it self, or in a *Lesser Circle*, parallel unto it: But the proper *Motion*, is about the *Poles* of the *Ecliptick*, either in the *Ecliptick* it self, or in a *Lesser Circle*, parallel unto it.

III. The Sun's *Center* keepeth always upon the *Ecliptick Line*, but the other *Planets* do go from the *Ecliptick* on both Sides 8 Deg. Hence the broad *Circle*, whose Middle is the *Ecliptick*, doth arise, and is called the *Zodiack*.

IV. The *Æquinoctial* is in the Heavens about that Streak, which the Sun doth make by his *Diurnal Motion* on the Days of the Two *Æquinoxes*, viz. the 10th of *March*, and the 12th of *September*.

V. The *Zodiack* is known by the Twelve *Asterisms* of *fixed Stars*, called, The *Twelve Signs*.

VI. The Two *Luminaries* are the *Sun* ☉, and the *Moon* ☾, the *Moon* cometh round by her proper *Motion*, in a *Month*, the *Sun* in a *Year*.

VII. The other *Planets* are either the *Superior*, as *Saturn* ♄, coming about in his proper Motion, once in 30 Years. *Jupiter*

Fig. XXV.

ter ♃ in 12, *Mars* ♂ in 2 Years. The *Inferior Planets* are *Venus* ♀ and *Mercury* ☿; *Venus* is 9 Months *Morning Star*, and other 9 Months *Evening Star*. These Two Planets keep always near to the *Sun*, so that *Mercury* ☿ is for the most part covered with its Beams.

VIII. The *fixed Stars* move also from *West* to *East*, either in the *Ecliptick*, or in a Parallel to the *Ecliptick*, but very slowly, *viz.* One Degree in 70 Years. Hence the *Signs* are distinguished in *Starred* and *Un-starred*. The *Starred Signs* are the Twelve *Asterisms* of the *Zodiack*; but the *Un-starred* are every one a Twelfth part of the *Ecliptick*. Now the *Starred Signs* left their former Places, and are preceded in some 1800 Years almost One whole *Sign*; so the *Starred Aries* ♈, stands now in the Place of the *Un-starred Taurus* ♉; and the *Starred Taurus* ♉, in the Place of the *Un-starred Gemini* ♊, &c.

IX. The *Æquinoctial* and *Ecliptick* are immutable, for there is never but One *Æquinoctial*, and One *Ecliptick*: But the *Horizon* and *Meridian* are mutable: For every Body walking upon the Superficies of the *Earth*, doth carry along with him his *Horizon*; So this *Circle* is as manifold as there are divers Points upon the Surface of the *Earth*. The *Horizon* is determined by the Eye of the Man turning about in an even open Field, where the Heaven seemeth to join with the *Earth*; and its Office is to shew the *Rising* and *Setting* of all *Heavenly Bodies*.

X. The *Meridian* is not alter'd by going on streight towards *South* or *North*, but only when you walk never so little towards the *East* or *West*, you have presently another *Meridian*. It is observable in the Heaven, by letting fall a Plummet or Perpendicular from the *Vertex*, by the *Sun* (or any *Star*) being at its highest.

XI. Every one of these Four *Circles* hath its *Poles*, which the *Circle* is just between, and every way equally distant from it, exactly dividing the *Sphere* into Two equal *Hemispheres*, and they divide each other into Two equal *Semicircles*: And by the *Poles* of each, there are described *Secondary Circles* (the *Meridian* only excepted) which *Secondary Circles* do cut their *Principal Circle* into Two Equal Parts, and at *Right Angles*.

XII. The

XII. The *Poles* of the *Æquinoctial* are the same with the *Poles* of the *World*; the one of which is called the *Artick Pole*, because it is near to the Two *Arktos* or *Bears*: The other opposite to it, is called *Antartick*: And the streight Line, which passeth between, through the Centre of the *Sphere* (from one *Pole* to the other) is called the *Axis* of the *World*. The *Æquinoctial* divideth the *Ecliptick* into Six *North*, and Six *South*, *Signs*: The *Secondary Circles* of the *Æquinoctial*, are called in the Heavenly Sphere *Circles* of *Declination*. Amongst these is one of chiefest Note, the *Meridian*; and besides it, Eleven *Hour-Circles*, passing by every 15th Degree of the *Æquinoctial*, to be reckoned from the *Meridian*, and so they divide the whole *Æquinoctial* into 24 equal *Hours*. There are also Two chief *Secondary Circles* of the *Æquinoctial*, which are called *Colures*; the one passing by the *Vernal* and *Autumnal Section*, is called the *Colure* of the *Æquinoxes*; the other passing by the Two Solstitial Points, *viz.* the beginning of *Cancer* ♋ and *Capricorn* ♑, is called the *Colure* of the *Solstice*. This latter, divideth the *Ecliptick* in *Ascending* and *Descending Signs*; because in the first the Sun doth ascend to our *Zenith* in *Capricorn*, *viz.* in ♑ ♒ ♓ ♈ ♉ ♊, which are also called the *Signs* of *Short Ascension*, because they rise in a short time equal to the *Shortest Day* of the Year: But in the *Descending Signs*, the Sun doth descend every Day more and more from our *Zenith*, and those are ♋ ♌ ♍ ♎ ♏ ♐: These are called also *Signs of Long Ascension*, because they Rise in a time, equal to the *Longest Day* in the Year. Both *Colures* together divide the *Ecliptick* into Four *Quadrants*; the *Vernal* containing *Aries* ♈, *Taurus* ♉, *Gemini* ♊; The *Summer* Quadrant, *Cancer* ♋, *Leo* ♌, *Virgo* ♍; The *Autumnal*, *Libra* ♎, *Scorpio* ♏, *Sagittarius* ♐; the *Winter* Quadrant, *Capricorn* ♑, *Aquaries* ♒, *Pisces* ♓.

Fig. XXV.

XIII. The Secondaries of the Terrestrial *Æquinoctial* are called *Meridians*, and they are 18, passing by every 10th Degree of the *Æquinoctial*, and in some *Globes*, through every 15th Degree of the *Æquinoctial*, which is equal to One Hour, in Time. The first of these doth pass by the Islands of *Azores*, in some *Globes*. But by the newest, and best made *English Globes*, at St. *Michael's* Island in the *Azores*.

XIV. The

Fig. XXV.

XIV. The *Parallels* to the *Æquinoctial*, are *Lesser Circles*, dividing the *Sphere* into Two *unequal* Parts: They are Two *Tropicks*, which the Sun describeth, the one in the *Longest Day*, the other in the *Shortest Day*. Their Name *Tropicks*, signifieth a returning, *viz.* of the *Sun*. And Two other *Lesser Circles* are described by the *Diurnal Revolution* of the *Poles* of the *Ecliptick*; the one called *Artick*, the other *Antartick*.

XV. The Two *Tropicks*, together with the Two *Polar Circles*, distinguisheth the whole Surface of the *Earth* into Five *Zones*; the *Hot* or *Torrid Zone* is between the Two *Tropicks*, but from the *Tropick* of *Cancer* ♋ to the *Artick Circle* is the North *Temperate Zone*, and from the *Tropick* of *Capricorn* ♑ to the *Antartick*, is the South *Temperate Zone*. What lyeth within the *Artick Circle*, is called the *North Cold or Frigid Zone*; and that which lyeth within the *Antartick*, the *South Frigid Zone*.

XVI. The *Inhabitants* of the *Torrid Zone* are called *Amphiscii*, because their *Mid-days Shadow* falleth now towards the *South*, and then towards the *North*: But the *Inhabitants* of the Two *Temperate Zones* are called *Heteroscii*, because their *Mid-days Shadow* falleth only towards the *North* in our *Hemisphere*, and only towards the *South* in the other *Hemisphere*: And lastly, the *Inhabitants* of the Two *Frigid Zones* are called *Periscii*, because their Shadow goeth round about them in 24 Hours.

XVII. There are *Parallels* also which distinguish the *Climates*; which *Climates* are, as it were, *Little Zones*: They are greatest near the *Æquinoctial*, and from thence they grow smaller and smaller towards both the *Poles*. Their Distance and Largeness is determined by the length of the *Longest Day*; for as often as the *Longest Day* gaineth *half an Hour* above 12 Hours, there is produced a new *Climate*: So the first natural *Climate* ought to have its midst where the *Longest Day* is 12 Hours and a half long: But the Ancients (perhaps knowing nothing of those Parts of the World so near to the *Æquinoctial*) left out this first natural *Climate*, and made their first *Climate*, where indeed the Second should have been, *viz.* where the *Longest Day* was 13 Hours; and because the midst of this did pass by the

the Island which the *Nilus* maketh, and is called *Meroe*; they named their first *Climate Dia Meries*, the second *Dia Syenes*, *Syene* being a Town lying under the Tropick of *Cancer*; the third they called *Dia Alexandria*, this being a Town situated upon the Month *Nilus*; the fourth, *Dia Rhodia*; the fifth, *Dia Romes*; the sixth, *Dia Ponta*; the seventh, *Dia Boristhenis*, where the Longest Day was 16 Hours; and here the Ancients left it: But the Modern have continued their *Climates* to the *Artick Circle*, where the Longest Day is 24 Hours, because the Sun doth not set there in the beginning of *Cancer*, but only toucheth the *Horizon* with his *Circle*. And now, there are as many *Climates* on the other Side of the *Æquinoctial*. The *Circle* of *Perpetual Apparition* is described by a Point touching the *North Cardinal*, being carried about by the daily Motion; and all the Stars that are within this *Circle* of *Perpetual Apparition* are always seen above our *Horizon*. Another *Circle* of *Perpetual Occultation* there is described by a Point touching the *South Cardinal*, and being carried about by the daily Motion; so all the *Stars* that are within this, do never *Rise* with us.

XVIII. The *Secondaries* of the *Ecliptick*, are called *Circles* of *Latitude*; there are Six of them upon the *Cælestial Globe*, dividing the Signs of the *Ecliptick*, and also the whole *Sphere* into *Twelve* equal Parts; by which Division all *Stars* are referred to that *Sign* which is between the Two next *Circles* of *Latitude*. Now Six Signs of the *Ecliptick* are always above the *Horizon*, and Six are beneath.

XIX. The *Poles* of the *Horizon*, are the *Zenith* and the *Nadir*; the *Secondary Circles* are called *Vertical* or *Azimuth*, or *Circles* of *Altitude*, amongst which, the chiefest are, (1.) The *Meridian*. (2.) The *Circle of the 90th Degree of the Ecliptick*, passing by the *Poles* of the *Ecliptick*, the *Zenith*, and the 90th Degree of the *Ecliptick*, being counted from the *Horizon*, either from the *East* or *West*. (3.) The *Vertical*, passing by the *East* and *West Cardinals*; where the Intersection is of the *Æquinoctial* and the *Horizon*; which Sections also, be Poles of the *Meridian*: And the *Poles* of this *Vertical* passing by *East* and *West*, are in the *South* and *North Cardinals*, where the *Meridian* doth divide the *Horizon* in the *Ortive* and *Occidental Semicircles*. Now the *Secondaries* of this said *Vertical*, which

passeth

Fig. XXV.

passeth by the true *East* and *West*, are the *Circles* of *Position*, passing by every 30th Degree of the *Æquinoctial*, reckoning from the *Meridian* or *Horizon*, and dividing the whole *Sphere* into *Twelve Houses*. The first *House* is called *Horoscope*, and is that which is next under the *Ortive Horizon*; from thence the other *Houses* do succeed under the *Earth* after the Succession of the *Signs*. Where every *Arch* of *Position* cutteth the *Ecliptick*, there is the *Cuspides* of the *Houses*. There are besides these *Circles* of *Position*, an infinite Number passing by every Point of the *Sphere*.

XX. The *Parallels* of the *Horizon* are called *Almacantarath*, and are described upon the *Astrolabe*, to shew the *Altitude* of the *Sun*, or of the *Stars* above the *Horizon*.

XXI. The *Meridian* is the *Original* of *Winds*; there are Four *Cardinals*, and Four *Mean*: The *North* is known by the *Flower-de-Luce*, and the *East* by the *Cross*: The *Mean* do compound their *Names* from the next adjacent *Cardinals*; being *North-East*, *North-West*, *South-East*, *South-West*: Now every one of the *Cardinals* and *Means* hath Two *Laterals*, bearing the same Name with their *Principals*; so they are called *North by East*; *North by West*; *North East by East*; *North East by North*, &c. Now these *Laterals* being 16, make, together with the 8 *Principals*, 24; and just in the midst, between every Two *Principals*, there are the Eight *Residual Winds*, bearing the same Name with the *Means* they are next unto; but taking a fore Name from the next *Cardinal*, so they are called *North-North East*, *East North-East*, &c.

XXII. These 32 *Winds* being continued upon the Surface of the *Earth*, do make as many *Rumbs*; that *Rumb* which passeth by the *South* or *North*, is always a *Meridian*; and that which passeth by *East* and *West*, is always either the *Æquinoctial*, or a *Parallel* unto it. The other *Rumbs* are *crooked Lines*, neither *Circular* nor *Elliptical*, and are *Seven* in every *Quadrant*, to be numbred both Ways from the *Meridian*. The general Propriety of all *Rumbs* is to cut all the *Meridians* they pass by into equal *Angles*. There may also, besides the said 32 *Rumbs*, pass one by every *Point* of the *Horizon*.

XXIII. The

XXIII. The Effects of all the foresaid *Circles*, both *Principal* and *Secondary*, are *Angles* and *Arks*. The *Angles* which they make are *Right*, when a *Secondary* meeteth with its own *Principal*, and the other *Angles* are altogether *Oblique*: So the *Angle* which the *Æquinoctial* maketh with the *Ecliptick* is always 23 Deg. 30 Min. as much as the greatest *Declination* of the *Sun*.

Fig. XXV.

XXIV. The general Way of *Measuring Angles* upon the *Sphere*, is to set one Foot of a Pair of *Callipers* in the Point where the *Angle* is made, and extend the other 90 Deg. from thence, and so describe an *Ark* between the Two *Legs*, (or *Circles* which make the *Angle*) for as many Degrees as that *Ark* containeth, so many Degrees also is the *Angle*. So the *Measure* of the *Angle* made by the *Æquinoctial* and *Ecliptick* is taken in the *Solsticial Colure*, between the beginning of ♋, and the 90th Degree of the *Æquinoctial*. Likewise the *Angle* which the *Æquinoctial* makes with the *Horizon*, is measured 90 Deg. from thence in the *Meridian*, between the *Horizon* and *Æquinoctial*; and that which is farther in the *Meridian*, from the *Æquinoctial* to the *Zenith*, is the *Latitude* of one *Place*; into which is, always, equal to the *Poles Height*, to be reckoned from the *North Cardinal* to the *Pole*. Sailing streight towards *South* or *North* One Mile, the *Æquinoctial* is raised or fallen One Minute; going 60 Miles, it riseth or falleth One Degree: So at length, coming to the middle Line of the World, the *Æquinoctial* will be raised to the very *Zenith*, both *Poles* lying in the *Horizon*, and all the *Parallels* to the *Æquinoctial* cut the *Horizon* at *Right Angles*; whence this *Position* of the *Sphere* is called *Sphæra Recta*; where all the *Stars* do rise, and abide as long above the *Horizon* as beneath; so there is a perpetual *Æquinoctial* all the Year: But as soon as one of the *Poles* doth *rise* above the *Horizon*, and the other cometh to be under the *Æquinoctial*, and all its *Parallels* make *Oblique Angles* with the *Horizon*: And for this Reason, such a *Position* of the *Sphere* is called *Sphæra Obliqua*; where, not all the *Stars* do rise, but some are always *above*, and some always *below*, the *Horizon*. When the *Pole* cometh to unite with the *Zenith*, then the *Æquinoctial* falleth wholly in the *Horizon*, and the *Parallels* of it, are also parallel to the *Horizon*; and this *Position* of the *Sphere* is

Fig. XXV.

is called *Sphæra Parallela.* The Sun moving in that part of the *Ecliptick*, which is above the *Horizon*, that is on one side of the *Æquinoctial*, doth never set, but is turned continually round about, and maketh a Day of Six Months: So likewise, running through the other Six *Signs* that are under the *Horizon*, it doth never *rise*, but maketh a Night of Six Months also.

XXV. The *Angle* which any Degree of the *Ecliptick* maketh with the *Right Horizon*, (that is, in *Sphæra Recta*) is equal to that *Angle* which the same Degree of the *Ecliptick*, maketh with the *Meridian*: But whether the *Ecliptick* make such an *Angle* with the *Right* or *Oblique Horizon*, the same *Angle* is always called the *Angle Orient*; that is, of the rising Degree of the *Ecliptick*; and its *Measure* is in the *Circle* of the 90th Degree, between the said 90th Degree and the *Horizon*.

XXVI. The *Angle* which the *Meridian* maketh in the *Pole* of the *World* with any *Circle* of *Declination*, taketh its Measure in the *Æquinoctial*, between the *Meridian* and the said *Circle* of *Declinatial*, and this they call, The *Distance of the Star from the Meridian.* So likewise the *Angle* which the *Meridian* maketh with the *Vertical Circle* at the *Zenith*, taketh its *Measure* in the *Horizon*, between the *Meridian* and the said *Vertical*; and this they call, The *Azimuth of the Sun* or *Star*.

XXVII. The *Arks* to be measured in every *Principal Great Circle*, or its *Secondaries*, have also their proper *Appellations*: So the *Ark* of the *Æquinoctial*, which is comprehended between the beginning of ♈, and the *Circle* of *Declination* passing by any *Star*, is called the *Right Ascension* of that *Star*: And the *Ark* which in the said Circle of *Declination*, is between the *Æquinoctial* and the *Star*, is called, The *Declination* of that *Star*. But the *Oblique Ascension* of a *Star*, is an *Ark* of the *Æquinoctial*, reckoned from the beginning of ♈, to the Point of the *Æquinoctial Rising* with that *Star*: So likewise, the *Oblique Descension* of a *Star* is, the *Ark* of the *Æquinoctial* reckoned from the beginning of ♈, to that Point of the *Æquinoctial*, which is a setting together with the *Star*. Now the Difference that is between the *Right* and *Oblique Ascension* of the *Sun*, or any *Star*, is called, *The Ascensional Difference.* Moreover, the *Ark* of the *Ecliptick*, which is between the beginning of ♈,

and

Lon-
le of
Stars
Globe.
le of
So
tial,
paſ-
re is,
tial

Fig. XXV.

n or
Point
mpli-
the
The
rcle,
tar:
Cir-

A2

Fig. XXV.

is
tl

a
t
r

XX
t

X

X

and *Circle* of *Latitude* passing by any *Star*, is called, The *Longitude* of that *Star*: And the *Ark*, which in the said *Circle* of *Latitude*, is between the *Ecliptick* and the *Star*, is the *Stars Latitude*: And all this is to be understood of the *Cœlestial Globe*. But upon the *Terrestrial Globe*, the *Longitude* and *Latitude* of any *Place* are referred to the *Æquinoctial* and *Meridian*: So the *Longitude* of an *Earthly Place* is an Ark of the *Æquinoctial*, intercepted between the *First Meridian*, and the *Meridian* passing by the same *Place*. And the *Latitude* of the same *Place* is, an Ark of the *Meridian*, to be reckoned from the *Æquinoctial* to the Place upon the *Globe*.

Fig. XXV.

XVIII. In the *Horizon* we reckon the *Amplitude* of the *Sun* or any *Star*, between the true *East* or *West* Points, and that Point where the *Sun* or *Star* doth Rise or Set: And the said *Amplitude* is either *North* or *South*, according to the Beaming of the *Sun* or *Star*, in respect of the true *East* or *West Points*. The *Altitude* of the *Sun* or a *Star*, is taken in the *Vertical Circle*, passing by the same, between the *Horizon* and the said *Star*: So the *Depression* of the *Star* is, An *Arch* of the *Vertical Circle*, between the *Horizon* and the said *Star*.

ANCILLA

ANCILLA MATHEMATICA.

VEL,

Trigonometria Practica.

SECTION III.

OF

GEOGRAPHY.

THE following *Geographical Problems* being first to be performed upon the *Terrestrial Globe*; upon which the *Spherical Triangle*, that resolves any *Question* is discovered, in order to the *Trigonometrical Calculation*: I conceive it necessary, in the first place, to insert this *General*

PROBLEM.

How to Measure *the* Sides *and* Angles, *of all* Spherical Triangles, *upon the Convex Superficies of the* Globe.

Fig. XVI.

THE *Sides* of all *Spherical Triangles* upon the *Globe*, are *Measured* by the *Degrees* of those *Great Circles*, that make (or constitute) the *Triangle*, contained between the Two *Angular Points*.

1. If the *Side*, or *Sides*, of the *Triangle* to be measured, do consist of such *Great Circles* as are actually divided into *Degrees* upon the *Globe*, or its Appendants; as the *Æquinoctial*, the *Colures*, the *Ecliptick*, the general *Meridian* or *Horizon*: Then, the number of *Degrees* contained in that *Great Circle*, contained between the Two *Angular Points*; is the Quantity of that *Side* of that *Triangle* in *Degrees*. But,

2. If

2. If the *Side* or *Sides* of the *Triangle* be composed of *Arches* of such *Great Circles* as are not actually divided (as all *Circles* of *Longitude*, and other *Oblique Great Circles*) then, take the Length of such Side in a Pair of *Calliper Compasses*, and apply it to any of the forementioned *Great Circles* (as the *Æquinoctial*, &c.) it shall thereupon shew you the Quantity of that *Side* in *Degrees*. —Or, the *Quadrant of Altitude* (but rather, a thin *Plate* of *Brass* longer than the *Quadrant of Altitude*, divided into *Degrees*, as the *Quadrant* is) applied to the *Side* to be *Measured*, between the Two *Angular Points*, shall give you the Quantity of the *Degrees* of that *Side* of the *Triangle*.

Fig. XXVI.

II. *For the* Angles.

The *Angles* of *Spherical Triangles* are *Measured* upon the Superficies of the *Globe*; by counting (or setting off) 90 Deg. from the *Angular Point*, of the *Angle* to be *Measured*, upon both the *Sides* which contains the *Angle* to be *Measured*: And at the Terminations of those 90 Deg. on both the *Sides*, make Two small *Marks* upon the *Globe*. Unto these Two *Marks*, apply the *Quadrant of Altitude*, or thin *Plate* of *Brass*; so the Number of the *Degrees* thereof, contained between the Two *Marks*, is the Quantity of that *Angle*.

Geographical Problems.

PROB. I.

To find the Longitude *of any* Place, *described upon the* Terrestrial Globe.

LOngitude is the Distance of a Place from the first *Meridian* reckoned in the Degrees of the *Equator*, beginning, as was said, in the *New Terrestrial Globe*, (made by Mr. *Morden*) as St. *Michael's* Island in the *Azores*.

Practice.] Bring the Place, (that is, the Mark of the Place) suppose *London*, to the *Brazen Meridian*; then count how many Degrees of the *Equator* are contained between the first *Meridian*, and that of *London* cut by the *Brazen Meridian*, which you will find to be 28 Deg. and that is the *Longitude* required. And in this manner you find

London

Fig. XXV.

		D.	M.
London	to be distant from the first *Meridian* by these *New Globes.*	28	0
Jerusalem		66	30
Jedo in *Japan*		167	0
Rio de la plata		32	0
Mexico		75	0
Charleton Isle		51	30

PROB. II.

To find the Latitude *of any Place.*

THE *Latitude* of a Place, is the Distance of the *Equator* from the Parallel of that Place, reckoned in the Degrees of the *Brass Meridian*; and is either *North* or *South*, according as it lyes between the *North* or *South Poles* of the *Equator*.

To find the *Latitude*, bring the Mark of the Place, suppose *London*, to be the *Brazen Meridian*; then count the Number of Degrees upon the *Meridian*, contained between the *Equator* and the Place ⊙. Thus you shall find the *Latitude*, by this new *Globe*, of *London*, to be 51 Deg. 30 Min. and of

	D.	M.		D.	M.
Labor in the *Mogul's* Country to be	31	30	By other *Globes* and *Maps.*	23	30
The *South* Part of the *Caspian* Sea to be	37	0		41	0
Astracan on the *North* Part of the *Caspian* Sea to be	46	0		49	0
The *North* Part of *China* to be	42	0		52	0
Delli in *India* to be	28	0		21	0

PROB. III.

Two Places, which differ only in Latitude, *to find their* Distance.

IN this there are Two Varieties of Position.

1. If both the *Places* lye under the same *Meridian*, and on one and the same Side of the *Æquinoctial*: *Substract the Lesser Latitude from the Greater, the Difference* (or Remainder) *reduced into Miles,* (by allowing 60 Deg. to One Mile) *shall give you the Distance.*

Exam-

Example. *London* and *Ribadio* lye both under the same *Meridian*, but differ in *Latitude:* For *London* hath 51 Deg. 30 Min. of *Latitude*, at L, and *Ribadio* hath 34 Deg. of *Latitude*, at R, both *North*; the *Difference* of *Latitude* is 17 Deg. 30 Min. equal to the *Arch* L R: And that Reduced into *Miles*, makes 1050 for their *Distance*.

Fig. XXVI.

2. If the Two *Places* lye under the same *Meridian*, but in different *Hemispheres*, *i. e.* one on the *North*, and the other on the *South* Side of the *Æquinoctial:* Then, *Add both the Latitudes together, and the Sum of them is their Distance.*

Example. *London*, and the Island *Tristan Dacunhu*, lye both under One *Meridian*, but *London* hath 51 Deg. 30 Min. of *North Latitude*, at L, and the Island hath 34 Deg. of *South Latitude*, at D; the Sum of these Two *Latitudes* is 85 Deg. 30 Min. equal to the *Arch* of the *Meridian* L Æ D; the which reduced into *Miles*, (by multiplying the Degrees by 60, and allowing for every Minute One Mile) makes 5130 Miles, for their Distance.

PROB. IV.

Two Places, which differ in Longitude *only; To find their Distance.*

IN this there are Two Varieties of Position.

1. If the Two Places lye both under the *Æquinoctial*, and have no *Latitude*; in this Case, *Their Difference of Longitude* (if it be less than 180 Deg) *is their Distance:* But, if the Difference exceed 180 Deg. Subtract it from 360 Deg. and the Remainder is their *Distance*, in Degrees.

Example. The Island *Samatra*, and Island St. *Thomas*, lye both under the *Æquinoctial*: St. *Thomas* having 22 Deg. 10 Min. of *Longitude* at T, and the Island *Samatra* 82 Deg. 10 Min. at S. Now, the Lesser Longitude 22 Deg. 10 Min. substracted from the Greater 82 Deg. 10 Min. leaves 60 Deg. equal to the Arch S T, for their Difference in Degrees: Which converted into Miles, makes 3600, and so many Miles are the Two Islands distant from each other.

2. But if the Two *Places* differ only in *Longitude*, and lye not in the *Æquinoctial*, but under some other intermediate *Parallel* of *Latitude:* As *Hierusalem* at H, and *Baldo* at B; both in the *Parallel* of 31 Deg. 40 Min. of North *Latitude*, but differing in *Longitude* 60 Deg. 15 Min. equal to the *Angle* H P B, to find the Distance of these Two Places.

Fig. XXVI.

I. *By the* Globe.

Apply the *Quadrant* of *Altitude*, or *Brass Plate*, to the Two Places, and the Number of Degrees thereof contained between the Two Places, is their Distance, which will be found to be 50 Deg. 32 Min.

II. *By* Trigonometrical Calculation.

The *Quadrant* of *Altitude* (or Brass Plate) applied to the Two Places, is represented by the *Arch* H B, and the *Arches* of the Two *Meridians*, which pass through the Two Places, are P B N, and P H M; and P B and P H, are equal to the Complement of of the *Latitude* of both the Places, *viz.* 58 Deg. 20 Min. So that now you have constituted upon the *Globe* an *Oblique Spherical Triangle* P B H; in which you have given, (1.) The Two Sides P B and P H, both equal to 58 Deg. 20 Min. the Complement of the *Latitude*. (2.) The *Angle* B P H 60 Deg. 15 Min. the *Difference* of the *Longitude* of the Two given *Places*. To find the *Side* B H, their *Distance*. For which this is

The Canon for Calculation. By *CASE* I. of R. *A. S.* T

As the Radius, Sine 90 Deg.
Is to the Co-sine of the *Common Latitude* (P H or P B) 58 D. 20 M.
So is the Sine of half the *Difference of Longitude*, (half B P H) 30 Deg. 07 $\frac{1}{2}$ Min.
To to Sine of half the Distance (half B H) 25 Deg. 16 Min.
The Double whereof, 50 Deg. 32 Min. is the *Distance* B H, which in Miles is 3032 Miles.

PROB. V.

Two Places, *which differ both in* Longitude *and* Latitude; *to find their* Distance.

IN this there are Three various *Positions*.

1. If one of the Places lye under the *Æquinoctial*, and so have no *Latitude*; and the other under some *Parallel* of *Latitude* between the *Æquinoctial*, and one of the *Poles*: As *London*, in 51 Deg. 30 Min. of North *Latitude* at L; and St. *Thomas* Island under the *Æquinoctial* at T, but differ in *Longitude* 18 Deg. For finding the *Distance* of these Two *Places*.

I. Upon

I. *Upon the* Terrestrial Globe.

Fig. XXVI.

Bring *London* to the *Brass Meridian*, and over it, fix the *Quadrant of Altitude:* The *Globe* being in this Position, bring the *Quadrant of Altitude* to lye just over St. *Thomas* Island, and you will find it cut the *Quadrant of Altitude*, in 54 Deg. 45 Min. for the *Distance* of the Two Places.

II. *By* Trigonometrical Calculation.

The *Globe* resting in the former Position, you will find constituted upon it a *Right-angled Spherical Triangle* L Æ T, composed of, (1.) L Æ, an *Arch* of the *Brass Meridian.* (2.) Æ T, an *Arch* of the *Æquinoctial.* And, (3.) L T, an *Arch* of a *Great Circle* (made by the *Quadrant of Altitude*) passing through both the Places: And in this *Triangle*, you have given, (besides the *Right Angle* at Æ) (1.) The *Perpendicular* L Æ, the *Latitude* of *London*, 51 Deg. 30 Min. (2.) The *Angle* at L, the *Difference* of *Longitude* 18 Deg. 10 Min. To find the Hypotenuse L T, the *Distance.*

The Canon for Calculation. By C A S E XIV. of *R. A. S. T.*

As Tang. of the Latitude L Æ, 51 Deg. 30 Min.
Is to the Radius;
So is the Co-sine of Æ L T, 18 Deg. 10 Min.
To the Co-tangent of L T, 52 Deg. 55 Min.
Which reduced into Miles, makes 3175 Miles, for *Distance* between *London*, and St. *Thomas* Island.

2. If both the *Places* proposed shall be without the *Æquinoctial*, but both of them, either on the *North* or *South* Side thereof: As *London* in 51 Deg. 30 Min. at L, and *Hierusalem* in 31 Deg. 40 Min. at H, both on the *North* Side of the *Æquinoctial*; and their Difference of *Longitude* 46 Deg: To find their *Distance*

Upon the Terrestrial Globe.

Bring one of the Places, as *London*, to the *Brass Meridian*, and over it screw the *Quadrant of Altitude*, and keep the *Globe* there fixed, then move the *Quadrant of Altitude*, till it lye over *Hierusalem*, and you shall find it to lye under 39 Deg. 32 Min. of the *Quadrant:* And that is the *Distance* of the Two Places.

By Trigonometrical Calculation.

Fig. XXVI.

The *Globe* being in the former Position, you will find upon it an *Oblique-angled Spherical Triangle*, composed of P L, an Arch of the *Brass Meridian:* Of P H, an Arch of the *Meridian* that passeth over *Hierusalem*; and of L H, an Arch of the *Quadrant of Altitude*. In which *Triangle* you have given, (1.) P L, the Complement of the *Latitude* of *London*, 38 Deg. 30 Min. (2.) The Side P H (the Complement of the *Latitude* of *Hierusalem* 59 Deg. 20 Min. (3.) The Angle L P H, the *Difference of Longitude* of the Two Places, 46 Deg. To find the Two Angles of *Position* at L and H: And, (2.) The Side L H, the *Distance* of the Two Places.

The Canons for Calculation.

By *CASE* III. and *CASE* IX. of *A.O.S.T.*

	D.	M.
The Side { P L / P H } is	38	30
	59	20
Their Sum is	96	50
Their Difference is	19	50
The Half Sum is	48	25
The Half Difference is	9	55

Being thus prepared, I say,

(1.) As the Sine of half the Sum of the Sides P L and P H, 48 Deg. 25 Min.

Is to the Sine of half their *Difference*, 9 Deg. 55 Min.

So is the Co-tangent of half the *Difference* of *Longitude* (*i. e.*) half the Angle L P H, 23 Deg.

To the Tangent of 28 Deg. 29 Min:

Which is the half *Difference* of the *Angles* P L H and P H L.

(2.) As the Co-sine of half the Sum of the Sides P L and P H, 41 Deg. 35 Min.

Is to the Co-sine of half their *Difference*, 80 Deg. 2 Min.

So is the Co-tangent of half the *Difference* of *Longitude*, *i. e.* half the Angle L P H 23 Deg.

To the Tangent of 74 Deg. 2 Min.

Which is the Tangent of half the Sum of the Two Angles P L H and P H L.

The

	D.	M.		
The Half Sum is	74	02		
The Half Difference is	28	29		
Their Sum	102	31	equal to ∢ P L H.	Fig. XXVI.
Their Difference	45	32	equal to ∢ P H L.	

(3.) For the Side L H, which is the *Distance*,

As the Sine of P H L, 45 Deg. 33 Min.

Is to the Sine of P L, 38 Deg. 30 Min.

So is the Sine of L P H, the Differ. of *Longitude*, 46 Deg.

To the Sine of L H, 30 Deg. 51 Min.

Which 38 Deg. 31 Min. reduced in Miles, makes 2331 Miles for the *Distance* of the Two Places.

3. If the Two Places proposed should be so situate, that one have *North*, and the other *South*, *Latitude*, and under different *Meridians*. As suppose *Constantinople*, lying in *North Latitude* 47 D. at C. and the *Cape* of *Good Hope*, lying in 35 Deg. of *South Latitude*, at V; and differing in *Longitude* 59 Deg. To find the *Distance* of these Two Places

Upon the Terrestrial Globe.

Bring one of the Places (as *Constantinople*) to the *Brass Meridian*, and there keep the *Globe*; then apply the *Quadrant of Altitude* (or rather a thin Plate of Brass divided as that is) to the Two *Places*, and you shall find 97 Deg. 42 Min. of the *Quadrant* (or *Brass Plate*) to be contained between them, and that is their *Distance*.

By Trigonometrical Calculation.

The *Globe* resting in its former Position, you will discover upon it an *Oblique-angled Spherical Triangle*, composed of P C, an Arch of the *Brass Meridian*: P V, an Arch of a *Meridian*, passing through the Place in *South Latitude*: And of C V, the *Quadrant* (or *Brass Plate*) representing the Arch of a *Great Circle* passing through both the *Places*: —In which *Triangle* you have given, (1.) The Side P C, the *Complement* of the *Latitude* of *Constantinople*, 43 Deg. (2.) The Side P V, the *South Latitude* of the Cape of *Good Hope*, with 90 Deg. added, which make 125 Deg. And, (3.) The Angle V P C, the *Difference* of *Longitude*, 59 Deg. To find, (1.) The Angles of *Position* P V C and P C V: And, (2.) The third Side V C, the *Distance* of the Two Places.

The

Fig. XXVI. *The Canons for Calculations, as in the last, by* CASE III. *and* IX. *of A. O. S. T.*

(1.) As the Sine of the *Half Sum* of the *Sides* P V and P C, 84 D.
Is to the Sine of half their *Difference*, 41 Deg.
So is the Co-tangent of half the *Difference of Longitude*, *i. e.* (half the Angle V P C) 29 Deg. 30 Min.
To the Tangent of 49 Deg. 23 Min.
Which is the *Half Difference* of the Angles P V C and P C V.

(2.) As the Co-sine of half the *Sum* of the *Sides*, P V and P C, 6 Deg.
Is to the Co-sine of half their *Difference*, 49 Deg.
So is the Co-tangent of half the *Difference of Longitude*, 61 Deg.
To the Tangent of 85 Deg. 32 Min.
Which is half the *Sum* of the Two Angles, P C V and P V C.

	D.	*M.*	
The half Sum is	85	32	
The half Difference is	49	23	
Their Sum	134	55	equal to ∠ P C V.
Their Difference	36	09	equal to ∠ P V C.

The Two *Angles* of *Position*.

(3.) For the Side V C, which is the *Distance* of the Two *Places*.
As the Sine of P V C, 36 Deg. 9 Min.
Is to the Sine of P C, 43 Deg.
So is the Sine of V P C, 59 Deg.
To the Sine of 82 Deg. 18 Min.
Whose Complement to 180 Deg. is 97 Deg. 42 Min. for the Side V C, which is the *Distance* of the Two *Places*. Which in Miles is 5862.

And these are all the Varieties of Positions that any Two *Places* upon the *Globe* can be situate.

ANCILLA

ANCILLA MATHEMATICA.
VEL,
Trigonometria Practica.

SECTION IV.

OF
ASTRONOMY.

ADVERTISEMENT.

WHereas, for the resolving of the following *Astronomical Problems* by *Trigonometrical Calculation*, it is absolutely necessary, that the true *Place* (or *Longitude*) of the *Sun* in the *Ecliptick* be first known; I have therefore inserted *Astronomical Tables* of the *Sun's Mean Longitude* and *Anomaly* in *Years*, *Months*, *Days* and *Hours*; and also of *Æquation*, whereby the Sun's true Place in the *Ecliptick* may be found at any time. It is also to be noted, That the several *Problems* in this *Section*, being wrought first by the *Cælestial Globe*, in order to their *Trigonometrical Calculation*; I have, therefore, to inform the Fancy, and ease the Memory, of the Reader (in all the *Schemes* relating to them, (which consist of *Great Circles* of the *Sphere*, and represent the *Triangle* to the *Eye*, which the *Problem* is to be resolved by) noted the *Circles*, *Lines* and *Points*, in all the *Schemes*, with the same *Letters* or *Characters*; and prefixed *Them*, and their *Significations*, next before the *Problems*.

Fig. XXVII.

TABLE

Fig. XXVII.

TABLE I.

The Sun's Mean Motion in Years.

	Longit. ☉.			Anomal. ☉				Longit. ☉			Longit. ☉		
Years.	S.	D.	M.	S.	D.	M.	Years.	S.	D.	M.	S.	D.	M.
B 1700	9	19	58	6	12	33	1731	9	20	27	6	12	30
1701	9	20	42	6	13	16	B 1732	9	20	12	6	12	14
1702	9	20	29	6	13	1	1733	9	20	57	6	12	58
1703	9	20	14	6	12	46	1734	9	20	43	6	12	43
B 1704	9	20	0	6	12	31	1735	9	20	29	6	12	27
1705	9	20	45	6	13	14	B 1736	9	20	14	6	12	12
1706	9	20	30	6	12	59	1737	9	20	59	6	12	56
1707	9	20	16	6	12	43	1738	9	20	45	6	12	40
B 1708	9	20	2	6	12	28	1739	9	20	30	6	12	25
1709	9	20	46	6	13	12	B 1740	9	20	16	6	12	10
1710	9	20	32	6	12	56	1741	9	21	1	6	12	53
1711	9	20	18	6	12	41	1742	9	20	46	5	12	38
B 1712	9	20	3	6	12	26	1743	9	20	32	6	12	23
1713	9	20	48	6	13	10	B 1744	9	20	18	6	12	7
1714	9	20	24	6	12	54	1745	9	21	2	6	12	51
1715	9	20	30	6	12	39	1746	9	20	48	6	12	36
B 1716	9	20	5	6	12	23	1747	9	20	34	6	12	20
1717	9	20	50	6	13	7	B 1748	9	20	20	6	12	5
1718	9	20	36	6	12	52	1749	9	21	4	6	12	49
1719	9	20	21	6	12	37	1750	9	20	50	6	12	33
B 1720	9	20	7	6	12	21	1751	9	20	36	6	12	18
1721	9	20	52	6	13	5	B 1752	9	20	21	6	12	2
1722	9	20	37	6	12	50	1753	9	21	6	6	12	46
1723	9	20	23	6	12	34	1754	9	20	52	6	12	31
B 1724	9	20	9	6	12	19	1755	9	20	38	6	12	16
1725	9	20	54	6	13	2	B 1756	9	20	23	6	12	0
1726	9	20	39	6	12	47	1757	9	21	8	6	12	44
1727	9	20	25	6	12	32	1758	9	20	54	6	12	29
B 1728	9	20	11	6	12	16	1759	9	20	39	5	12	13
1729	9	20	55	6	13	0	B 1760	9	20	25	6	11	58
1730	9	20	41	6	12	45	1761	9	21	10	6	12	42

TABLE

TABLE II.

The Sun's Mean Motion in Months and Days.

Days in Years.		JANUARY.						Days in Years.		FEBRUARY.					
		Longit. ☉			Anom. ☉					Longit. ☉			Anom. ☉		
Com.	Bis.	S.	D.	M.	S.	D.	M.	Com.	Bis.	S.	D.	M.	S.	D.	M.
1	0	0	0	59	0	0	59	1	0	1	1	32	1	1	32
2	1	0	1	58	0	1	58	2	1	1	2	32	1	2	31
3	2	0	2	57	0	2	57	3	2	1	3	31		3	31
4	3	0	3	56	0	3	56	4	3	1	4	30		4	30
5	4	0	4	56	0	4	56	5	4	1	5	29	1	5	29
6	5	0	5	55	0	5	55	6	5	1	6	28	1	6	28
7	6	0	6	54	0	6	54	7	6	1	7	27	1	7	27
8	7	0	7	53	0	7	53	8	7	1	8	26	1	8	26
9	8	0	8	52	0	8	52	9	8	1	9	25	1	9	25
10	9	0	9	51	0	9	51	10	9	1	10	25	1	10	24
11	10	0	10	50	0	10	50	11	10	1	11	24	1	11	24
12	11	0	11	50	0	11	50	12	11	1	12	23	1	12	23
13	12	0	12	49	0	12	49	13	12	1	13	22	1	13	22
14	13	0	13	48	0	13	47	14	13	1	14	21	1	14	21
15	14	0	14	47	0	14	47	15	14	1	15	20	1	15	20
16	15	0	15	46	0	15	46	16	15	1	16	19	1	16	19
17	16	0	16	45	0	16	45	17	16	1	17	19	1	17	18
18	17	0	17	44	0	17	44	18	17	1	18	18	1	18	18
19	18	0	18	44	0	18	43	19	18	1	19	17	1	19	17
20	19	0	19	42	0	19	43	20	19	1	20	16	1	20	16
21	20	0	20	42	0	20	42	21	20	1	21	15	1	21	15
22	21	0	21	41	0	21	41	22	21	1	22	14	1	22	14
23	22	0	22	40	0	22	40	23	22	1	23	13	1	23	13
24	23	0	23	39	0	23	39	24	23	1	24	13	1	24	12
25	24	0	24	38	0	24	38	25	24	1	25	12	1	25	12
26	25	0	25	38	0	25	37	26	25	1	26	11	1	26	11
27	26	0	26	37	0	26	37	27	26	1	27	10	1	27	10
28	27	0	27	36	0	27	35	28	27	1	28	9	1	28	9
29	28	0	28	35	0	28	35	29	28	1	29	8	1	29	8
30	29	0	29	34	0	29	34								
31	30	1	0	33	1	0	33								

TABLE II.

The Sun's Mean Motion in Months and Days.

		March.								*April.*					
		Longit. ☉			*Anom.* ☉					*Longit.* ☉			*Anom.* ☉		
Com.	*Bis.*	*S.*	*D.*	*M.*	*S.*	*D.*	*M.*	*Com.*	*Bis.*	*S.*	*D.*	*M.*	*S.*	*D.*	*M.*
1	0	1	29	8	1	29	8	1	0	3	29	42	2	29	41
2	1	2	0	7	2	0	7	2	1	3	0	41	3	0	40
3	2	2	1	6	2	1	6	3	2	3	1	40	3	1	40
4	3	2	2	6	2		6	4	3	3	2	39	3	2	39
5	4	2	3	5	2	3	5	5	4	3	3	38	3	3	38
6	5	2	4	4	2	4	4	6	5	3	4	37	3	4	37
7	6	2	5	3	2	5	3	7	6	3	5	36	3	5	36
8	7	2	6	2	2	6	2	8	7	3	6	36	3	6	35
9	8	2	7	1	2	7	1	9	8	3	7	35	3	7	34
10	9	2	8	0	2	8	0	10	9	3	8	34	3	8	34
11	10	2	9	0	2	8	59	11	10	3	9	33	3	9	33
12	11	2	9	59	2	9	58	12	11	3	10	32	3	10	32
13	12	2	10	58	2	10	58	13	12	3	11	31	3	11	31
14	13	2	11	57	2	11	57	14	13	3	12	30	3	12	30
15	14	2	12	56	2	12	56	15	14	3	13	30	3	13	29
16	15	2	13	55	2	13	55	16	15	3	14	29	3	14	28
17	16	2	14	54	2	14	54	17	16	3	15	28	3	15	27
18	17	2	15	54	2	15	53	18	17	3	16	27	3	16	27
19	18	2	16	53	2	16	53	19	18	3	17	26	3	17	26
20	19	2	17	52	2	17	52	20	19	3	18	25	3	18	25
21	20	2	18	51	2	18	51	21	20	3	19	24	3	19	24
22	21	2	19	50	2	19	50	22	21	3	20	23	3	20	23
23	22	2	20	49	2	20	49	23	22	3	21	23	3	21	22
24	23	2	21	48	2	21	48	24	23	3	22	22	3	22	21
25	24	2	22	49	2	22	47	25	24	3	23	21	3	23	21
26	25	2	23	47	2	23	46	26	25	3	24	20	3	24	20
27	26	2	24	46	2	24	46	27	26	3	25	19	3	25	19
28	27	2	25	45	2	25	45	28	27	3	26	18	3	26	18
29	28	2	26	44	2	26	44	29	28	3	27	17	3	27	17
30	29	2	27	43	2	27	43	30	29	3	28	17	3	28	16
31	30	2	28	42	2	28	42	31	30	3	29	16	3	29	15

TABLE II.

The Sun's Mean Motion in Months and Days.

		MAY.								*JUNE.*					
		Longit. ☉			*Anom.* ☉					*Longit.* ☉			*Anom.* ☉		
Com.	*Bis.*	*S.*	*D.*	*M.*	*S.*	*D.*	*M.*	*Com.*	*Bis.*	*S.*	*D.*	*M.*	*S.*	*D.*	*M.*
1	0	3	29	16	3	29	15	1	0	4	29	49	4	29	49
2	1	4	0	15	4	0	14	2	1	5	0	48	5	0	48
3	2	4	1	14	4	1	14	3	2	5	1	47	5	1	47
4	3	4	2	13	4	2	13	4	3	5	2	46	5	2	46
5	4	4	3	12	4	3	12	5	4	5	3	46	5	3	45
6	5	4	4	11	4	4	11	6	5	5	4	45	5	4	44
7	6	4	5	10	4	5	10	7	6	5	5	44	5	5	43
8	7	4	6	10	4	6	9	8	7	5	6	43	5	6	43
9	8	4	7	9	4	7	8	9	8	5	7	42	5	7	42
10	9	4	8	8	4	8	8	10	9	5	8	41	5	8	41
11	10	4	9	7	4	9	7	11	10	5	9	40	5	9	40
12	11	4	10	6	4	10	6	12	11	5	10	40	5	10	39
13	12	4	11	5	4	11	5	13	12	5	11	39	5	11	38
14	13	4	12	4	4	12	4	14	13	5	12	38	5	12	37
15	14	4	13	4	4	13	3	15	14	5	13	37	5	13	36
16	15	4	14	3	4	14	2	16	15	5	14	36	5	14	36
17	16	4	15	2	4	15	2	17	16	5	15	35	5	15	35
18	17	4	16	1	4	16	1	18	17	5	16	34	5	16	34
19	18	4	17	0	4	17	0	19	18	5	17	33	5	17	33
20	19	4	17	59	4	17	59	20	19	5	18	33	5	18	32
21	20	4	18	58	4	18	58	21	20	5	19	32	5	19	31
22	21	4	19	58	4	19	57	22	21	5	20	31	5	20	30
23	22	4	20	57	4	20	56	23	22	5	21	30	5	21	29
24	23	4	21	56	4	21	55	24	23	5	22	29	5	22	28
25	24	4	22	55	4	22	55	25	24	5	23	28	5	23	28
26	25	4	23	54	4	23	54	26	25	5	24	28	5	24	27
27	26	4	24	53	4	24	53	27	26	5	25	27	5	25	26
28	27	4	25	52	4	25	52	28	27	5	26	26	5	26	25
29	28	4	26	51	4	26	51	29	28	5	27	25	5	27	24
30	29	4	27	51	4	27	50	30	29	5	28	24	5	28	24
31	30	4	28	50	4	28	49	31	30	5	29	23	5	29	23

TABLE II.

The Sun's Mean Motion in Months and Days.

		JULY.								*AUGUST.*					
		Longit. ☉			*Anom.* ☉					*Longit.* ☉			*Anom.* ☉		
Com.	*Bis.*	S.	D.	M.	S.	D.	M.	*Com.*	*Bis.*	S.	D.	M.	S.	D.	M.
1	0	5	29	23	5	29	23	1	0	6	29	56	6	29	56
2	1	6	0	22	6	0	22	2	1	7	0	56	7	0	55
3	2	6	1	21	6	1	21	3	2	7	1	55	7	1	54
4	3	6	2	21	5	2	20	4	3	7	2	54	7	2	53
5	4	6	3	20	6	3	19	5	4	7	3	53	7	3	52
6	5	6	4	19	6	4	18	6	5	7	4	52	7	4	52
7	6	6	5	18	5	5	17	7	6	7	5	51	7	5	51
8	7	6	6	17	5	6	17	8	7	7	6	50	7	6	50
9	8	6	7	18	5	7	16	9	8	7	7	50	7	7	49
10	9	6	8	16	5	8	15	10	9	7	8	49	7	8	48
11	10	6	9	15	6	9	14	11	10	7	9	48	7	9	47
12	11	6	10	14	6	10	13	12	11	7	10	47	7	10	46
13	12	6	11	13	6	11	12	13	12	7	11	46	7	11	45
14	13	6	12	12	6	12	11	14	13	7	12	45	7	12	45
15	14	6	13	11	6	13	11	15	14	7	13	44	7	13	44
16	15	6	14	10	6	14	10	16	15	7	14	44	7	14	43
17	16	6	15	9	6	15	9	17	16	7	15	43	7	15	42
18	17	6	16	9	6	16	8	18	17	7	16	42	7	16	41
19	18	6	17	8	6	17	7	19	18	7	17	41	7	17	40
20	19	6	18	7	6	18	6	20	19	7	18	40	7	18	39
21	20	6	19	6	6	19	5	21	20	7	19	39	7	19	39
22	21	6	20	5	6	20	5	22	21	7	20	38	7	20	38
23	22	6	21	4	5	21	4	23	22	7	21	38	7	21	37
24	23	6	22	3	6	22	3	24	23	7	22	37	7	22	36
25	24	6	23	3	6	23	2	25	24	7	23	36	7	23	35
26	25	6	24	2	5	24	1	26	25	7	24	35	7	24	34
27	26	6	25	1	5	25	0	27	26	7	25	34	7	25	33
28	27	5	26	0	6	25	59	28	27	7	26	33	7	26	33
29	28	6	26	59	6	26	59	29	28	7	27	32	7	27	32
30	29	6	27	58	6	27	58	30	29	7	28	31	7	28	31
31	30	6	28	57	6	28	57	31	30	7	29	31	7	29	30

TABLE

TABLE II.

The Sun's Mean Motion in Months and Days.

		SEPTEMBER.								OCTOBER.					
		Longit. ☉			Anom. ☉					Longit. ☉			Anom. ☉		
Com.	Biss.	S.	D.	M	S.	D.	M	Com.	Biss.	S.	D.	M.	S.	D.	M
1	0	8	0	30	8	0	25	1	0	9	0	4	9	0	3
2	1	8	1	29	8	1	28	2	1	9	1	3	9	1	2
3	2	8	2	28	8	2	28	3	2	9	2	2	9	2	2
4	3	8	3	27	8	3	27	4	3	9	3	1	9	3	
5	4	8	4	26	8	4	26	5	4	9	4	1	9	4	[illegible]
6	5	8	5	25	8	5	25	6	5	9	5	0	9	4	59
7	6	8	6	25	8	6	24	7	6	9	5	59	9	5	58
8	7	8	7	24	8	7	23	8	7	9	6	58	9	6	57
9	8	8	8	23	8	8	22	9	8	9	7	57	9	7	56
10	9	8	9	22	8	9	21	10	9	9	8	56	9	8	55
11	10	8	10	21	8	10	21	11	10	9	9	55	9	9	54
12	11	8	11	20	8	11	20	12	11	9	10	54	9	10	54
13	12	8	12	19	8	12	19	13	12	9	11	54	9	11	53
14	13	8	13	19	8	13	18	14	13	9	12	53	9	12	52
15	14	8	14	18	8	14	17	15	14	9	13	52	9	13	51
16	15	8	15	17	8	15	16	6	15	9	14	51	9	14	50
17	16	8	16	16	8	16	15	7	16	9	15	50	9	15	49
18	17	8	17	15	8	17	15	8	17	9	16	49	9	16	48
19	18	8	18	14	8	18	14	9	18	9	17	48	9	17	48
20	19	8	19	13	8	19	13	20	19	9	18	48	9	18	47
21	20	8	20	13	8	20	12	21	20	9	19	47	9	19	46
22	21	8	21	12	8	21	11	22	21	9	20	46	9	20	45
23	22	8	22	11	8	22	10	23	22	9	21	45	9	21	44
24	23	8	23	10	8	23	9	24	23	9	22	44	9	22	43
25	24	8	24	9	8	24	8	25	24	9	23	43	9	23	42
26	25	8	25	8	8	25	8	26	25	9	24	42	9	24	42
27	26	8	26	7	8	26	7	27	26	9	25	42	9	25	41
28	27	8	27	7	8	27	6	28	27	9	26	41	9	26	40
29	28	8	28	6	8	28	5	29	28	9	27	40	9	27	39
30	29	8	29	5	8	29	4	20	29	9	28	39	9	28	38
31	30	9	0	4	9	0	3	31	30	9	29	38	9	29	37

TABL.

TABLE II.

The Sun's Mean Motion in Months and Days.

Co.	Bi.	NOVEMBER. Longit. ☉ S.	D.	M.	Anom. ☉ S.	D.	M.
1	0	10	0	37	10	0	36
2	1	10	1	36	10	1	35
3	2	10	2	36	10	2	35
4	3	10	3	35	10	3	34
5	4	10	4	34	10	4	33
6	5	10	5	33	10	5	32
7	6	10	6	32	10	6	31
8	7	10	7	31	10	7	30
9	8	10	8	30	10	8	29
10	9	10	9	30	10	9	29
11	10	10	10	29	10	10	28
12	11	10	11	28	10	11	27
13	12	10	12	27	10	12	26
14	13	10	13	26	10	13	25
15	14	10	14	25	10	14	24
16	15	10	15	24	10	15	23
17	16	10	16	23	10	16	23
18	17	10	17	23	10	17	22
19	18	10	18	22	10	18	21
20	19	10	19	21	10	19	20
21	20	10	20	20	10	20	19
22	21	10	21	19	10	21	18
23	22	10	22	18	10	22	17
24	23	10	23	17	10	23	16
25	24	10	24	17	10	24	16
26	25	10	25	16	10	25	15
27	26	10	26	15	10	26	14
28	27	10	27	14	10	27	13
29	28	10	28	13	10	28	12
30	29	10	29	12	10	29	11
31	30	11	0	11	11	0	10

Co.	Bi.	DECEMBER. Longit. ☉ S.	D.	M.	Anom. ☉ S.	D.	M.
1	0	11	0	11	11	0	10
2	1	11	1	11	11	1	10
3	2	11	2	10	11	2	9
4	3	11	3	9	11	3	8
5	4	11	4	8	11	4	7
6	5	11	5	7	11	5	6
7	6	11	6	6	11	6	5
8	7	11	7	5	11	7	4
9	8	11	8	5	11	8	4
10	9	11	9	4	11	9	3
11	10	11	10	3	11	10	2
12	11	11	11	2	11	11	1
13	12	11	12	1	11	12	0
14	13	11	13	0	11	12	59
15	14	11	14	0	11	13	58
16	15	11	14	58	11	14	57
17	16	11	15	58	11	15	57
18	17	11	16	57	11	16	56
19	18	11	17	56	11	17	55
20	19	11	18	55	11	18	54
21	20	11	19	54	11	19	53
22	21	11	20	53	11	20	52
23	22	11	21	52	11	21	51
24	23	11	22	52	11	22	51
25	24	11	23	51	11	23	50
26	25	11	24	50	11	24	49
27	26	11	25	49	11	25	48
28	27	11	26	48	11	26	47
29	28	11	27	47	11	27	46
30	29	11	28	46	11	28	45
31	30	11	29	46	11	29	45

TABLE III.

The Sun's Mean Motion in Hours.

Hours.	Longitude. Mi.	Aphelion. Mi.
1	2	2
2	5	5
3	7	7
4	10	10
5	12	12
6	15	15
7	17	17
8	20	20
9	22	22
10	25	25
11	27	27
12	30	30
13	32	32
14	34	34
15	37	37
16	39	39
17	42	42
18	44	44
19	47	47
20	49	49
21	52	52
22	54	54
23	57	57
24	59	59

TABLE

TABLE IV.

A Table of the Sun's Æquation.

	Sign 0.		Sign 1.		Sign 2.		Sign 3.		Sign 4.		Sign 5.		
	Æ. Subf.		Æ. Subf.		Æ. Subf.		Æ. Subf.		Æ. Subf.		Æ. Subf.		
	D.	M.	D.	M.	D.	M.	D.	M.	D.	M.	D.	M.	
0	0	0		8	1	41	1	57	1	43	0	0	30
1	0	2	1	0	1	42		7		42	0	58	29
2	0	4	1	1	1	43	1	57	1	41	0	56	28
3	0	6	1	3	1	44	1	57	1	40	0	55	27
4	0	8	1	5	1	45		57	1	39	0	53	26
5	0	10	1	6	1	46		57	1	38	0	51	25
6	0	12	1	8	1	47		57	1	36	0	49	24
7	0	14	1	10	1	47		57	1	35	0	47	23
8	0	16	1	11	1	48		57	1	34	0	45	22
9	0	18	1	13	1	49	1	56	1	33	0	43	21
10	0	20	1	5	1	50	1	56	1	32	0	41	20
11	0	22	1	6	1	50	1	56	1	30	0	39	19
12	0	24	1	18	1	51	1	56	1	29	0	37	18
13	0	26	1	19	1	52	1	55	1	27	0	35	17
14	0	28	1	21	1	52	1	55	1	26	0	33	16
15	0	30	1	22	1	53	1	54	1	25	0	31	15
16	0	32	1	23	1	54	1	54	1	23	0	29	14
17	0	34	1	25	1	54	1	53	1	22	0	27	13
18	0	36	1	26	1	54	1	52	1	20	0	25	12
19	0	38	1	28		55	1	52	1	19	0	23	11
20	0	39	1	29	1	55	1	51	1	7	0	21	10
21	0	41	1	30	1	56	1	51	1	5	0	19	9
22	0	43	1	32	1	56	1	50	1	14	0	17	8
23	0	45	1	33	1	56	1	49	1	12	0	15	7
24	0	47	1	34	1	57	1	48	1	10	0	13	6
25	0	49	1	35	1	57	1	48	1	9	0	10	5
26	0	51	1	36	1	57	1	47	1	7	0	8	4
27	0	52	1	38	1	57	1	46	1	5	0	6	3
28	0	54	1	39	1	57	1	45	1	3	0	4	2
29	0	56	1	40	1	57	1	44	1	2	0	2	1
30	0	58	1	41	1	57	1	42	1	0	0	0	0
	Æ. Add.		Æ. Add.		Æ. Add.		Æ. Add.		Æ. Add.		Æ. Add.		
	11 Sign.		10 Sign.		9 Sign.		8 Sign.		7 Sign.		6 Sign.		

The Uſe of the Tables.

TO find the *Sun*'s true Place in the *Ecliptick* at any time, by theſe *Tables*, this is the

RULE.

From the ſeveral *Tables*, collect the *Years*, *Month*, and *Day* of the *Month*, and alſo the odd *Hours*, if any be, and ſet them down, one under another, with the reſpective *Longitudes* and *Anomalies* anſwering thereto: Then add the ſeveral *Longitudes* and *Anomalies* together, and with the *Sum* of the *Anomalies*, enter (*Table* IV.) of *Æquations*, which *Æquation*, *Add to*, or *Subſtract from*, the *Sum* of the *Longitudes*, and the *Sum* or *Remainder*, ſhall be the *Sun*'s true Place in the *Ecliptick* for the time propoſed.

How Time is to be computed.

Any *Day* begins upon its own *Noon*; ſo that the Firſt Day of *January*, at 12 at *Noon*, is the common *Term* of the *Old* and *New Years*.

Example.

Let the true Place of the *Sun* in the *Ecliptick* be required, for the 16th Day of *May*, in the Year of our Lord 1703. at 5 in the *Afternoon*.

Firſt, Set down the Year 1703. then the Month and Day, *May* 16; and *laſtly*, the Hours, 5, as is here due.

Secondly, Look for 1703. (in *Table* I.) againſt which ſtands 9 S. 10 D. 14 M. for the ☉ *Longitude*; and 6 S. 12 D. 46 M. for the ☉ *Anomaly*; both which ſet down as here you ſee.

Thirdly, Look for *May* 16. (in *Table* II.) againſt which ſtands 4 S. 14 D. 3 M. for the *Long.* And 4 S. 12 D. 2 M. for the *Anomaly*; both which ſet under the former, as you ſee.

	Long. ☉ S. D. M.	*Anom.* ☉ S. D. M.
Year 1703.	9 20 14	06 12 46
May, Day 19.	4 14 03	04 12 02
Hour 5	12	12
☉ Mean Lon.	1 04 29	10 25 00
Æquat. Add.	01 06	
☉ true Place.	2 05 35	♊ *Gemini.*

Fourthly, Look for 5 Hours (in *Table* III.) againſt which ſtands 12 Minutes, both for the *Longit.* and *Anomal.* both ſet down, as you ſee.

Fifthly,

Fifthly, Add all the *Longitudes* together, and they make 14 Sig. 04 Deg. 29 Min. (from which abate 12 *Signs*, and there remains only Two *Signs*.) —Alſo *Add* the *Anomalies* together, and they make 10 Sig. 25 Deg. 00 Min.

Sixthly, (in *Table* IV.) Look for 10 *Signs* at the *Bottom* of the *Table*, and 25 Deg. in the laſt Column towards the *Right Hand*, ſo againſt it, over 10 *Signs*, you ſhall find 1 Deg 6 Min. to be *Added*. Set them under the *Mean Longitude*, and add them to it, ſo will the *Sum* be 2 Sig. 05 Deg. 35 Min. And that is the true *Place* of the Sun in the Ecliptick, which is 35 Deg. diſtant from the *Æquinoctial Point* ♈ *Aries*. Note, that

Signs	0	1	2	3	4	5	6	7	8	9	10	11
Is	♈	♉	♊	♋	♌	♍	♎	♏	♐	♑	♒	♓

Other Examples.

	Longit. ☉ S. D. M.	*Anom.* ☉ S. D. M.
Year 1720.	9 20 07	6 12 21
Febr. Day 29	1 28 09	1 28 09
Hours 6	15	15
Mean Motion	11 18 31	8 10 45
Æquat. Add.	1 52	
☉ true Place.	11 20 23	♓ *Piſces.*

In this *Example*, it being *Leap-Year*, I take the Day of the Month out of the Column that hath *Biſſex.* at the Head of it.

	Longit. ☉ S. D. M.	*Anom.* ☉ S. D. M.
Year 1713	9 20 48	6 13 10
April 23	3 21 23	3 21 22
At Noon	0 00 00	0 00 00
Mean Motion	1 12 11	10 04 32
Æqua. Add.	01 35	
☉ true Place.	1 13 46	♉ *Taurus.*

	Longit. ☉ S. D. M.	*Anom.* ☉ S. D. M.
Year 1761	9 21 10	6 12 42
Jan. Day 12	11 50	11 50
Hours 16	39	39
Mean Motion	10 3 39	6 25 11
Æquat. Add.	51	
☉ true Place.	10 4 30	♒

Note alſo, That if the *Sum* of the Sun's *Anomalies* be leſs than Six *Signs*, they will be found at the *Head* of the *Æquation Table*, and the *Degrees* in the *Firſt Column* towards the *Left Hand*, and the *Æquation* muſt (always) be *Subſtracted* from the *Mean Longitude*; whereas in all theſe *Examples* it hath been *Added*.

In

In all the *Problems* in this *Section*, it is to be understood, That (in all the *Spherical Schemes*, or *Figures* following.

	Represents the
P	*Pole* of the World.
Æ Q	*Æquinoctial* Circle, or *Æquator*.
P P	*Parallels* (or smaller Circles) of *Declination* of the *Sun*, or of a *Star*.
F	*Pole* of the *Ecliptick*:
E C	*Ecliptick*.
F F	*Parallels* (or lesser *Circles*) of *Altitude* of the *Sun*, or of a *Star*, or the *Latitude* of a *Place* or *Country*.
♈	*Vernal* Intersection of the *Ecliptick* and *Æquator*.
♎	*Autumnal* Intersection of the *Ecliptick* and *Æquator*.
A	*Ascension* of any Point of the *Ecliptick*.
D	*Descension* of any Point of the *Ecliptick*.
Z	*Zenith*.
N	*Nadir*.
H O	*Horizon*.
Z N	*Prime Vertical* Circle, or *Azimuth* of *East* and *West*.
O *r*	*East* *Pole* of the *Meridian*: Or the Place in the *Horizon*, where the *Sun*, or a *Star*, *Rises* or *Sets*.
O *c*	*West* *Pole* of the *Meridian*: Or the Place in the *Horizon*, where the *Sun*, or a *Star*, *Rises* or *Sets*.
R	*Right Angle*.
☉	*Sun*.
✱	*Star*.
S	*Side*, *Sun* (and sometimes) *Star*.
b	*North* *Latitude*, *Declination*, *Amplitude*.
m	*South* *Latitude*, *Declination*, *Amplitude*.

R. A. P. T. Right-angled Plain Triangles.
O. A. P. T. Oblique-angled Plain Triangles.
R. A. S. T. Right-angled Plain Triangles.
O. A. S. T. Oblique-angled Plain Triangles.

Astrono-

Aſtronomical Problems.

PROBLEM I.

The Longitude, *or* Place *of the* Sun, *in the* Ecliptick, *being given; To find,*

1. *The Sun's* Right Aſcenſion.
2. *The Sun's* Declination.
3. *The* Angle of Poſition *made by the Interſection of the* Meridian *and the* Ecliptick.

IN this *Problem* there is given the *Sun's* or a *Star's Place,* in Reſpect of the *Ecliptick*; and his Place, in reſpect of the *Æquinoctial,* is required,

I. *By the* Cœleſtial Globe.

Example. Let the Place of the Sun be in 29 Deg. of *Taurus* ♉, (that is, 59 Deg. from the beginning of *Aries* ♈, which is the neareſt *Æquinoctial Point.*)

The *Globe* being in any *Poſition,* (for in this *Problem* there is no regard to be had to the *Latitude*) count the Sun's Place in the *Ecliptick* upon the *Ecliptick Circle,* from ♈; and bring that Point to the Graduated Side of the *Braſs Meridian.* Then, (1.) Will the *Braſs Meridian* cut the *Æquinoctial Circle* in 56 d. 46 m. counted alſo from ♈; and that is the *Right Aſcenſion.* And, (2.) The Number of Degrees of the *Braſs Meridian,* comprehended between the *Æquinoctial* and *Ecliptick Circles,* will be 20 d. which is the *Sun's Declination.*—And, (3.) The *Angle* made by the Interſection of the *Braſs Meridian,* and the *Ecliptick Circle* in the Point of the Sun's Place, will be 77 d. 23 m. which is the *Angle of Poſition,* in reſpect of the *Meridian* and *Ecliptick.*

II. *By* Trigonometrical Calculation.

The *Globe* being in this *Poſition,* there is repreſented upon the *Superficies* thereof, Two *Right-angled Spherical Triangles,* ſuch as are expreſſed in the *Diagrams*; and are there noted with ⊙ R ♈ and ⊙ R ♎; in which, the Sides ⊙ ♈ and ⊙ ♎ are Arches of the *Ecliptick Circle,* and is the *Sun's Longitude:* *Fig.* XXVII. XXVIII

The

Fig. XXVII. XXVIII. The Sides ♈ R and ♎ R are Arches of the *Æquinoctial*, and is the Sun's *Right Ascension*; and the Sides ☉ R are Arches of the *Brass Meridian*, and is the Sun's *Declination*.—Also, the Angle at ♈ or ♎, is the Angle of the greatest *Obliquity* of the *Ecliptick*, (and is equal to the Sun's greatest *Declination* 23 d. 31 m.) The Angle R is a *Right Angle*, and the Angle ☉ is the Angle at the *Sun's Position*, in respect of the *Ecliptick* and *Meridian Circles*.

To resolve this *Problem Trigonometrically*, you are to consider the *Quadrant* of the *Ecliptick*, in which the Sun is, which in the *Figures* are signified by one of these Numbers, 1. 2. 3. 4. Of which, the first is of the *Spring*, from the beginning of *Aries* ♈, to the beginning of *Cancer* ♋, &c. Then in the Triangle R ♈ ☉ or R ♎ ☉.

There is given, besides the *Right Angle* at R,
1. The Angle at ♈ or ♎, 23 d. 31 m.
2. R ♈ or R ♎: The Sun's Distance from the next *Æquinoctial Point*; to be numbred from ♈ or ♎, unto the Degrees of the Sun's *Longitude* or *Place* in the *Ecliptick* given.

And,

There is required,
1. R ☉, The Sun's *Declination* for that Sign which he is in; whether *North* or *South*.
2. The Arch ♈ R, or ♎ R, which if it be in Quadrant,
   1. The Degrees found by the Canon, are the Degrees of
   2. The Degrees found, must be substracted from 180, and the remainded are the Degrees of
   3. The Degrees found must be added to 180, and the Sum are the Degrees of
   4. The Degrees found must be substracted from 360 d. and the Remainder are the Degrees of
   } *Right Ascension*.
3. The Angle ☉, or the Angle of *Position*, in respect of the *Meridian* and *Ecliptick Circles*.

The Canons for Calculation.

The Sun being in 29 d. of *Taurus* ♉, which is 59 d. from *Aries* ♈.

1. For

1. For the *Right Ascension* ♈ or ♎ R. By *Case* III. of R. A. S. T. | Fig. XXVII. XXVIII.

As the Sine of 90 d.

Is to the Co-sine of the greatest Obliquity of the Ecliptick 66 d. 29 m.

So is the Tangent of the Sun's Distance from ♈ or ♎, 59 d.

To the Tangent of 56 d. 46 m. Which is the Sun's *Right Ascension*; because his Place given was in the First *Quadrant*:—But if the Sun had been in 1 d. of *Leo* ♌, or 29 d. of *Scorpio* ♏, or 1 d. of *Aquarius* ♒. All which Points are 59 d. distant from ♈ or ♎; the *Right Ascension* would be found the same, as before, *viz.* 56 d. 46 m. But by the Rule before given in this *Problem.*

			D.	M.
If the Sun's Place given had been in	1 d. of *Leo*, in Quad. 2.	The R. *Ascension* would have been	123	14
	29 d. of *Scorpio*, in Qu. 3.		236	46
	1 d. of *Aquarius*, in Qu. 4.		303	15

2. For the *Sun's Declination* R ☉. By *Case* II. of R. A. S. T.

As the Sine of 90 d.

Is to the Sine of the greatest *Obliquity* of the *Ecliptick* 23 d. 31 m.

So is the Sine of the Sun's Distance from ♈ or ♎ 59 d.

To the Sine of 20 d. the Sun's present *Declination.*

Which is *North*, because he is in a *Northern Sign.*

3. For the Angle of the Sun's *Position* ☉. By *Case* III. of R. A. S. T.

As the Sine of the Sun's present *Declination* 20 d.

Is to the Sine of greatest *Declination* 23 d. 31 m.

So is the Sine of the *Sun's* R. *Ascension* 56 d. 46 m.

To the Sine of 77 d. 23 m. The Sun's *Angle of Position*, made by the *Meridian* and *Ecliptick Circles.*

PROB. II.

The Right Ascension, *or* Declination *of the Sun given*; *To find his* Longitude (*or* Place) *in the* Ecliptick.

THIS is the Converse of the foregoing *Problem*; for in this, the Sun's *Place*, in respect of the *Æquinoctial Circle*, is given: And his *Place*, in respect of the *Ecliptick Circle*, is required: And may be resolved as followeth.

1. By

I. *By the* Cœleſtial Globe.

Fig. XXVII. XXVIII.

Example. Let the Sun's *Right Aſcenſion* be 303 d. 14 m. and his *Declination* 20 d. Southward: And let his *Longitude* (or *Place* in the *Ecliptick*) be required.

Count 303 d. 14 m. the *Right Aſcenſion* given, upon the *Æquinoctial Circle*, from the *Vernal Æquinoctial* Point *Aries* ♈, and bring that Point to the graduated Side of the *Braſs Meridian:* Then will the *Braſs Meridian* cut the *Ecliptick Circle* in 1 d. of *Aquarius* ♒, and in that Sign and Degree the Sun is at *Noon*, when his *Right Aſcenſion* is 303 d. 14 m.

But to find his Place by his *Declination*, Count 20 d. (the *Declination* given) upon the *Braſs Meridian*, downwards towards the South Pole (becauſe the *Declination* given was *Southerly*) and turn the *Body* of the *Globe* about till 20 d. of the *Meridian* do cut the *Ecliptick Circle*; which it will do in Two Points, *viz.* One in 29 d. of *Scorpio* ♏, in the third *Quadrant*; and the other in 1 d. of *Aquarius* ♒, in the fourth *Quadrant*; in both which Points the Sun being, he hath 20 d. of *South Declination*, and which of thoſe Points you ſeek is determined by the Degrees of *Right Aſcenſion* given, as here 303 d. 14 m. and the *Globe* being in this Poſition, will ſign out the ſame Triangles, mentioned in the foregoing *Problem.*

II. *By* Trigonometrical Calculation.

Conſider the *Quadrant* of the *Ecliptick* in which the Sun is, (which the Degrees of *Right Aſcenſion* given, will determine) and in this Example will be the fourth *Quadrant:* But the given *Declination* is indifferent in all the Four *Quadrants.*

Wherefore, in the Triangle R ♈ ☉.

There is given, beſides the *Right Angle* at R.

1. The Angle at ♈ or ♎, 23 d. 31 m. the *Sun's greateſt Declination.*
2. The Side R ☉, the Sun's preſent *Declination* 20 d. And alſo either of the other Sides R ♈ (or R ♎) an Ark of the *Equator*, to be numbred from the neareſt *Æquinoctial* Point, to the Degrees of *Right Aſcenſion given.*

And

And there is required	The Side ☉ ♈ (or ☉ ♎) the Sun's Distance in the *Ecliptick Circle*; from the nearest *Æquinoctial Point*, from which Point the *Longitude* required is to be numbred upon the *Ecliptick Circle*.	Fig. XXVII. XXVIII.

So the *Right Ascension* 303 d. 14 m. given, it being in the *Fourth Quadrant*, will give in the *Ecliptick* 1 d. of *Aquarius* ♒.

The Canons for Calculation.

1. By the *Right Ascension* given.

The *Right Ascension* given being 303 d. 14 m. are found in the *Fourth Quadrant*, and therefore must be substracted from 360 d. and the Remainder will be 56 d. 46 m. which must be made use of in the Calculation, instead of 303 d. 14 m. And then in the Triangle ♎ ☉ R,

1. By the *Right Ascension* given.

As the Tangent of the Side ♎ R, 56 d. 46 m. the *Right Ascension*,
Is to the Sine of 90 d. or Radius:
So is the Co-sine of the Angle at ♎ 66 d. 29 m.
To the Co-tangent of ♎ ☉, 59 d. 00 m. Sun's *Longitude*.
Now, because the *Right Ascension* given was in the *Fourth Quadrant*, the *Longitude* 56 d. 46 m. thus found, must be in the *Fourth Quadrant* also, which will be in 1 d. of *Aquarius* ♒.

2. By the *Declination* given.

As the Sine of the Sun's *greatest Declination*, 23 d. 31 m. the Angle at ♎.
Is to the Sine of the Side ☉ R, 20 d. the Sun's *present Declination* given:
So is the Radius, Sine 90 d.
To the Sine of the Side ♎ ☉, 59 d. from *Libra* ♎.
And being in the *Fourth Quadrant*, it gives the Longitude of the Sun to be in 1 d. of *Aquarius* ♒, as before.

PROB.

PROB. III.

Fig. XXIX.

The Latitude *of the Place*, 51 *d.* 30 *m. and the Sun's Place in the* Ecliptick, 29 *d. of* Taurus, *given, to find*,

1. *The Sun's* Declination.
2. *The Sun's* Amplitude.
3. *The Hour from Midnight or Ascension all* Difference.
4. *The* Time *of the Sun's* Rising *and* Setting.
5. *The* Length *of the* Day *and* Night.
6. *The* Angle *of the Sun's* Position *at the time of his* Rising *or* Setting.

I. *By the* Cœlestial Globe.

BRing the 29 d. of *Taurus* to the *Brass Meridian*, and you will find 29 of *Taurus* to lye under 20 d. of the *Meridian*, and that is the Sun's *Declination* North, the Sun being in a *Northern* Sign. Then, set the *Index* of the *Hour-Circle* to the *South* 12, and turn the Body of the Globe Eastward, till 29 d. of *Taurus* do just touch the *Horizon*, and there fix it. Then shall you find 29 d. of *Taurus* to cut the *Horizon* in 33 d. 18 m. counted from the *East*, which is the *Amplitude* of the Sun's *Rising* from the true *East* Point of the *Horizon* Northward: And also, the *Index* of the *Hour-Wheel*, will point out 11 m. after 4 in the Morning, at which time the Sun *Riseth*. —And if you turn the Body of the Globe about Westward, till 29 d. of *Taurus* doth touch the *West* Side of the *Horizon*; then shall the *Index* point at 49 m. after 7 at Night, at which Time the Sun *Setteth*.

Again, Turn the Globe about, till 29 d. of *Taurus* touch the *East* Side of the *Horizon* (as before) and set the *Index* of the *Hour-Circle* to the *North* (or undermost) 12: And then turning it Westward till 29 d. of *Taurus* touch the *Horizon* on the *West* Side; and then shall the *Index* point at 3 ho. 38 m. more than 12 Hours, from the *South* 12: Which shews that the *Day* is then 15 ho. and 38 m. *Long*.—And if you count the Hours between the *North* 12 and the *Index*, you shall find them to be 8 ho. and 22 m. and that is the *Length* of the Night.

And then, for the Angle of the *Sun's Position*, that is to be found, as is directed in the 24th *Element*, and in the *Problem* to this Second Part, and will be found to be 56 d. 23 m.

The

The Globe being in this Position, there is represented upon the Superficies of it a Spherical Triangle P ☉ O, as in *Fig.* XXIX. compounded of the Arches of such *Great Circles* of the *Sphere*, as are ingredient in the *Proposition*, *viz.* Of P O, an Arch of the *Brass Meridian*; of ☉ O, an Arch of the *Horizon*; and of P ☉, an Arch of that *Meridian*, which cutteth the *Ecliptick* in that Point in which the Sun is. —And in this Triangle, the Angle P O ☉ is a *Right Angle*: ☉ P O, is the *Hour* from *Midnight*: And the *Angle* P ☉ O, is the Angle of *Position* made by the *Meridian* and *Horizon*, at the time of the *Sun's Rising*.

II. *By* Trigonometrical Calculation.

In the Triangle P O ☉, you have given, (besides the *Right Angle* at O). (1.) The Hypotenuse P ☉, the Complement of the Sun's *Declination*, (70 d.) (2.) The Perpendicular P O, the *Latitude* 51 d. 30 m. —To find, (1.) The Base ☉ O, the Sun's *Amplitude* at his Rising. (2.) The Angle ☉ P O, the *Hour* from *Midnight*: And, (3.) The Angle P ☉ O; the Angle of the *Sun's Position* at his *Rising*.

The Canons for Calculation.

1. For the *Amplitude*, ☉ O. By *Case* VI. of *R. A. S. T.*

As the Co-sine of P O, (the Latitude) 38 d. 30 m.
Is to the Radius, 90 d.
So is the Co-sine of P ☉, (the Sun's *Declination*) 20 d.
To the Co-sine of ☉ O, (the *Amplitude*) 33 d. 20 m.
Which is equal to the Arch of the *Horizon*. *Or* ☉, which is the *Amplitude* of his *Rising* from the *East* Northward, because the Sun was in a *Northern* Sign: For if he had been in a *Southern* Sign, as in 59 d. of *Scorpio*, the *Amplitude* would have been the same, but from the *East* Southward.

2. For the *Hour* from *Midnight* ☉ P O. By *Case* V. of *R. A. S. T.*

As the Radius,
Is to the Tangent of P O (the *Latitude*) 51 d. 30 m.
So is the Co-tangent of P ☉ (the Sun's Distance from the Pole) 20 d.
To the Co-sine of ☉ P O (the *Hour from Six*) 27 d. 14 m.
Whose Complement, 62 d. 46 m. is the *Hour* from *Midnight*; which turned into *Time*, (by allowing 15 d. for 1 h. and 4 d. for 1 m. of *Time*) makes 4 h. and 11 m. Which is the *Hour* from *Midnight*.

Fig. XXIX.

3. For the *Ascensional Difference.*

The Hours from *Midnight*, 4 h. 11 min. substracted from 6 h. there remains 1 h. 49 m. which is the *Ascensional Difference.* Or, The *Time* that the *Sun Rises* before *Six* in the *Summer*, or after *Six* in the *Winter.*

4. and 5. For the *Time* of the *Sun's Rising* and *Setting*, and (consequently) the *Length* of the *Day* and *Night.*

The *Ascensional Difference*, 1 h. 49 m. being added to 6 h. gives 7 h. 49 m. for the *Semidiurnal Ark* in *Summer*; or substracted from 6 h. gives 4 h. 11 m. for the *Semidiurnal Ark* in *Winter.*

The *Semidiurnal Ark*, 7 h. 49 m. substracted from 12 h. there remains 4 h. 11 m. for the time of the *Sun's Rising:* And that substracted from 12 h. leaves 7 h. 49 m. for the time of the *Sun's Setting.* And,

The *Semidiurnal Ark*, 7 h. 49 m. doubled, gives 15 h. 38 m. for the *Length* of the *Day:* —And 15 h. 38 m. substracted from 24 h. there will remain 8 h. 22 m. for the *Length* of the *Night.*

6. For the *Angle* P ☉ O, of the Sun's *Position*, at the time of his *Rising.* By *Case* IV. of R. A. S. T.

As the Sine of P ☉ (the *Co-Declination*) 70 d.
Is to the Radius:
So is the Sine of P O (the *Latitude*) 51 d. 30 m.
To the Sine of P ☉ O (the *Angle* of *Position* at the *Sun's Rising*) 56 d. 23 m.

PROB. IV.

To find the Sun's Meridian Altitude, *and his* Depression *at* Midnight, *he being in any Point of the* Ecliptick.

LET the Place of the Sun be the same, as in the former *Problem*, *viz.* in 29 d. of *Taurus.* Then,

I. *By the* Cœlestial Globe.

Turn the Globe about, till the 29 d. of *Taurus* be just under the *Meridian*; then shall you find the Number of Degrees of the *Meridian*, which are comprehended between that Point and the Horizon to be 58 d. 30 m. which is the *Meridian Altitude.* And if you bring the 29 d. of *Scorpio*, which is the opposite Point of the

the *Ecliptick* in which the *Sun* is, to the *Meridian*, the Number of the Degrees of the *Meridian* between that Point and the *Horizon* will be found to be 18 d. 30 m. which is the *Sun*'s Depression at *Midnight*.

Fig. XXIX.

In like manner,

When the *Sun* is in	D.	M.		You shall find the *Meridian Altitude.*	D.	M.	And the *Depres.*	D.	M.
	16	00	♌		54	36		22	24
	25	00	♍		19	25		57	25
	29	00	♒		26	38		50	22
	13	00	♈		43	39		33	21

PROB. V.

To know when Twilight *Begins and Ends.*

TWilight begins, when the Sun is 18 d. below the *Horizon* before its *Rising*: And endeth, when the Sun cometh to be 18 d. below the *Horizon*, after its *Setting*.

The Globe Rectified, and the *Sun* in 29 d. of *Taurus*, find the opposite Point thereunto, which is the 29 d. of *Scorpio*; and bring that Point, as also the *Quadrant of Altitude*, both of them on the West-side of the *Meridian*; and then move both the Body of the *Globe*, and the *Quadrant of Altitude* also, till the 29 d. of *Scorpio* lye directly under 18 d. of the *Quadrant of Altitude*: Which done, keep them both together, and then see how many Hours the *Index* is removed from 12, which you shall find to be 1 h. and 8 m. So that *Twilight* begins at 8 m. after 1 in the Morning. And this being taken from 4 h. 11 m. the time of the *Sun*'s *Rising* that Day, there will remain 3 h. 3 m. which is the Length or Continuance of the *Twilight*. Also if you double the time of the beginning of *Twilight* 1 h. 8 m. you shall have the length of dark Night, which will be but 2 h. 16 m.

In like manner, if you would know when the *Twilight* endeth after *Sun-setting*, you must bring the 29 d. of *Scorpio* (the Point opposite to the *Sun*) on the East-side of the *Meridian*, making it, and 18 d. of the *Quadrant of Altitude*, to meet, then the *Index* will shew 10 h. 52 m. and till that time of Night doth *Twilight* continue.

Fig. XXIX.

And so;

The *Sun* in	D.		Day-break will be at	H.	M.		Twilight last till	H.	M.	
	29	♋		0	10	aft. Midn.		11	50	at Night.
	8	♐		5	52			6	8	
	2	♎		4	6			7	54	
	0	♉		2	41			9	19	

And if you go about to find the time of the beginning and end of *Twilight*, all the Time that *Sun* is passing from 2 d. of *Gemini* to 30 d. of *Cancer*, which is from about the 12th of *May* to the 12th of *July*, you shall find that there will be no *Twilight* at all, but all that Time continual Day: For all that Space of Time the *Sun* never descendeth so much as 18 d. under the *Horizon*, in the Latitude of 51 d. 30 m.

The former of these Two last *Problems* needeth no *Trigonometrical Calculation:* But, if to the Sun's *Declination* for the Day (or *Place* of the *Sum*) given, 20 d. you add the Complement of the *Latitude* given, 38 d. 30 m. the *Sum* of them 58 d. 30 m. is the *Meridian Altitude* for that Day that the Sun is in 58 d. of *Taurus.*

But if the Sun had been in 29 d. of *Scorpio*, a Southern Sign, then the *Declination* 20 d. substracted from 38 d. 30 m. the Complement of the *Latitude*, the Remainder 18 d. 30 m. will be the *Meridian Altitude.*

For the *Depression* of the *Sun*, he being upon the *Meridian* at Midnight, it is the same with the *Meridian Altitude*, when the Sun is in the opposite *Sign.*

So the *Depression* at *Midnight* when the Sun is in

29 d. of *Taurus*, } is { 18 d. 30 m.
29 d. of *Scorpio*, } is { 58 d. 30 m.

PROB.

PROB. VI.

The Latitude *of* Place, *and the* Sun's Place *in the* Ecliptick *(and consequently his* Declination) *being given; To find*

1. *What* Altitude *the Sun shall have at Six a Clock.*
2. *What* Azimuth *he will be upon, at Six a Clock, Morning or Evening.*
3. *The* Angle *of the* Sun's Position, *made by the Intersection of the* Vertical Circle, *and the* Meridian *which passes by the* Sun *at the time of the* Question.

ALL these *Problems* are only in use, when *Sun* is in a *Northern Sign*, and so hath *North Declination:* For the *Sun* is never above the *Horizon* (in either of the *Temperate Zones*) when he is in a *Southern Sign*, or hath *South Declination*. Fig. XXIX.

I. *By the* Cœlestial Globe.

1. For the *Altitude* at Six. Bring the 29 d. of *Taurus* to the *Meridian*, and set the *Index* of the *Hour-Circle* to 12; then turn the *Globe* Eastward, till the *Index* of the *Hour-Circle* come just to Six a Clock: Then holding the *Globe* there, lay the *Quadrant of Altitude* just over the 29 d. of *Taurus*, and there you shall find it to cut 15 d. 30 m. of the *Quadrant*. And such *Altitude* shall the *Sun* have at Six a Clock in the Morning, and the same at Six at Night.

And so,

	D. M.		D. M.
When the *Sun* is in	16 00 ♌	The *Sun's Altitude* at Six will be found to be	12 32
	..		
	13 00 ♈		4 2

2. For the *Azimuth* at Six. Bring the 29 of *Taurus* to the *Meridian*, and set the *Index* of the *Hour-Wheel* to 12; then move the *Globe* till the *Index* lye upon Six; and holding the *Globe* there, lay the *Quadrant of Altitude* just over 29 d. *Taurus:* Then shall you find, that there are 77 d. 14 m. of the *Horizon* contained between the Intersection of the North Part of the *Meridian*, and the *Quadrant of Altitude* which is the *Azimuth* from the North; or 12 d. 46 m. from the East, which is the *Azimuth* from the East, or 102 d. 46 m. from the South, which is its *Azimuth* therefrom.

In

Fig. XXIX.

In like manner,

Sun in		D. M. North	D. M. East or West	D. M. South
D.M. 16 0 ♌	The *Sun's Azimuth* at 6 will be found to be from the	79 49	10 11	100 11
13 0 ♈		86 47	3 13	93 13

II. *By* Trigonometrical Calculation.

The *Globe* being in this Position, you may discover upon the Superficies thereof a *Spherical Triangle*, such as in the *Figure*, is noted with Z P ☉; in which, Z P is an Arch of the *Meridian*, P ☉ an Arch of the *Hour-Circle* of Six, and Z ☉ an Arch of an *Azimuth*, or *Vertical Circle:* In which *Triangle* there is given, (besides the *Right Angle* at P). (1.) The Perpendicular Z P, 38 d. 30 m. equal to the Complement of the *Latitude.* (2.) The Base ☉ P, 70 d. equal to the Complement of the Sun's *Declination:* To find, (1.) The Hypotenuse Z ☉, the Complement of the Sun's *Altitude* at Six. (2.) The *Angle* ☉ Z P, the Sun's *Azimuth* at Six a Clock: (3.) Z ☉ P, the *Angle* of the Sun's *Position.*

Fig. XXX.

The Canons for Calculation.

1. For the *Hypotenuse* Z ☉, (the Complement of the Sun's *Altitude* at Six) by *Case* XIV. of *R. A. S. T.*

As the Radius,
Is to the Co-sine of Z P, 51 d. 30 m.
So is the Co-sine of ☉ P, 20 d.
To the Sine of 15 d. 32 m. (the Complement of Z ☉.)
And such *Altitude* the Sun will have, at Six a Clock, (Morning or Evening) when he is in 29 d. of *Taurus.*

2. For the *Angle* P Z ☉, the Sun's *Azimuth* at Six a Clock, by *Case* XIII. of *R. A. S. T.*

As the Tangent of ☉ P (the *Co-Declination*) 70 d.
Is to the Radius:
So is the Sine of Z P (the *Co-Latitude*) 38 d. 30 m.
To the Tangent of 12 d. 46 m.
Whose Complement 77 d. 14 m. is the *Angle* ☉ Z P, the Sun's *Azimuth* from the North Part of the *Meridian.*

3. For

3. For the *Angle* of the Sun's *Position*, Z ⊙ P: By *Case* XI. of *R. A. S. T.* *Fig.* XXVII.

As the Sine of ⊙ P, 70 d.
Is to the Sine of ⊙ Z P, 77 d. 14 m.
So is the Sine of Z P, 31 d. 30 m.
To the Sine of Z ⊙ P, 40 d. 15 m.
Which is the *Angle* of the Sun's *Position.*

PROB. VII.

Having the Latitude, 51 *d.* 30 *m. the Sun's* Place, 59 *of* Taurus, *(or* Declination 20 *d.) Given, as in the foregoing* Problem; *To find,*

1. *At what* Hour *the* Sun *shall be upon the* East *or* West Azimuth.
2. *What* Altitude *the Sun shall have when he is upon the* East *or* West Azimuth.
3. *The* Angle *of the Sun's* Position.

I. *By the* Cœlestial Globe.

BRing the 29 d. of *Taurus* to the *Meridian*, and the *Index* to 12 of the Clock. Also bring the beginning of the Degrees of the *Quadrant of Altitude* to the East Point of the *Horizon*, and turn the Globe about till the 29 d. of *Taurus* do touch the Degrees of the *Quadrant of Altitude*; then shall the *Index* point at 7 m. past Seven; at which time, in the Morning, will the Sun be exactly upon the East *Azimuth*, or Point of the Compass. And if you carry the *Quadrant of Altitude* to the West Point of the *Horizon*, and turn the *Globe* about till 29 d. of *Taurus* touch the edge of Degrees thereof, the *Hour-Index* will point at Four of the Clock, and 53 m. at which time, in the Afternoon, will the Sun be upon the West *Azimuth* or Point of the Compass.

In the same manner,

	D. M.		D. M.		D. M.
When the Sun is in	16 00 ♌	It will be due *East* at	6 57	*West* at	5 03
	13 00 ♈		6 17		5 43

Then for the Sun's *Altitude* when he will be upon the *East* or *West Azimuth.*

Bring 29 d. of *Taurus* to the *Meridian*, and the *Quadrant of Altitude* to the East or West Points of the *Horizon*: Then turn the

Fig. XXX. *Globe* about, till the 29 d. of *Taurus* touch the *Quadrant of Altitude*, and you shall find it to touch at 25 d. 55 m. of the *Quadrant of Altitude*; and such *Altitude* hath the Sun, when he is upon the *East* or *West Azimuth*.

In like manner,

	D.	M.			D.	M.
When the Sun is in	16	00	♌	its *Altitude*, when *East* or *West*, will be found,	20	19
	13	00	♈		6	36

II. *By* Trigonometrical Calculation.

The *Globe* being in any of the former *Positions*, you will find represented upon the Superficies of it a *Spherical Triangle*, Z P ☉, (such as in the *Figure*) *Right-angled* at Z, which is composed of Z P, an Arch of the *Brass Meridian*, equal to the *Complement* of the *Latitude* 38 d. 30 m. Z ☉, an Arch of the *Quadrant of Altitude*: And of P S, an Arch of that *Meridian* (or *Hour-Circle*) which cutteth the *Ecliptick* in the Place the Sun is in.—And in this *Triangle* there is given (besides the *Right Angle* at Z, (1.) The *Perpendicular* P Z, the *Complement* of the *Latitude* 38 d. 30 m. (2.) The *Hypotenuse* P ☉, the *Complement* of the Sun's *Declination* 70 d. To find, (1.) The *Angle* Z P ☉, the *Hour* at which the Sun will be upon the *East* or *West Azimuth*. (2.) The *Side* Z ☉, the *Complement* of the *Altitude* he shall then have: And, (3.) The *Angle* Z ☉ P of his *Position*.

Fig. XXXI.

The Canons for Calculation.

1. For the *Angle* Z P ☉, (the *Hour* from *Noon*, that the Sun will be due *East* or *West*; by *Case* V. of *R. A. S. T.*

As the Radius,
Is to the Tangent of Z P, 31 d. 30 m.
So is the Co-tangent of P ☉, 20 d.
To the Co-sine of Z P ☉, 73 d. 10 m.
Which converted into Time, is 4 h. 53 m. Whose Complement is 7 h. 7 m. at which Time the Sun will be due *East* or *West*.

2. For

2. For the Base Z ☉, the Complement of the Sun's *Altitude*, when due *East* or *West* : By *Case* VI. of *R. A. S. T.*

Fig. XXXI.

As the Co-sine of Z P, (the *Co-Latitude*) 51 d. 30 m.
To the Radius:
So is the Co-sine of P ☉ (the *Declination*) 20 d.
To the Co-sine of Z ☉ (the *Altitude*) 25 d. 55 m.
And that is the Sun's *Altitude*.

3. For the Angle Z ☉ P, of Position: By *Case* IV.

As the Sine of P ☉, 70 d.
To the Radius:
So Sine Z P, 38 d. 30 m.
To the Sine of Z ☉ P, 41 d. 29 m.
Which is the Angle of Position.

PROB. VIII.

The Latitude *of the* Place, *the Sun's* Place *in the* Ecliptick *(or his* Declination) *being given, to find,*

1. *What* Altitude *the Sun shall have at any time of the* Day.
2. *What* Azimuth *he shall then have. And,*
3. *The Sun's* Angle *of* Position.

Let the Latitude *given be 51 d. 30 m.* North, *the Sun's* Place *in the* Ecliptick *29 d. of* Taurus, *(and so his* Declination *20 d.) And let the* Hour *be 9 in the* Morning, *or 3 in the* Afternoon.

I. *By the* Cœlestial Globe.

BRing the 29 d. of *Taurus* to the *Meridian*, and set the *Hour-Index* to 12 a Clock. Then turn about the *Globe*, till the *Hour-Index* point to the given Hour, (suppose 9 in the Morning, or 3 in the Afternoon:) There keep the *Globe*; and laying the *Quadrant of Altitude* over the 29 d. of *Taurus*, you shall find 43 d. cut thereby; and such *Altitude* shall the Sun have at 9 in the Morning, or 3 in the Afternoon. And at the same time you will find the *Quadrant of Altitude* to cut the *Horizon* in 24 d. 37 m. counted from the *East* or *West* Points thereof; and such *Azimuth* will the *Sun* have at 9 a Clock in the Morning, or 3 in the Afternoon, when he is in 29 d. of *Taurus*. And by this *Problem* the Sun's *Altitudes* in any Sign or Degree of the *Ecliptick* at all Hours may be found, as in these following *Synopsis* or *Tables*.

Fig. XXXI. A Table, *shewing what* Altitude *the Sun shall have* Hour *of the* Day, *in the beginning of every of th* Signs *of the* Zodiack.

So the *Sun* being in the beginning of	*Cancer*		*Gemi.* or *Leo.*		*Taur.* or *Virgo.*		*Aries* or *Libra.*		*Scorp.* or *Pisces.*		*Aqua* or *Sagi*
At the H.	*D.*	*M.*	*D.*	*M.*	*D.*	*M*	*D.*	*M*	*D.*	*M.*	*D.*
XII.	62	00	58	42	50	00	38	30	27	1	18
XI. I.	59	43	56	34	48	12	36	58	25	40	17
X. II.	53	45	50	55	43	12	32	37	21	51	13
IX. III.	44	42	43	3	36	0	26	7	15	58	8
VIII. VI.	35	41	34	13	27	31	18	8	8	33	1
VII. V.	27	17	24	56	18	18	9	17	0	6	
VI.	18	11	15	40	9	0					
V. VII.	9	32	6	50							
IV. VIII.	1	32									

His *Altitude* will be

'able, *shewing what* Altitude *the Sun shall have, he being on every Tenth* Azimuth *from the South, in the beginning every of the Twelve Signs.* Fig. XXXI.

(*Azimuth* from the *South*, viz. ... will be found to be)

; *Sun* in the begin-ing of	*Cancer.*		*Leo* or *Gemi.*		*Virgo* or *Taur.*		*Libra* or *Aries.*		*Scorp.* or *Pisces*		*Sagitt.* or *Aqua.*		*Capric.*	
d. *Az.*	D.	*M.*	D.	*M.*	D.	*M.*	D.	*M.*	D.	*M.*	D.	*M.*	D.	*M.*
0	62	0	58	42	50	0	38	30	27	0	28	18	15	0
10	61	43	58	24	49	38	38	4	26	30	17	45	14	25
20	60	51	57	28	48	33	36	46	25	10	16	5	12	41
30	59	52	55	52	46	40	34	34	22	27	13	15	9	45
40	57	20	53	29	43	51	31	21	18	48	9	14	5	34
50	54	3	50	12	40	11	27	5	13	58	3	57	0	6
60	49	56	45	53	35	23	21	41	8	0				
70	44	40	40	25	29	27	15	13	1	0				
80	28	11	33	46	21	29	7	52						
90	30	38	26	10	14	25								
100	22	27	18	2	6	45								
110	14	14	9	58										
120	6	34	2	30										

.hese Tables are of good use for the making of *Cylinders, drants*, and other Instruments, that give the *Hour* and *Azi-*, by the height of the Sun, and also to insert the Tropicks other Signs of the *Zodiacks* and *Azimuths* into Sun-Di- *&c.*

II. *By* Trigonometrical Calculation.

he *Globe* continuing in the same Position you left it, you discover upon it an *Oblique Spherical Triangle* Z P ☉, in the Figure) constituted by the Intersections of Z P, Arch of the *Brass Meridian*, Z ☉, a Part of the *Quadrant Altitude:* And of P ☉, an Arch of a *Meridian*, passing ugh the 29 d. of *Taurus* in the *Ecliptick*, or *Hour-Circle* or 3 a Clock. Fig. XXXII.

In which *Triangle* there is given,

The Side Z P, (the Complement of the *Latitude*) 38 d. 30 m.

 2. The

Fig. XXXII.

2. The Side P ☉ (the Complement of the Sun's *Declination*, or his Distance from the Pole) 70 d.
3. The Angle Z P ☉, (the Hour-Circle of 9 or 3 a Clock) 45 d.

To find,

1. The Side Z ☉ (the Complement of the Sun's *Altitude.*)
2. The Angle ☉ Z P, (the Sun's *Azimuth* from the *North* Part of the *Meridian.*)
3. The Angle of Position, Z ☉ P.

The Canons for Calculation.

By *Case* III. and *Case* IX. of *O. A. S. T.*

(1.) As the Sine of half the *Sum* of the Two given Sides, Z P, 38 d. 30 m. and P ☉ 70, *viz.* 54 d. 15 m.
Is to the Sine of half the *Difference* of those Sides, *viz.* 15 d. 54 m.
So is the Co-tangent of half the given *Angle*, Z P ☉, *viz.* 22 d. 30 m.
To the Tangent of half the *Difference* of the unknown *Angles*, at Z and ☉, *viz.* 38 d. 48 m.

Then,

(2.) As the Co-sine of half the *Sum* of the given Sides Z P and P ☉, *viz.* 35 d. 45 m.
Is to the Co-sine of half the *Difference* of those Sides, *viz.* 74 d. 15 m.
So is the Co-tangent of half the given *Angle* Z P ☉, *viz.* 22 d. 30 m.
To the Tangent of half the *Sum* of the Two unknown *Angles* P Z ☉ and Z ☉ P, *viz.* 75 d. 49 m.

Then,	*D.*	*M.*
The half Sum before found is	75	49
The half Difference before found is	38	48
The Sum of them is	114	37
The Difference of them is	37	01

The one equal to the greater *Angle* ☉ Z P 114.37, which is the Sun's *Azimuth* from the *North* Part of the *Meridian*: From the *South* Part 65 d. 23 m. and from the *East* or *West* 24 d. 37 m.

The other, equal to the lesser *Angle* Z ☉ P 37 d. 1 m. which is the *Angle* of the *Sun's Position*.

PROB.

Fig. XXXII.

PROB. XI.

The Latitude *of the* Place *(51 d. 30 m.) the* Sun's Place *in the* Ecliptick *(29 d. of* Taurus; *or 29 d. of* North Declination) *and the* Sun's Altitude *(12 d.) being given; To find,*

1. *The Sun's* Azimuth, *from the* Eaſt, Weſt, North *or* South *Points of the* Horizon:
2. *The* Hour *of the* Day, *from* Noon *or* Midnight.
3. *The* Angle *of the* Sun's Poſition.

I. *By the* Cœleſtial Globe.

1. *For the* Sun's Azimuth.

THE *Globe* being Rectified, *&c.* and the *Quadrant of Altitude* fixed, and brought to the *Horizon*; turn 29 d. of *Taurus* toward the *Eaſt*, if in the Morning; or towards the *Weſt*, if in the Evening, till it come to lye juſt under 12 d. of the *Quadrant of Altitude*; and then note at what Degree in the *Horizon* the *Quadrant of Altitude* reſteth; which will be at 17 d. 8 m. from the *Eaſt*, if in the Morning, or 17 d. 8 m. from the *Weſt*, if in the Afternoon, which is the *Azimuth* from the *Eaſt* or *Weſt* towards the *North*. And this *Azimuth*, if reckoned by the Points of the Compaſs upon the *Horizon*, will be *E.* by *N.* 5 d. 35 m. Northward, if in the Morning; or *W.* by *N.* 5 d. 53 m. Northward, if in the Evening, when the *Sun* is in 29 d. of *Taurus*, and hath 12 d. of *Altitude*. Now if you count the Degrees of the *Horizon* between the *Quadrant of Altitude*, and the *North* Part of the *Meridian*, you ſhall find them to be 72 d. 52 m. which is the *Azimuth* from the *North*: And if you count them from the South Part of the *Meridian*, you ſhall find them to be 107 d. 8 m. which is the *Azimuth* from the *South*.

In like manner,

	D.	M.
The *Latitude* being	51	30
The *Sun's* Place	1	00 *Aquarius*.
The *Sun's Altitude*	12	00

Then will the *Azimuth* be found to be 56 d. from the *Eaſt* or *Weſt*, towards the *South*; which (by the Points of the Compaſs upon the *Horizon*) will appear to be S. E. by S. if in the Morning, or S. W. by S. if in the Evening.

2. For

2. *For the Hour of the Day.*

Fig. XXXII. Being 29 d. of *Taurus* to the *Meridian*, and set the *Index* to 12 a Clock: Then, if it be in the Forenoon, set the *Quadrant of Altitude* on the *East* Side of the *Meridian*; but on the *West* Side, if it be in the Afternoon. And turn the *Globe* about, till the 29 d. of *Taurus* meet with 12 d. of the *Quadrant of Altitude*, and then shall the *Index* of the *Hour-Circle* point at 5 a Clock, and 36 m. if it be in the Morning; or at 24 m. after 6 of the Clock, if it be at Night. And that is the true Hour of the Day.

In like manner,

	D.	M.	
The *Latitude* being	51	30	
The *Sun*'s Place	1	00	*Taurus.*
The *Altitude*	36	00	

Then will the Hour of the Day be found to be either 9 in the Morning, or 3 in the Afternoon. And which of these Hours it is, may best be known by a second Observation of the *Altitude*: For if the *Altitude* do increase, it is the Forenoon; but if it decrease, it is the Afternoon.

Again,

	D.	M.
The *Latitude* being	52	30
The *Sun*'s Place in the beginning of *Taurus.*		
The *Sun*'s *Altitude*	25	56

The Hour of the Day would be found to be either 8 m. past 4 in the Afternoon. Or if it were in the Forenoon, 52 m. after 7 in the Morning.

3. *For the* Angle *of the* Sun's Position.

This may be found by the 26th Element, and as is taught in the Proeme before this Second Part: And in this Example will be found to be 39 d. 16 m.

II. *By* Trigonometrical Calculation.

Fig. XXXIII. The *Globe* resting in the same *Position* you left it, when you wrought these *Problems* upon it, you will discover upon it an *Oblique-angled Spherical Triangle*, (such as in the *Figure*, is noted with Z ☉ P: Which *Triangle* consists of, (1.) Z P, an Arch of the *Brass Meridian*, equal to the Complement of the given *Latitude*, 38 d. 30 m. (2.) Of Z ☉, an Arch of the *Quadrant of Altitude*

217

Fig.
XXXIII.

;
;

d

oe
of
d,
le
of
m.
id-
om

For

Fig.
XXXII

Fig.
XXXIII

Altitude, croſſing the *Ecliptick* in 29 d. of *Taurus*, where the Sun is 12 d. high, and is equal to the *Complement* of the Sun's *Altitude* 78 d. (3.) An Arch of a *Meridian* (or *Hour-Circle*) P ☉, paſſing through 29 d. of *Taurus*, and croſſing the *Quadrant of Altitude* in 12 d. thereof, and is equal to the *Complement* of the *Sun's Declination* (or his Diſtance from the *Pole*) 70 d. So that the *Three Sides* of this *Triangle* are *Given*, and the *Three Angles* are *Required*.

Fig. XXXIII.

The Canons for Calculation.

1. For the Sun's *Azimuth*, the *Angle* ☉ Z P. By *Caſe* IX. of *O. A. S. T.*

	D.	M.
The Side ☉ P, the *Co-declination*, is	70	00
The Side ☉ Z, the *Co-altitude*, is	78	00
The Side Z P, the *Co-latitude*, is	38	30
Their *Sum* is	186	30
Their half *Sum* is	93	15
The *Difference* between the half *Sum*, and 70 d. ☉ P	23	15

Being thus prepared, the Proportions are,

(1.) As the Radius, Sine 90 d.
Is to the Sine of Z P (one of the *Sides* containing the enquired *Angle* Z) 38 d. 30 m.
So is the Sine of Z ☉ (the other Side containing the enquired *Angle* Z) 78 d.
To the Sine of 37 d. 31 m.

(2.) As this Sine of 37 d. 31 m.
Is to the Sine of the half *Sum*, 93 d. 15 m. (or 86 d. 45 m.)
So is the Sine of the *Difference*, 23 d. 15 m.
To the Sine of 40 d. 20 m.

(3.) To this Sine of 40 d. 20 m. add the Radius, and it will be 19.811063; the half whereof 9.905531 is the Sine of 53 d. 34 m. Whoſe Complement 36 d. 26 m. doubled, makes 72 d. 52 m. And that is the Quantity of the *Angle* ☉ Z P: Or the Sun's *Azimuth* from the *North* Part of the *Meridian*. And that taken from 90 d. leaves 17 d. 8 m. for the Sun's *Azimuth* from the *Eaſt* or *Weſt*. And that added to 90 d. gives 107 d. 8 m. for the Sun's *Azimuth* from the *South*.

2. For

Fig. XXXIII.

2. For the *Hour*, the *Angle* Z P ☉. By *Case* I. of *O. A. S. T.*

As the Sine of the Side P ☉ (70 d.)
Is to the Sine of the *Angle* ☉ Z P, (72 d. 52 m.)
So is the Sine of the Side Z ☉ (78 d.)
To the Sine of the *Angle* Z P ☉, (84 d. 7 m.)
And that is the *Hour*, counted from the *North* Part of the *Meridian* (which in Time is 5 h. 36 m. that is, 36 m. after 5 in the Morning, or 95 d. 53 m. (or 6 h. 24 m.) from the *South*.

3. For the *Angle* Z ☉ P, the *Angle* of *Position*. By *Case* I. of *O. A. S. T.*

As the Sine of the Side P ☉, (70 d.)
Is to the Sine of the *Angle* ☉ Z P, (72 d. 52 m.)
So is the Sine of Z P, (38 d. 30 m.)
To the Sine of the *Angle* Z ☉ P, 39 d. 16 m.
Which is the *Angle* of the *Sun's Position*.

PROB. X.

To find the Longitude *and* Latitude *of any* Star.

THE *Longitude* of any *Star*, is an *Arch* of the *Ecliptick*, contained between the beginning of *Aries*, and the Intersection of an *Arch* of a great Circle, which passeth through both the Poles of the *Ecliptick*, and also through the Body of that *Star*.

The *Latitude* of a *Star* is that Part of an *Arch* of a great Circle, which passeth through both the Poles of the *Ecliptick*, and through the Body of the *Star*, and is contained between the *Ecliptick* Line and that *Star*.

I. For the *Longitude* ☉, Skrew the *Quadrant of Altitude* over that Pole of the *Ecliptick* which is nearest to the *Star*, whose *Longitude* you seek. Then laying the *Quadrant* just over the Centre of the Star, look what Degrees of the *Ecliptick*, are cut by the (counting them from the beginning of *Aries*) *Quadrant of Altitude*, and those Degrees are the Degrees of the *Star's Longitude*. So the *Quadrant of Altitude* skrewed over the *North* Pole of the *Ecliptick*, and laid upon the bright Star *Capella*, the *Quadrant* shall cut 77 d. 16 m. of the

the *Ecliptick* Circle, counted from the beginning of *Aries*; and that is that *Star's Longitude*.

Fig. XXXIII.

Note, That the Poles of the *Ecliptick* are distant from the Poles of the World 23½ d. on either side.

For the *Latitude*, the *Quadrant* fitted as before, and laid over the Centre of *Capella*, the *Star* shall cut 22 d. 50 m. of the *Quadrant of Altitude*; and such is the *Latitude* of that *Star*, *North*, for that it lyes on the *North* Side of the *Ecliptick* Line.

This needs no Trigonometrical Calculation.

PROB. XI.

To find the Right Ascension *and* Declination *of a* Star.

THE *Right Ascension* of a *Star* is that Arch of the *Æquinoctial*, which is contained between the beginning of *Aries*, and that Point which comes to the *Meridian* with that *Star*.

The *Declination* of a *Star* is an Arch of the *Meridian* contained between the *Æquinoctial* and any *Star*.

For the *Right Ascension*, (the *Globe* being rectified) bring *Capella* to the *Meridian*, and then shall you find 73 d. 7 m. of the *Æquinoctial* contained between the beginning of *Aries* and the *Meridian*; and that is the *Right Ascension* of *Capella*.

For the *Declination*, bring *Capella* to the *Meridian*, so shall you find 45 d. 37 m. of the *Meridian* contained between the *Æquinoctial* and *Capella*; and that is the *Declination* of that *Star*. And in this manner you may find the *Longitude*, *Latitude*, *Right Ascension*, and *Declination*, of any other *Star* upon the *Cœlestial Globe*, as in this following Table of the principal Fixed Stars of the first Magnitude you shall find.

This needs no Trigonometrical Calculation.

Fig. XXXIII.

Stars Names.	Longit. D.	M.	Latit. D.	M.		R. Ascen. D.	M.	Declin. D.	M.	
Arcturus	119	35	31	2	B	210	13	20	58	B
Lucida Lyra	280	43	61	47	B	276	27	38	30	B
Algol	51	37	22	22	B	41	46	39	39	B
Capella	77	16	22	50	B	73	7	45	37	B
Aldebaran	65	12	35	31	A	64	17	15	48	B
Regulus	145	17	0	26	B	147	43	13	33	B
Cauda Leonis	167	3	12	18	B	173	4	16	25	B
Spica Virgin.	199	16	1	59	A	196	56	9	31	A
Antares	245	13	4	27	A	242	23	25	37	A
Fomahant	329	11	21	00	A	339	46	31	17	A
Regel	72	17	1	11	A	74	44	8	37	A
Syrius	99	33	39	30	A	97	42	16	14	A
Procyon	111	18	15	57	A	110	34	6	3	B

PROB. XII.

To find the Distance of Two Stars.

1. IF the Two *Stars* be both of them under the same *Meridian*, Bring them under the General (or Brass) *Meridian*, and see what Degrees of the *Meridian* are contained between them, for that is their Distance.

2. If they lye not under the same *Meridian*, but have the same Declination, or lye in the same Parallel, Bring one of them to the *Meridian*, and see what Degrees of the *Æquinoctial* are cut thereby: Then bring the other *Star* to the *Meridian*, and count what Degrees of the *Æquinoctial* are contained between the *Meridian* and the Degrees before found; for that is the Distance of those Two *Stars*.

3. If the Two *Stars* do neither lye under the same *Meridian*, nor in the same Parallel, Then lay the *Quadrant of Altitude* (it being loose) to both the *Stars*, and the Degrees of the *Quadrant* contained between the Two *Stars* is their Distance. And if the *Quadrant* be too short, you may use the Circle of Position, or take their Distance with a pair of Calope-Compasses, and measure their Distance upon the *Æquinoctial*, or any other great Circle.

Thus,

Thus,

The Right Shoulder of *Auriga*, and the Right Shoulder of *Orion*, being under the same *Meridian*, their Distance will be found to be 37 d. 38 m.

Fig. XXXIII.

Also,

Arcturus and the Lion's Neck, being near in the same Parallel, their Distance will be found to be 57 d.

Likewise,

Lyra the Harp, and *Marchad* in the Wing of *Pegasus*, will be found, to be distant 63 d.

The finding of the Distance of *Stars* upon the *Cælestial Globe* is the same as the finding of Distance of Places upon the *Terrestrial*, and consequently the *Triangles* made upon the *Globe* are the same also: So that the Canons for Calculation for finding them will be the same also, and needs not be here again repeated.

PROB. XIII.

To know what Stars *will be upon the* Meridian *at any Hour of the Night.*

THE *Sun* being in 29 d. of *Taurus*, what *Stars* will be upon the *Meridian* at 10 a Clock, and 12 m. at Night. Bring 29 d. of *Scorpio* (which is the opposite Sign to *Taurus*) to the *Meridian*, and set the *Index* of the Hour-Circle to 12. Then turn the Globe about Westward till the *Index* point at 12 m. after 10 a Clock, and there hold the *Globe*; and all those *Stars* which lye under the *Brass Meridian* are then upon the *Meridian*, of which *Arcturus* is the Chief.

This needs no Trigonometrical Calculation.

PROB. XIV.

To know what Day in the Year any Star *shall be upon the* Meridian *at 12 a Clock at Night.*

BRing the *Star* to the *Meridian*, and mark what Degree of the *Ecliptick* is just under the *Meridian* at the same time: Then find that Degree of the *Ecliptick* in the *Horizon*, and note what Day of the Year standeth against it, for that Day of the Year will that *Star* be upon the *South* Part of the *Meridian* at 12 at Night: And when the Sun is in the opposite Point of

Fig. XXXIII. the *Ecliptick*, the ſame *Star* will be upon the *North* Part of the *Meridian* at 12 at Noon.

PROB. XV.

The Sun's *Place, and the* Altitude *of a known* Star *given*; *To find the Hour of the Night.*

THE *Sun* being in 21 d. of *Capricorn*, the *Altitude* of the *Great Dog* 14 d. I demand the Hour of the Night.

The *Globe* Rectified, &c. Bring 21 d. of *Capricorn* to the *Meridian*, and the *Index* to 12 a Clock. Then move the *Globe* and *Quadrant of Altitude* ſo together, that the *Great Dog* meet with 14 d. of the *Quadrant*; and then ſhall the *Index* point at 8 of the Clock, and 22 m. which is the true Hour of the Night.

And thus,

When the *Sun* is in	D.		and the Alti- tude of	D.	The ho. will be	D.	M.
	20	♐	*The Bull's Eye*	39		7	12
	20	♏	*The Bull's Eye*	30		9	2
	5	♊	*Arcturus*	50		11	3

PROB. XVI.

The Altitude *of* Aldebaran *(or any other Star) being given, in a known* Latitude; *To find the* Star's Azimuth.

THE *Quadrant of Altitude* being fixed in the *Zenith*, move it and the *Globe* till the Degrees of *Altitude* given do meet with the Centre of the *Star*; then ſhall the end of the *Quadrant of Altitude* ſhew you upon the *Horizon* the *Azimuth* in which the *Star* then is. And thus, if you bring the *Quadrant of Altitude* on the *Eaſt* Side of the *Globe*, moving it and the *Globe* both, till the Centre of *Aldebaran* do meet juſt with 42 d. of the *Quadrant*, you ſhall then find the *Quadrant of Altitude* to reſt at 33 d. of the *Horizon*, counted from the *Eaſt*; or at 57 d. if you count them from the *South*: And that is the *Azimuth* of *Aldebaran* when he hath 42 d. of *Altitude*; and that is near the S. E. by E. Point of the Compaſs.

The

The Latitude *of the* Place, (51 *d.* 30 *m.*) *and the* Declination *of a* Star, (*suppose the* Bull's Eye, Aldebaran) *given: To find*

Fig. XXXIII.

PROB. XVII.

Its Right Ascension.

THE *Globe* Rectified to the *Latitude*, &c. Bring *Aldebaran* to the *Meridian:* Then count how many Degrees of the *Æquinoctial* are contained between the *Meridian* and the beginning of *Aries*; which will be 64 d. 17 m. and that is the *Right Ascension* of that *Star*; which in time (by allowing 15 d. for an Hour, and 1 d. for 4 m. of time) in 4 h. 16 m. its *Right Ascension* in time.

And in the same manner may you find

The Right Ascension of		D.	M.		H.	M.
	Arcturus	210	13	in time	14	1
	Syrius	97	42		6	30
	Algol	39	39		2	38

The same Trigonometrical Calculation, as for the *Sun*'s *Right Ascension*, as in *Prob.* III. serves for this also.

PROB. XVIII.

Its Ascensional Difference.

BRing the *Star* to the *Meridian*, and the *Hour-Index* to 12. Then bring the *Star* either to the *East* or *West* Side of the *Horizon*, and there you shall find 1 h. and 27 m. contained between the *Index* and 6 a Clock: And such is the *Ascensional Difference* of *Aldebaran*.

In like manner you may find

The *Ascensional Difference* of		D.	M.		H.	M.
	Arcturus	28	40	or in time	1	55
	Syrius	21	28		1	26
	Algol	——			——	

Algol his *Declination* being more than the Complement of the *Latitude*, never rises nor sets, but is always above the *Horizon*.

The same Trigonometrical Calculation, as for the *Sun*.

PROB.

Fig. XXXIII.

PROB. XIX.

Its Amplitude.

BRing *Aldebaran* to the *Horizon* on either Side of the *Globe*, and you shall find it to touch the *Horizon* at 25 d. 56 m. from the *East* or *West* Northward; which is the *Amplitude* of the *Bull's Eye* rising or setting. And according to the Points of the Compass it riseth E. N. E. 2 d. 26 m. Northerly, and sets W. N. W. 2 d. 26 m. Northerly.

And thus may you find

				D.	M.
That	*Arcturus*	Riseth from the *East* or *West*	Northward	35	6
	Syrius		Southward	26	41
	Algol		Never rises or sets.		

The same Trigonometrical Calculation as for the *Sun*.

PROB. XX.

The Semidiurnal Arch, *and the time that* Aldebaran *(or any other* Star) *continues above the* Horizon.

BRing the *Aldebaran* to the *Meridian*, and set the Hour-Circle to 12. Then turn the *Globe* Westward, till *Aldebaran* touch the *Horizon*; then shall the Hour-Index point at 7 h. 27 m. And so long time is *Aldebaran* above the *Horizon*, before he comes to the *Meridian*, and continues so many Hours and Minutes above the *Horizon*, after he hath past the *Meridian*, and sets in the *West*. And those 7 h. and 27 m. is the *Semidiurnal Arch* of that *Star*; which doubled, is 14 h. 54 m. And so long doth that *Star* continue above the *Horizon* after the time of his rising.

And in this manner you find

			D.	M.		H.	M.
The Semidiurnal Arch of	*Arcturus*	to be	7	55	And his Continuance above the *Horizon*	15	50
	Syrius		4	34		9	8
	Algol		12	00		24	00

The same Trigonometrical Calculation as for the *Sun*.

PROB.

PROB. XXI.

Fig. XXXIII.

At what Hour (any time of the Year) Aldebaran *comes to the* Meridian.

LET the time be the First of *January*, at which time the *Sun* is in 22 d. of *Capricorn*. Bring 22 d. of *Capricorn* to the *Meridian*, and set the Hour-Index to 12. Then turn the *Globe* about till *Aldebaran* be under the *Meridian*, and then you shall find the Index to point at 42 m. after 8 of the Clock, at which time *Aldebaran* will be upon the *Meridian* that Night.

In like manner you may find that

Upon			will be upon the *Meridian* at	*H.*	*M.*
October	28	*Arcturus*		11	10
January	21	*Syrius*		9	33
January	1	*Algol*		7	12

The same Trigonometrical Calculation as for the *Sun*.

PROB. XXII.

At what Hour (at any time of the Year) Aldebaran *(or any other* Star) *riseth or setteth.*

LET the time be *January* 1. By the last before-going, you found that *Aldebaran* came to the *Meridian* at 8 h. 42 m. And by the last but one you found his Semidiurnal Arch to be 7 h. 27 m. This being taken from 8 h. 42 m. the time of his being *South*, leaveth 1 h. 15 m. the time of its rising; so that upon the First of *January Aldebaran* did rise at 15 m. after 1 in the Afternoon. Again, if you add his Semidiurnal 7 h. 27 m. to the time of its being *South* 8 h. 42 m. the Sum will be 16 h. 9 m. from which take 12 h. and the Remainder will be 4 h. 9 m. So that *Aldebaran* did set at 9 m. after 4 of the Clock the next Morning.

And in like manner you may find that

Upon				Rises at *H.*	*M.*	Sets at *H.*	*M.*
October	28	*Arcturus*		3	8	7	12
January	21	*Syrius*		5	3	2	3
January	1	*Algol*		———		———	

Algol never Rises nor Sets.

The same Trigonometrical Calculation as for the *Sun*.

PROB.

Fig. XXXIII.

PROB. XXIII.

At what Horary Distance from the Meridian, Aldebaran *will be due East or West: And what* Altitude *he shall then have.*

BRing *Aldebaran* to the *Meridian*, and the Hour-Index to 12, and the *Quadrant of Altitude* to the *West* Point of the *Horizon*: Then turn the *Globe* Eastward, till the Centre of the *Star* be just under the Edge of the *Quadrant*; then shall the Index point at 5 h. and 40 m. So that when *Aldebaran* is due *East* or *West*, he will be 5 h. 40 m. of time short of, or gone beyond, the *Meridian*.

And when the Centre of *Aldebaran* is just under the Edge of the *Quadrant of Altitude*, you shall find it to touch 20 d. 21 m. And such is the *Altitude* of *Aldebaran* when he is upon the *East* or *West Azimuth*.

In like manner may you find that

		H. M.		D. M.
Arcturus	will be upon the *East* or *West*	4 49	and his	27 13
Syrius	*Azimuth*, when he is di-	5 04	*Altit.*	20 56
Algol	stant from the *Meridian*.	3 15	will be	54 37

The same Trigonometrical Calculation as for the *Sun*.

PROB. XXIV.

What Altitude *and* Azimuth Aldebaran *(or any other* Star*) shall have, when Six Hours distant from the* Meridian.

BRing *Aldebaran* to the *Meridian*, and the Index to 12. Then turn the *Globe* about till the Index point at 6; then lay the *Quadrant of Altitude* over the Centre of the *Star*, and you shall find it to lye under 12 d. 18 m. of the *Quadrant*: And such is the *Altitude* of *Aldebaran*. At the same time look what Degrees of the *Horizon* are cut by the *Quadrant of Altitude*, and you shall find 8 d. between it and the *East* or *West* Points Northwards. And such is the *Azimuth* of *Aldebaran*. And according to this Rule you shall find that when

		Altit. D. M.	*Azim.* D. M.	
Arcturus	is 6 h. distant from	16 15	76 36	from the
Algol	the *Meridian* his	12 38	79 43	*North*.

Syrius

Syrius is never 6 h. distant from the *Meridian*, nor any other *Star* that hath *South* Declination. Fig. XXXIII.

The Trigonometrical Calculation as for the *Sun*.

PROB. XXV.

To find what Altitude *and* Azimuth *any Star hath when he is at any horary Distance from the* Meridian.

THIS is no other than the last. For having brought the *Star* to the *Meridian*, and the Index to 12, move the *Globe* till it come to the designed Hour. Then the *Quadrant of Altitude* being laid over the *Star*, shall at the same time shew you both the *Altitude* and *Azimuth* thereof as before. This needeth no Example.

The same Trigonometrical Calculation as for the *Sun*.

PROB. XXVI.

Having the Azimuth *of a* Star, *to find at what Horary Distance that* Star *is from the* Meridian, *and what* Altitude *that* Star *then hath.*

BRing the *Star* to the *Meridian*, the *Index* to 12, and the *Quadrant of Altitude* to the *Given Azimuth*; then turn the *Globe* about till the Centre of the Star lye just under the *Quadrant of Altitude*; the *Index* at that time shall give the *Horary Distance*, and *Quadrant* the *Altitude* of the *Star*.

Example. Aldebaran being seen upon 80 d. of *Azimuth* from the North-Westward, that is, near upon the W. by N. Point of the Compass; the *Star* brought to the *Meridian*, and the *Quadrant of Altitude* to 80 d. and the *Hour-Index* to 12. If you bring the *Star* to the *Quadrant of Altitude*, you shall find the *Index* to point at 6 h. which is the *Star's Horary Distance* from the *Meridian*. And the *Quadrant of Altitude* will shew 12 d. 18 m. the *Altitude* of *Aldebaran* at that time.

These Twelve last *Problems* may be resolved by Trigonometrical Calculation in all Respects as those of the *Sun*. if (in all of them) instead of the Word [*Sun*] there, read [*Star*] here. And the Canons for Calculation will be the same in both.

Fig. XXXIII.

PROB. XXVII.

The Longitude *and* Latitude *of a* Star, *which is situate out of the* Ecliptick, *being given; To find the Right Ascension and Declination of that* Star.

HERE is given the Situation of the *Star,* in respect to the *Ecliptick,* altho' its Situation be out of the *Ecliptick;* and his Place, in respect of the *Æquinoctial,* is required

I. *By the* Cœlestial Globe.

Screw the *Quadrant of Altitude* over the Pole of the *Ecliptick,* and lay it to the Degree of the *Star's Longitude* in the *Ecliptick.* —Then count the *Star's Latitude* given upon the *Quadrant of Altitude:* And observe what *Meridian* passeth through that Point: For the Degrees of that *Meridian,* comprehended between that Point, and the Pole of the World, are the Degrees of the *Star's Declination:* And the same *Meridian* continued, will cross the *Æquinoctial* Circle in the Point of the *Star's Right Ascension.*

II. *By* Trigonometrical Calculation.

When the *Longitude* of the *Sun* falls to be in any Sign {Ascending, as in Fig. XXXIV. / Descending, as in Fig. XXXV.} make choice of the *Triangle* PFS, for the *Latitude* of the given *Star* if it be {North, / South,} the which the Letters *b* and *m* written within them will determine. In which *Triangle* there is

Fig. XXXIV. XXXV.

Given,
1. The Side PF, the Distance of the Pole of the *Æquator* from the Pole of the *Ecliptick,* 23 d. 30. m.
2. The Angle F, whose Measure is the Arch of the *Ecliptick,* intercepted between the Sides FP and FS, from whom the *Longitude* of the *Star* is produced, and the Solstitial Point found out.
3. The Side FS, the Complement of the given *Star's Latitude.*

Required,

Required,
1. The Side P S, the Complement of the *Star's Declination.*
2. The Angle P, which the Ark of the *Æquator* intercepted between the Sides P F and P S; of which, this *Right Ascension* of the *Star* is made, whither from the Degree of the *Ecliptick* 270, or 90 d.

Fig. XXXIV. XXXV.

Example. Let there be given the *Head of Andromeda*, which is a *Star* of the Second Magnitude; whose *Longitude* is in 9 d. 38 m. ♈, and *Latitude* 25 d. 42 m. *b.* The *Longitude* of the *Star* falls out to be in an Ascending Sign *Aries*; therefore, make choice of the Triangle P F S, Fig. XXXV. the *Latitude* being North, in which there is given, (1.) P F, the Distance of the Pole of the *Æquator* from the Pole of the *Ecliptick*, 23 d. 30 m. (2.) The Angle F, which the Ark of the *Ecliptick* intercepted between the Sides F P and F S, being produced, measureth, and gives the *Longitude* of the *Star* to be 9 d. 38 m. ♈, so that the Solstitial Point ♋ is nearest; substract therefore 9 d. 38 m. ♈ from 90 d. the beginning of ♋, the Remainder will be 80 d. 22 m. which is the Angle at F. (3.) F S, the Complement of the *Latitude* of the given *Star*, 64 d. 18 m. —And let there be sought, (1.) P S, the Complement of the *Declination* 62 d. 45 m. therefore the *Declination* of the *Star* is 27 d. 15 m. (2.) The Angle P, 87 d. 47 m. Which added to 270 d. makes the *Right Ascension* of the *Star* to be 357 d. 47 m.

The Canons for Calculation.

I. For the Angles S and P.

(1.) As the Sine of half the Sum of the Sides S F and F P, 43 d. 54 m.
Is to the Sine of half the Difference of those Sides, 20 d. 24 m.
So is the Co-tangent of half the Angle at F, 40 d. 11 m.
To the Tangent of half the Difference of the Angles at S and P, 30 d. 46 m.

(2.) As the Co-sine of half the Sum of the Sides F S and F P, 46 d. 6 m.
Is to the Co-sine of half the Difference of those Two Sides, 39 d. 36 m.

 So

Fig. XXXIV. XXXV.

So is the Co-tangent of half the Angle at F, 40 d. 11 m.
To the Tangent of half the Sum of the Angles at S and P, 57 d. 1 m.
Which 57 d. 1 m. added to the aforefound Difference, gives 87 d. 47 m. for the greater Angle at P; and 30 d. 46 m. the Difference aforefound substracted from 57 d. 1 m. leaves 26 d. 15 m. for the lesser Angle at S.

Then,

(3.) As the Sine of the Angle at S, 26 d. 15 m.
Is to the Sine of the Side F P, 23 d. 30 m.
So is the Sine of the Angle at F, 80 d. 22 m.
To the Sine of the Side S P, 62 d. 45 m.
Whose Complement, 27 d. 15 m. is the *Star's Declination*.

PROB. XXVIII.

The Declination *and* Right Ascension *of a* Star *being given: To find the* Longitude *and* Latitude *of that* Star.

THIS is the Converse of the former *Problem*, and may be resolved as followeth.

I. *By the* Cœlestial Globe.

Screw the *Quadrant of Altitude* over the Pole of the *Ecliptick:* Then count upon the *Æquinoctial* Circle the Degrees of *Right Ascension*; and upon that *Meridian* which passes through that Point, count the *Star's Declination* from the *Æquinoctial*, towards the *Pole*, and to that Point bring the *Quadrant of Altitude*; then will the *Quadrant* cut the *Ecliptick* in the Degrees of the *Star's Longitude*; and the Degrees of the *Quadrant*, comprehended between the *Ecliptick* Circle and its *Pole*, will give the Degrees of the *Star's Latitude*.

II. *By* Trigonometrical Calculation.

When the *Right Ascension* of the *Star* falls out to be in the *First* or *Fourth Quadrant* of the *Æquinoctial*; as in *Fig.* XXXIV. or in the *Second* or *Third:* As in *Fig.* XXXV. make use of the Triangle P F S for the *Star's Declination* North or South. In which there is

Given,

Fig. XXXIV. XXXV.

Given,
1. P F, the Diſtance of the Pole of the *Ecliptick* from the Pole of the *World*, 23 d. 30 m.
2. P S, the Complement of the *Star's Declination*, 64 d. 13 m.
3. The Angle P, whoſe Meaſure is the Arch of the *Æquator* intercepted between the Sides P F and P S, 87 d. 47 m. from whence the *Right Aſcenſion* of the *Star* is produced, and will be found to be 357 d. 47 m.

Required,
1. The Angle F, whoſe Meaſure is the Arch of the *Ecliptick* intercepted between the Sides F P and F S, from whence the *Longitude* of the *Star* from the next *Solſtitial* Point is found.
2. F S; the Complement of the *Star's Latitude*.

The Canons for Calculation.

(1.) As the Sine of half the Sum of the given Sides P F and P S, 43 d. 8 m.
Is to the Sine of half the Difference of thoſe Sides, 19 d. 37 m.
So is the Co-tangent of half the given Angle F, 43 d. 54 m.
To the Tangent of half the Difference of the Angles at F and S, *viz.* 27 d. 3 m.

(2.) As the Co-ſine of half the Sum of the Sides F S and F P, 43 d. 8 m.
Is to the Co-ſine of half the Difference of thoſe Sides; 19 d. 37 m.
So is the Co-tangent of half the given Angle at F 43 d. 54 m.
To the Tangent of half the Sum of the Angles at F and S; *viz.* 53 d. 18 m.

The half Difference before found 27 d. 3 m. added to the half Sum now found, gives 80 d. 21 m. for the Angle at F, the Complement of the *Star's Longitude*. And ſubſtracted from the half Sum 53 d. 18 m. leaves 26 d. 15 m. for the Angle at S.

Then ſay,

(3.) As the Sine of the Angle at F; 80 d. 21 m.
Is to the Sine of the Side S P, 62 d. 45 m.
So is the Sine of the Angle at P, 87 d. 47 m
To the Sine of the Side S F, 64 d. 18 m.
Whoſe Complement 25 d. 42 m. is the *Star's Latitude*.

A:

A Table of the Right Ascension and Declination of 100 Eminent Fixed Stars, Calculated for the Year of Christ 1700, compleated according to Ricciolus. *And their Difference of Right Ascension and Declination in 100 Years.*

Names of the STARS.	*Right Ascension for the Year 1700.*			*Declination for the Year 1700.*				*Diff. of Ascen. in 100 Y.*		*Declin. in 100 Years.*	
	De.	*Mi.*	*Se.*	*De.*	*Mi.*	*Se.*		D.	M.	*Mi.*	
Head of *Andromeda.*	358	14	08	27	27	06	B	1	17	34	A
Girdle of *Andromeda.*	13	11	20	34	2	49	B	1	23	33	A
South Foot of *Andromeda.*	26	21	51	40	52	30	B	1	29	30	A
Fomalhant	340	11	00	31	08	10	A	1	25	31	S
Right Shoulder } of ♒.	327	56	55	1	43	44	A	1	20	29	S
Left Shoulder	318	55	54	6	48	46	A	1	21	26	S
Left Hand	307	45	54	10	33	44	A	1	26	19	S
Bright Star in the Eagle.	294	2	47	8	6	32	B	1	27	13	A
First in Horn of ♈.	24	17	2	17	48	24	B	1	23	31	A
Second in ♈ Horn.	24	30	3	16	19	24	B	1	22	31	A
Bright one in *Aries.*	27	35	58	22	1	30	B	1	25	30	A
Goat of *Auriga.*	73	35	56	45	40	00	B	1	49	10	A
Right Shoulder of *Auriga.*	84	29	42	44	51	30	B	1	58	4½	A
Arcturus in *Bootes.*	210	33	2	20	48	2	B	1	11	26½	S
Left Shoulder of *Bootes.*	215	2	33	39	35	12	A	1	2	27	S

Names

Names of the STARS.	*Right Ascension for the Year* 1700.			*Declination for the Year* 1700.				*Diff. of Ascen. in* 100 *Y.*		*Declin. in* 100 *Years.*	
	De.	*Mi.*	*Se.*	*De.*	*Mi.*	*Se.*		*D.*	*M.*	*Mi.*	
Præsepe in *Cancer.*	125	46	2	20	43	4	B	1	28	19	S
Northern *Asinego* in ♋.	126	26	0	22	3	0	B	1	30	20	S
Southern *Asinego* in ♋.	126	54	3	19	15	0	B	1	27	20	S
Great Dog *Cyrius.*	97	57	6	16	18	6	A	1	7		A
Little Dog *Porcyon.*	110	54	33	5	59	12	B	1	20	12	S
Upper Horn of ♑.	300	24	34	13	22	6	A	1	25	16	S
Lower Horn of ♑.	301	7	29	15	38	2	A	1	27	17	S
First in the Tail of ♑.	320	56	29	17	54	21	A	1	26	26	S
Second in the Tail of ♑.	322	43	40	17	22	22	A	1	25	27	S
Bright one in *Cassiopea*'s Chair.	358	14	33	57	32	16	B	1	15	34	A
Scheder. or the Breast of *Cassiopea.*	5	56	0	54	55	16	B	1	22	34	A
In the Flexture of *Cassiopea.*	9	45	58	59	7	36	B	1	27	34	A
In *Cassiopea*'s Knee.	16	36	0	58	40	22	B	1	35	33	A
Cephus his Girdle.	321	6	20	69	17	10	B	1	22	26	A
Bright one in the *Whale*'s Jaw.	41	38	7	2	50	50	B	1	15	25	A
Northern in the *Whale*'s Belly.	24	12	0	11	44	50	A	1	15	31	S
Southern in the *Whale*'s Belly.	7	5	8	19	35	40	A	1	18	34	S
Northern in the *Whale*'s Tail.	1	4	12	10	24	54	A	1	17	34	S

Names

Names of the STARS.	*Right Ascension for the Year* 1700.			*Declination for the Year* 1700.				*Diff. of Ascen. in* 100 *Y.*		*Declin. in* 100 *Years.*	
	De.	*Mi.*	*Se.*	*De.*	*Mi.*	*Se.*		*D.*	*M.*	*M.*	
Bright one in the North. Crown.	230	39	0	27	45	20	B	1	5	21	S
In the Beak of the Swan.	289	39	48	27	22	40	B	1	1	11	A
In the Swan's Breast.	302	55	52	39	20	5	B	0	53½	18	A
In the Swan's Tail.	307	47	17	44	14	51	B	0	51½	20½	A
Upper Wing of the Swan.	293	56	2	44	26	21	B	0	48	14	A
Lower Wing of the Swan.	308	29	10	32	51	24	B	0	0	21	A
Bright one of the *Dragon.*	267	25	20	51	35	2	B	0	35	2	S
In the Head of *Castor.*	108	50	46	32	30	26	B	1	44	11	S
In the Head of *Pollux.*	111	43	36	28	43	2	B	1	34	12	S
Bright one in the Twins Foot.	95	3	32	16	37	32	B	1	28	2	S
Hercules Head.	255	21	37	14	46	48	B	1	8	8	S
Hercules Right Shoulder.	244	19	35	22	41	40	B	1	5	15	S
Hercules Left Shoulder.	255	31	33	25	15	48	B	0	52	8	S
Heart of *Hydra.*	138	12	22	7	21	30	A	1	15	25	A
Lion's Heart *Regulus.*	148	4	15	13	25	16	B	1	22½	28½	S
Lion's Tail.	175	25	34	16	14	4	B	1	19	24	S
Bright one in *Juba Leonis.*	150	48	47	21	21	0	B	1	25½	29	S
Bright one in *Lumbis Leonis.*	164	32	20	22	7	44	B	1	27	34	S
Uppermost in the Neck.	149	58	52	24	53	54	B	1	28	29	S
Lowest in the Neck.	147	47	52	18	13	33	B	1	28	28	S

Names

Names of the STARS.	Right Ascension for the Year 1700.			Declination for the Year 1700.				Diff. of Ascen. in 100 Y.		Declin. in 100 Years.	
	De.	M.	Se.	De.	Mi.	Se.		D.	M.	M.	
Thigh of the Hare.	78	51	30	20	59	0	A	1	5	7	S
Northern Scale of *Libra.*	225	15	26	8	14	46	A	1	$21\frac{1}{2}$	24	A
Southern Scale of *Libra.*	218	38	12	14	45	18	A	1	25	27	A
Bright one in *Lyra.*	276	39	32	38	32	16	B	0	50	4	A
Head of *Ophiucus.*	260	15	38	12	49	22	B	1	11	7	S
Left Hand of *Ophiucus.*	239	47	37	2	52	2	A	1	23	18	A
Right Knee of *Ophiucus.*	252	39	40	9	53	30	A	0	50	10	A
Left Knee of *Ophiucus.*	245	11	37	15	16	30	A	1	23	15	A
Right Shoulder of *Ophiucus.*	162	8	38	4	44	40	B	1	13	5	S
Uppermost in the Head of *Orion.*	79	41	10	9	34	38	B	1	22	7	A
Right Shoulder of *Orion.*	84	43	4	7	18	20	B	1	22	4	A
Left Shoulder of *Orion*	97	16	40	6	3	2	B	1	19	8	A
Foot of *Orion Rigel.*	75	2	50	8	33	42	A	1	$15\frac{1}{2}$	$9\frac{1}{2}$	S
1 In the Belt of *Orion.*	79	9	48	0	32	50	A	1	$17\frac{1}{2}$	7	S
2 In the Belt of *Orion.*	80	12	54	1	25	46	A	1	17	6	S
3 In the Belt of *Orion.*	81	18	25	2	8	20	A	1	16	5	S

Names of the STARS.	*Right Ascension for the Year* 1700.			*Declination for the Year* 1700.				*Diff. of Ascen. in* 100*Y.*		*Declin. in* 100 *Years.*	
	De.	*Mi.*	*Se.*	*De.*	*Mi.*	*Se.*		D.	*M.*	*M.*	
Mouth of *Pegasus.*	322	27	36	8	32	14	B	1	18	26	A
Sa'd, Alpharius in th Leg.	342	20	36	26	28	38	B	1	12	32	A
Marcab in the joining of the Wing.	342	28	10	13	35	58	B	1	15	32	A
End of *Pegasus* Wing.	359	27	25	13	32	56	B	1	16	34	A
Bright one in the Side of *Perseus.*	45	32	18	48	44	54	B	1	28	12	A
Ras Al Gol of *Perseus.*	42	12	42	39	46	30	B	1	37	25	A
The hindermost in the Head of the Southern Fish.	345	24	5	1	40	2	B	1	17	33	A
In the Knot in the Line of ♓.	26	38	5	1	19	0	B	1	18	30	A
Bright one in the Head of ♐.	283	1	5	21	22	48	A	1	31	8	S
Antares, Heart of *Scorpius.*	242	47	28	25	39	54	A	1	32	16	A
Northern Front of *Scorpius.*	236	58	15	18	53	36	A	1	28	16	A
Middlemost *Scorpius.*	236	14	34	21	41	40	A	1	30	20	A
Southern Front of *Scorpius.*	235	18	0	25	9	54	A	1	$37\frac{1}{2}$	21	A
Brightest in the Neck of *Scorpius.*	232	24	0	7	24	36	B	1	15	21	S
Aldebaran, or Southern Eye of ♉.	64	41	35	15	52	10	B	1	$26\frac{1}{2}$	15	A
In the Northern Horn of ♉.	76	51	18	28	19	2	B	1	37	8	A
Southern Horn of ♉.	79	55	20	20	55	58	B	1	31	7	A
Northern Eye of ♉.	62	43	36	18	30	28	B	1	24	17	A
Lowest of the *Hyades* ♉.	60	39	35	14	55	38	B	1	25	17	A
Bright one of the *Pleyades* ♉.	52	27	35	23	9	24	B	1	29	21	A

Names

Fig. XXXIV. XXXV.

Names of the STARS.	Right Ascension for the Year 1700.			Declination for the Year 1700.				Diff. of Ascen. in 100 Y.		Declin. in 100 Years.	
	De.	Mi.	Se.	De.	Mi.	Se.		D.	M.	M.	
Spica Virginis.	197	22	55	9	33	30	B	1	$19\frac{1}{2}$	$32\frac{1}{2}$	A
Girdle of *Virgo.*	190	10	22	5	2	54	B	1	18	34	S
Vindemiatrix in *Virgo.*	191	52	20	12	34	58	B	1	17	33	S
The Bright one in the Shoulder of the Greater Bear.	161	17	5	63	22	2	B	1	41	32	S
The Bright one in its Side.	160	52	20	57	59	2	B	1	37	32	S
The Bright one in the hindermost Thigh.	174	23	34	55	23	42	B	1	23	34	S
On the Back near the Tail.	180	8	2	58	41	42	B	1	20	34	S
The First in the Tail.	190	7	56	57	36	58	B	1	9	33	S
The Second in the Tail.	197	55	2	56	30	52	B	1	3	32	S
The Last in the Tail.	203	53	50	56	50	56	B	1	2	31	S
The last in the Tail of the lesser Bear, now the Pole Star.	9	52	10	87	42	51	B	3	10	$34\frac{1}{2}$	A
The Bright one in the Shoulder, heretofore called *Cynosura.*	222	39	20	75	37	30	B	1	15	$2\frac{1}{2}$	A

PROB.

PROB. XXIX.

The Latitude *of a* Place, *(or Elevation of the* Pole) *and* Declination *of the* Sun, *or of a* Star, *being given: To find the* Ascensional Difference *of that* Star: *And from thence, (the* Right Ascension *of the same* Star *being known) to find its* Oblique Ascension: *The* Semidiurnal Ark, *the Continuance of it above and below the* Horizon: *Its* Ortive *and* Occasive Amplitude: *And the* Oriental Angle, viz. *that of the Point of the* Ecliptick.

I. *By the* Cœlestial Globe.

Fig. XXXVI. THE *Globe* being rectified to the *Latitude*, and the Place of the *Sun* brought to the *Meridian*; and the *Quadrant of Altitude* screw'd to the *Zenith*, and brought to the *East* or *West* Points of the *Horizon*; the several Arches and Angles made by the Positions and Intersections of these Circles being measured or counted upon the *Globe*, will resolve the several Demands: But with some additional Works; which I here omit. Wherefore,

II. *By* Trigonometrical Calculation.

Here the Situation of a *Star* is given, in respect of the *Æquinoctial*; and Enquiry is made where its *Place* should be, in respect of an *Oblique Horizon* given.

The *Globe* continuing in the same Position as when you wrought the *Problem* upon it, you will find constituted upon it *Two Spherical Triangles*; such as in the *Figure* are noted with R *Or* S, for the *Declination* of the *Star* Northern or Southern. In either of which *Triangles*, there is (besides the *Right Angle* at R)

Given,
1. The *Angle Or*, which is the *Angle* made by the *Æquator* and the *Horizon* of *London*, 38 d. 30 m.
2. The *Perpendicular* R S, the *Star's Declination* North.

And

Fig. XXXVI.

And there is Required,

1. *Or* R, the *Ascensional Difference*; which, when a *Star* declineth towards

An *Elevated Pole*:
- *Substract* the *Decl.* from the R. *Ascension*, it gives the *Obl. Ascension.*
- Add the *Decl.* to the R. *Ascension*, it gives the *Obl. Descension.*
- Add 90 d. it gives the *Semidiurnal Ark.*

A *Depressed Pole*,
- Add the *Decl.* to the R. *Ascension*, it gives the *Obl. Ascension.*
- Substract the *Decl.* from the R. *Ascension*, it gives the *Obl. Ascension.*
- Substract 90 d. it gives the *Semidiurnal Ark.*

The *Semidiurnal Ark* being turned into *Hours*, gives half the Time of the *Stars* Continuance above the *Horizon*: Whose Complement to 12 *Hours* is its *Seminocturnal Ark*, or half his Continuance below the *Horizon*.

2. *Or* S, the *Amplitude* of the *Star*, at his *Rising* or *Setting*, having either *North* or *South* *Declination*.

3. The *Angle* S; which *Substracted* in *Signs*,
- *Ascending*, from the *Angle* of the *Ecliptick* and *Meridian*,
- *Descending*, from the Complement of the *Angle* of the *Ecliptick*; adding 180 d.

Gives the *Oriental Angle*

EXAMPLE I. *Let there be given an Elevation of the Pole (as at* London*) 51 d. 30 m. And the Sun's Declination, 0 d. 50 m. South.*

In

Fig. XXXVI. In the *Triangle* R *Or* S, (the *Declination* being Southward) there is given, (besides the *Right Angle* at R,) (1.) The *Angle* R *Or* S, the Complement of the given *Latitude*, 38 d. 30 m. (2.) The *Side* R S, the *Declination* of the *Sun*, 1 d. 2 m. by which may be found,

(1.) The *Side Or* R, the *Ascensional Difference*, 1 d. 2 m. Which, when the *Sun* declines (as here) towards the *Depressed Pole*, being added to the *Right Ascension* before found to be 358 d. 3 m. gives the *Oblique Descension*, to be 359 d. 7 m. But being taken from the said *Right Ascension*, 358 d. 3 m. gives the *Oblique Descension* to be 356 d. 59 m. And being taken likewise from 90 d. gives 88 d. 56 m. for the *Semidiurnal Ark:* Which being turned into *Time*, gives 5 h. 55 m. for the half continuance of the *Sun* above the *Horizon*, whose Complement to 12 h. is 6 h. 5 m. the half continuance of the Sun below the *Horizon:* But double the continuance of the Sun above the *Horizon*, the *Length* of the *Day* will be 11 h. 50 m. And double the continuance below, and the *Length* of the *Night* will be 12 h. 10 m.

(2.) The *Side Or* S, the *Ortive Amplitude* of the *Sun*, 1 d. 20 m.

(3.) The *Angle* R S *Or*, 51 d. 30 m. the given *Latitude*.

Which being substracted (because, in this *Example*, the *Sign* is *Ascending*, *viz.* ♓) from the *Angle* of the *Ecliptick* and *Meridian*, before found, to be 66 d. 31 m. gives 15 d. 1 m. for the *Oriental Angle* of the *Point* of the *Ecliptick*.

This *Example* was of the *Sun*, I will here give another *Example* of a Fixed Star, *viz.* The Head of *Andromeda*

EXAMPLE II. *The* Declination *of the* Head *of* Andromeda *is 27 d. 15 m. represented by the* Side R S.

The Complement of the given *Latitude* is 38 d. 30 m. represented by the *Angle* R *Or* S.

These being given, we are to find,

1. The *Side* R *Or*, the Star's *Ascensional Difference*.
2. The *Side Or* S, the *Amplitude* of the Star.
3. The *Angle* at S.

III. *The Canons for Calculation.*

1. For the *Side* R *Or*, by *Case* VII. of *R. A. S. T.*

As the Radius,

Is to the Tangent of R S, the Star's *Declination*, 27 d. 15 m.

So

So is the Co-tangent of the *Angle Or*, 38 d. 30 m.
To the Sine of the Side R *Or*. The Star's *Ascensional Difference*, 40 d. 21 m.

Fig. XXXVI.

2. For the *Side Or* S. By *Case* IX. of R. *A. S. T.*

As the Sine of the Angle R *Or* S, 38 d. 30 m.
Is to the Radius:
So is the Sine of the Side R S, 27 d. 15 m.
To the Sine of the Side S R, (the Star's *Amplitude* at his *Rising*) 47 d. 22 m.

3. For the *Angle* R S *Or*. By *Case* VIII. of R. *A. S. T.*

As the Co-sine of the Side R S, 27 d. 15 m.
To the Radius:
So is the Sine of the Angle R *Or* S, 38 d. 30 m.
To the Sine of the Angle R S *Or*, 61 d. 41 m.

By which the *Oriental Angle* may be found.

	D.	*M.*
The *Right Ascension* of *Andromeda*'s Head was found by (*Pr.* 27.) to be	357	47
And his *Declination* given *North*	27	15
His *Ascensional Difference* (by the (1.) hereof) is found to be	40	21
Which substracted from the Star's *Right Ascension*, gives the *Oblique Ascension* to be	317	26
And the *Declination* added to the *Right Ascension*, gives	398	08
From which substract the whole Circle	360	00
There will remain for the *Oblique Descension*	38	08
To which add	90	00
And it gives the *Semidiurnal Arch* to be	128	08

	H.	*M.*
Which turned into Time, is	8	32
Which doubled, is	17	04

And so long the *Star* continues above the *Horizon* of *London*.
And that substracted from 24 h. leaves 6 h. 56 m. for the Star's continuance below the *Horizon* of *London*.

By the (2.) hereof, the *Amplitude* of the Star's Rising from the *East* Northward, is found to be	47	22
And by the (3.) hereof, the *Angle* at S,	61	41

Which { Substracted in Signs Ascending, from / Added in Signs Descending, to } the *Angle* of the *Ecliptick* and *Meridian*, gives the *Oriental Angle*.

PROB.

Fig. XXXVI.

PROB. XXX.

The Declination *of the* Sun, *and the* Length *of a Day being given, to find the* Latitude *or* Height *of the* Pole.

I. *By the* Cœleftial Globe.

BRing the *Solftitial Colure* to the *Meridian*, and count the *Sun's Declination* upon it, either *North* or *South*, (fuppofe 23 d. 30 m. *North*) then let the *Sun's Declination* be 23 d. 30 m. *North*, and the Length of the Day 16 h. 26 m.

Bring the *Æquinoctial Colure* to the *Meridian*, and upon it count 23 d. 30 m. *North*, and there fix the Body of the *Globe*. Then, upon the Parallel of 23 d. 30 m. of *Declination*, count 8 h. 13 m. the half Length of the Day, and there make a Mark. Then move the *Brafs Meridian* up and down in the Frame, till the Mark you made do juftly touch the *Horizon*; for then, the Degrees of the *Brafs Meridian*, comprehended between the *Horizon* and the *Pole*, will be the *Latitude* fought, *viz.* 31 d. 30 m.

II. *By* Trigonometrical Calculation.

This being (in effect) but the Converfe of the foregoing *Problem*, the *Spherical Triangle* will be the fame upon the *Globe* as in that; namely, The *Triangle* R *Or* S: In which there is given, (befides the *Right Angle* at R) (1.) The *Side* R S, the *Sun's Declination*, 23 d. 30 m. *North*. (2.) The *Side* R *Or*, the *Afcenfional Difference*, which the *Sun* hath when the Length of a Day given is either *Greater* or *Lefs* than 12 Hours.

1. Let the *Sun's Declination* be 23 d. 30 m.
2. The longeft Day at *London*, 16 h. 26 m.
3. The *Afcenfianal Difference*, 2 h. 13 m. which is half the Time by which the longeft Day exceeds 12 h. *viz.* 4 h. 26 m. the half whereof, 2 h. 13 m. turned into Degrees, is 33 d. 52 m.

III. *The Canons for Calculation.* By *Cafe* XIII. of *R. A. S. T.*

As the Tangent of the Side R S, (the *Sun's Declination* given) 23 d. 30 m.

Is to the Radius:

So is the Sine of the Side R *Or*, (the *Afcenfional Difference*) 33 d. 10 m.

To the Tangent of 51 d. 30 m. the *Latitude* fought. Whofe Complement, 38 d. 30 m. is the Angle R, *Or* S. This

This *Problem* is of good use in *Geography*, where the Length of the *Longest Day*, of every *Clime* is given, to find what *Elevation* of the *Pole* answers thereunto. Fig. XXXVI.

PROB. XXXI.

The Latitude *of a* Place, *or Elevation of the* Pole, *and the* Oblique Ascension *of a* Star *being given: To find the* Cosmical, Achronical *and* Heliacal Rising *of that* Star.

THE *Cosmical Rising* or *Setting* of a *Star*, is that which happens in the *Morning*, but the *Achronical* in the *Evening*; the one commencing at the *Sun's Rising*, the other at his *Setting*. The *Heliacal Rising* is, when a *Star* begins to get from under the *Sun Beams*; the *Setting*, when it begins to go under them.

But for the better seeing of the *Stars*, there must be a certain Depression of the *Sun* below the *Horizon*, which is called The *Arch of Vision*; for the Light of a *Star* is *Greater* or *Less*. In *Ptolomy's* Opinion, *Venus* should have an *Arch* of 5 Degrees, tho' it is visible even in the Day-time, at a very great Distance.

			D.	M.
Jupiter and *Mercury*			10	00
Saturn			11	00
Mars			11	30
The Fixed Stars of the	First	Magnitude	12	00
	Second		13	00
	Third		14	00
	Fourth		15	00
	Fifth		16	00
	Sixth		17	00
	Nubilus		18	00

But in this and the following *Problem*, the Situation of a *Star* is given in respect of the *Horizon* and *Æquator* together; and his Situation is sought for in respect of the *Sun's* being in, or near, the *Horizon*, and in the *Ecliptick*.

When therefore the *Oblique Ascension* of a *Star* falls in the	First	Quadrant of the *Ecliptick*, as in	Fig. XXXVII.
	Second		Fig. XXXVIII.
	Third		Fig. XXXIX.
	Fourth		Fig. XL.

Fig. XXXVII. XXXVIII. XXXIX. XL.

I. *By the* Cœleſtial Globe.

1. For the *Coſmical Riſing* and *Setting.*

Rectifie the Globe to your Latitude, and bring the Place of the Sun (ſuppoſe 11 d. of ♒) to the *Eaſt* Part of the *Horizon:* Then look what *Stars* are about the Verge of the *Eaſtern Semicircle* of the *Horizon*; for all thoſe *Stars* do at that time *Riſe Coſmically.* And thoſe *Stars* which touch, or are near the Rim of the *Weſt Semicircle* of the *Horizon*, do *Set* at that time *Coſmically.* So ſhall you find

May 27.	*Aldebaran* or the *Bull's Eye*, with divers other ſmaller *Stars*,	Riſing, and	The Right Leg of *Serpentarius*, and ſeveral other ſmaller *Stars*,	Setting *Coſmically.*

2. For the *Acronical Riſing* and *Setting.*

Rectifie the Globe to the *Latitude*, bringing the Place of the *Sun*, ſuppoſe 5 d. of *Scorpio* to the Weſt Part of the *Horizon*, then ſhall all thoſe *Stars* which you ſee on the Verge of the Eaſt Side of the *Horizon*, be Riſing *Acronically.* And all thoſe that are about the Verge of the Weſtern Part of the *Horizon*, are then Setting *Acronically.* And ſo upon the forementioned Time, you ſhall find

May 27.	A *Star* in the *Whale's Tail*, and ſeveral other ſmaller *Stars*	Riſing, and	The *Tail* of the *Lion*, the South Ballance, and ſeveral other ſmaller *Stars*	Setting *Acronically.*

3. For the *Heliacal Riſing* and *Setting.*

Rectifie the *Globe* to the *Latitude*, and the *Quadrant of Altitude* in the *Zenith*; then bring the given *Star* (ſuppoſe *Regulus*, or the *Lion's Heart*) to the *Eaſt Side* of the *Horizon*, and the *Quadrant of Altitude* to the *Weſt Side*; then *Regulus* being a *Star* of the *Firſt Magnitude* (by the Rule of the Ancients) may be ſeen when the *Sun* is but 12 d. below the *Horizon*; wherefore, ſee what Degree of the *Ecliptick* doth cut the *Quadrant of Altitude* in 12 d. which you will find to be 9 d. of *Piſces*; the oppoſite Degree to which is 9 d. of *Virgo.* To which Sign and Degree, when the Sun cometh (which will be about the 23th of *Auguſt*) then will *Regulus*, or the *Lion's Heart*, *Riſe Heliacally.* Then,

Then, for his *Heliacal Setting*, bring the *Star* to the *West Side* of the *Horizon*, and turn the *Quadrant* of *Altitude* to the *East Side*, and see what Degree of the *Ecliptick* is elevated upon the *Quadrant*, as the *Magnitude* of the *Star* you are to deal with doth require. For when the *Sun* comes to the opposite Degree of the *Ecliptick*, that *Star* shall *Set Heliacally*. So,

Fig. XXXVII. XXXVIII. XXXIX. XL.

The { *Pleiades* / *Aldebaran* / *Arcturus* } shall *Rise Heliacal-ly* upon { *June* 4. / *June* 26. / *Sept.* 26. } And *Sets Heliacal-ly* upon { *April* 20. / *April* 22. / *Novem.* 19.

And so for others, as in the following *Table*.

A Table shewing the Time of the Year when 50 Eminent Fixed Stars do Rise, both Cosmically and Achronically.

Names of the Stars.	*Cosmical Rising.*	*Achronical Rising.*
Marchab. Pegasi.	*January* 1	*March* 9
Right Shoulder of *Aquarius*	6	*Febr.* 14
Extream Star in the Wing of *Pegasus*	28	19
Following Tail of the Goat	*February* 5	*Janu.* 26
Bright Star in the Ram's Head	*March* 2	*April* 20
The following Horn of the Ram	5	15
The former Horn of the Ram	*April* 10	30
North Tail of the Whale	12	*March* 19
Brightest of the Pleiades	22	*May* 10
The Knot in the Net of *Pisces*	*May* 1	*April* 25
North Horn of the Bull	15	*June* 7
North Eye of the Bull	20	*May* 12
In the Belly of the Whale	23	*March* 16
The Bull's Eye, *Aldebaran*	27	*May* 11
North Horn of the Bull	*June* 5	*June* 17
Lower Head of *Gemini*	21	*July* 20
Bright Foot of *Gemini*	25	*June* 5
Middle Star in *Orion*'s Girdle	*July* 9	*May* 4
North Asellus	12	*July* 25
Presepe	14	19
South Asellus	17	17
Lesser Dog, *Porcyon*	18	*June* 6
Great Dog, *Palilicium*	30	*May* 3
Lion's Heart	*August* 8	*August* 9

Fig. XXXVII. XXXVIII. XXXIX. XL.

Names of the Stars.	*Cosmical Rising.*	*Achronical Rising.*
Lion's Back	*August* 10	*Octob.* 19
Hydra's Heart	20	*June* 16
Lion's Tail	22	*Octob.* 16
Hare's Thigh	*Septem.* 9	*April* 14
Vindemiatrix	13	*Nov.* 8
Arcturus	15	*Decem.* 13
Virgin's Girdle	19	*Octob.* 18
Bright Star of the Crown	28	*Jan.* 7
Virgin's Spike	*Octob.* 3	*Sept.* 24
Right Shoulder of *Hercules*	6	*Jan.* 9
Left Shoulder of *Hercules*	10	21
Head of *Hercules*	21	8
North Ballance	21	*Nov.* 17
South Ballance	23	*Octob.* 24
Swan's Bill	31	*Febr.* 16
Right Shoulder of *Ophiucus*	*Novem.* 5	*January* 4
Left Knee of *Ophiucus*	6	*Decem.* 6
Lower Wing of the Swan	4	*March* 11
Vulture's Tail	11	*Jan.* 27
Right Knee of *Ophiucus*	16	*Decem.* 6
Scorpion's Heart	22	*Novem.* 4
The Eagle	25	*Decem.* 31
Pegasus Scheat	*Decem.* 11	*March* 24
Andromeda's Girdle	22	*April* 29
Andromeda's Head	25	6
Upper Horn of the Goat	25	*Jan.* 14

II. *By* Trigonometrical Calculation.

The *Latitude* of *London* is 51 d. 30 m.

The *Oblique Ascension* of the Head of *Andromeda* is 317 d. 26 m. and therefore in the Fourth *Quadrant* of the *Ecliptick*. Wherefore in the *Oblique-angled Spherical Triangles* ♈ Or. A, or ♎ Or. A.

In

In which there is Given:

1. The Angle at ♈ or ♎, (the greatest *Declination* of the *Sun*, or *Obliquity* of the *Ecliptick*, 23 d. 30 m.
2. The Angle at *Or.* (which is at the Intersection of the *Æquator*, and the *Horizon* of *London*) 38 d. 30 m.
3. The Side *Or.* ♈, or *Or.* ♎, (the Distance of the *Oblique Ascension* from the nearest *Æquinoctial Point*, 42 d. 34 m.

Fig. XXXVII. XXXVIII. XXXIX. XL.

To find: 1. The Side ♈ A, or ♎ A; for, A is that Point of the *Ecliptick*, which *Ascends* with a Star.

When therefore the Sun is in that Point, a *Star Riseth Cosmically*: But when the Sun is in that Point of the *Ecliptick* that is opposite to A, a *Star Riseth Achronically*.

To find: 2. The *Oriental Angle* at the Point A, *viz.* the Angle ♈ A *Or.* or ♎ A *Or.*

Moreover, in the Triangle R A ☉, (besides the Right Angle at R)

There is Given:

1. The Side R ☉, the *Ark of Vision*, agreeing with the Star, *viz.* 13 d. for the second Magnitude.
2. The Angle R A ☉ (the same as ♈ A *Or.* or ♎ A *Or.* before found, 26 d. 50 m.

To find the Side A ☉, the Arch which must be added to the Point A, to find out the Point ☉; which, when the Sun is in, a *Star Rises Heliacally.*

The Canons for Calculation.

1. For the Sides A *Or.* and *Or.* ♈. By *Case* III. of *O. A. S. T.*

The *Angles* Given are A *Or.* ♈, 141 d. 30 m. and *Or.* ♈ A, 23 d. 30 m. And the Given Side is *Or.* ♈ comprehended by them, 42 d. 34 m.

The *Sum* of the Given Angles is 164 d. Their *Difference* 118 d. The *half Sum* 82 d. The *half Difference* 59 d. And half the Given *Side* is 21 d. 17 m.

Then,

(1.) As the Sine of half the *Sum* of the Angles A *Or.* ♈, and *Or.* ♈ A, 82 d.

Is to the Sine of half the *Difference* of those Angles:

So is the Tangent of half the Given *Side Or.* ♈, 21 d. 17 m.

To the Tangent of 18 d. 38 m. Which is half the *Difference* of the Two unknown *Sides Or.* ♈, and *Or.* A.

(2.) As

Fig. XXXVII. XXXVIII. XXXIX. XL.

(2.) As the Co-sine of half the *Sum* of the Angles A *Or.* ♈, and *Or.* ♈ A. 82 d.

Is to the Co-sine of half the *Difference* of those Angles:

So is the Tangent of half the Given *Side Or.* ♈, 21 d. 17 m.

To the Tangent of 55 d. 19 m. Which is half the Sum of the Two unknown *Sides Or.* A, and *Or.* ♈. —The half Difference 18 d. 38 m. substracted from 55 d. 19 m. (the half Sum) leaves 36 d. 41 m. for the *Lesser Side* A *Or.* —And the half Sum 55 d. 19 m. added to the half Difference, gives 73 d. 57 m. for the *Greater Side* A ♈, but (because the Angle *Or.* opposite thereto, is above a *Quadrant*) add 90 d. to 73 d. 57 m. and the Sum 163 d. 57 m. is the Side A ♈.

2. For the Angle *Or.* A ♈. By *Case* II. of R. A. S. T.

(3.) As the Sine of the Side A *Or.* 36 d. 41 m.

Is to the Sine of *Or.* ♈ A, 23 d. 30 m.

So is the Sine of the Side *Or.* ♈, 42 d. 34 m.

To the Sine of the Angle *Or.* A ♈, 26 d. 50 m.

Then in the Triangle A R ☉, Right-angled at R, to find the Side A ☉; by *Case* IX. of *R. A. S. T.*

(4.) As the Sine of the Angle at A, 26 d. 50 m.

Is to the Sine of the Side R ☉ (the Angle of *Vision*, 13 d. it being a Star of the Second Magnitude).

So is the Radius,

To the Sine of the Side A ☉, 29 d. 53 m. Which added to the Side A ♈, 163 d. 57 m. the Sum is 193 d. 50 m. from which substract 180 d. there will remain 13 d. 50 m.

Wherefore, the Head of *Andromeda* Riseth *Cosmically* at *London*, with 13 d. 50 m. of ♑ *Capricorn*, in which Sign and Degree the Sun is about the 24th of *December*: And *Achronically* with 13 d. 50 m. of ♋ *Cancer*, in which Sign and Degree the Sun is about the 25th Day of *June*. —The *Oriental Point* of *Rising* is at A, or the Angle ♈ A *Or.* 25 d. 50 m. And the Head of *Andromeda Riseth Heliacally*, when the Sun is in 13 d. 50 m. of ♒ *Aquarius*, in which Sign the Sun is about the 22th of *February*, and *Sets Heliacally* when in the opposite Point, *viz.* 13 d. 50 m. of ♌ *Leo*, that is about the 26th of *August*.

PROB.

Fig. XLI. XLII. XLIII. XLIV.

PROB. XXXII.

The Elevation of a Pole, and the Oblique Descension of a Star given; To find the Cosmical, Acronical *and* Heliacal Setting *of that Star.*

I. *By the* Cœlestial Globe.

THE Practical Performance of this and the following *Problems* is shewed before in *Prob.* XXXI. and needs not be here again recited. Wherefore,

II. *By* Trigonometrical Calculation.

When the Oblique Descension of a Star falls

In the	First Second Third Fourth	Quadrant of the Eclyptick, as in	Figure XLI. Figure XLII. Figure XLIII. Figure XLIV.	Then,

By Trigonometrical Calculation.

In the Oblique-angled Triangle ♈ *Oc.* D, or ♎ *Oc.* D, there is given,

(1.) The Angle ♈ *Oc.* ♎, 23 d. 30 m.

(2.) The Angle *Oc.* the Angle of the *Æquator* and *Horizon* of *London*, 38 d. 28 m. its Comp. to 180.

(3.) ♈ *Oc.* or ♎ *Oc.* the Distance of the Oblique Descension, from the nearest Part of the *Æquinoctial.*

To find, 1. The Side ♈ D or ♎ D.

For, D is the Point of the Ecliptick, which descends with the Star: When therefore the Sun is in that Point, where D is, a Star Sets *Achronically:* But when the Sun is in the Point of the Ecliptick, opposite thereto, the Star then Sets *Cosmically.*

2. The Angle ♈ D *Oc.* or ♎ D *Oc.*

Moreover, in the Triangle R D ☉ (besides the Right Angle at R) there is given,

(1.) The Side R ☉, the Arch of Vision.

(2.) The Angle R D ☉, the same with the Angle ♈ D *Oc.* or ♎ D *Oc.* before found.

To find the Side D ☉, the Arch which must be taken from the Point D in the Ecliptick, to get the Point ☉ in the Ecliptick; to which Point, when the Sun comes, the Star Sets *Heliaeally.*

EXAM-

EXAMPLE.

Fig. XLI. XLII. XLIII. XLIV.

The Head of *Andromeda* (at London) sets *Achronically* with 26 d. 46 m. of ♈: — *Cosmically*, with 26 d. 46 m. of ♎. —But, the Angle ♈ D *Oc.* is found to be 121 d. 32 m. and its Complement to 180 d. the Angle R D ⊙, 58 d. 28 m. Wherefore, the Head of *Andromeda* sets Heliacally, when the Sun is in 11 d. 31 m. ♈.

PROB. XXXIII.

The Latitude *of a Place, the Oblique Ascension or Descension of Star, and the* Longitude *of the Sun being given: To know whether a Star may be seen* Heliacally.

I. *By the* Cœlestial Globe.

TO perform this by the Globe, see before *Prob.* XXXI.

II. *By* Trigonometrical Calculation.

This is but the Converse of the Two foregoing *Problems*: And the Spherical Triangles upon the Globe will be the same; and therefore, in the Triangle R A ⊙ or R D ⊙

There is given, besides the Right Angle at R,

1. The Oriental Angle at the Point A, or the Occidental at the Point D: The one being found in the Triangle ♈ *Or.* A, or ♎ *Or.* A, as in the *Problem*: The other in the Triangle ♈ *Oc.* D, or ♎ *Oc.* D, as in the *Problem.*
2. The Side A ⊙, or D ⊙; the Arch which is intercepted between the *Longitude* of the Sun, and the Ascending or Descending Point of the Ecliptick.

To find out the Arch R ⊙.

Which Arch, if it be greater, or equal, to the visible Arch of the Given Star, that Star doth appear: But, if it be less, it lyes hid under the Sun's Beams.

[*Note.*] Sometimes the Oriental Angle is less than the Arch of Vision; and then (even at Midnight) such a Star cannot be seen. But, if you substract the Arch of Vision from the greatest Depression of the *Æquator* under the *Horizon* (which is the Complement of the Latitude (or Elevation of the Pole at *London* 38 d. 30 m.) there remains the Northern Declination of a Degree of the Ecliptick, unto which the Sun must first come, before that Star will begin to appear at Midnight. As

As for Example. All Stars of the Fifth Magnitude, have an Arch of Vision of 16 d. therefore substract 16 from 38 d. 30 m. the Depression of the *Æquator*, and there will remain 22 d. 30 m. the Northern Declination of the Sun, *viz.* as much as is required that it be depressed 16 d. below the *Horizon*; that the Stars of the Fifth Magnitude may appear. But the Sun hath as great Declination in 13 d. 26 min. of *Gemini* ♊, and in 16 d. 33 m. of *Cancer* ♋: Therefore, whilst the Sun goes through that Arch of the Ecliptick (which will be between the 24th of *May*, and the 28th of *June*) the Stars of the Fifth Magnitude (are not visible at *London* (even at Midnight) because the Twilight hinders.

Fig. XLI. XLII. XLIII. XLIV.

Here needs no Canons for Calculation.

PROB. XXXIV.

The Elevation of the Pole, and the Longitude of the Sun: To find the Beginning of the Morning, and End of the Evening, Twilight.

I. *By the* Cœlestial Globe.

IN this and the following *Problem*, the Situation of the Sun is given, in respect of the *Æquator*, and its Situation is sought for in respect of the Meridian.

But, because, at every Cessation of the Twilight, there is commonly required a Depression of the Sun, of 18 d. below the *Horizon.*

Now, in the Latitude of 48 d. 30 m. the Sun being in the beginning of *Cancer* ♋, at Midnight, is depress'd below the *Horizon*, precisely 18 d. Hence it follows, that in Places, having a greater Elevation of the Pole than 43 m. the Sun being in the *Tropick*, must (at Midnight) be depressed less than 18 d. from whence it follows, that Twilight will neither Begin nor End, but will last all Night. Therefore, to know when this *Problem* is in Use, substract 18 d. from the greatest Depression of the *Æquator*, (that is, at *London*, 38 d. 30 m.) and there remains the Northern Declination 20 d. 30 m. (as by the *Problem* conversed): So the Sun, when he hath 20 d. 30 m. of North Declination, which he will have when he is in 1 d. 29 m. of *Gemini* ♊, and in 28 d. 40 m. of *Cancer* ♋: Therefore, whilst the Sun goes through that Arch of the Ecliptick (which is between the 12th Day of *May*, and

 the

the 11th Day of *July*) the Twilights lasts all Night at *London*, all which Time there is no Use of this *Problem* there.

At other Times of the Year it will be in Use, and may be resolved as followeth.

II. *By* Trigonometrical Calculation.

Fig. XLV. Upon the *Globe*, the *Triangle* resolving this *Problem*, is the *Oblique-angled Spherical Triangle* P N ⊙ in Fig. XLV. for the Declination of the *Sun*, Northern or Southern. In which there is

Given,
1. The Side P N, the Distance of the *Pole* of the *Æquator*, and the *Horizon* of *London*, 38 d. 30 m.
2. The Side N ⊙, the Complement of the Depression of the Sun under the *Horizon*, which the Sun must have when the *Twilight Begins* or *Ends*, *viz.* 72 d.
3. The Side P ⊙, the Complement of the *Declination* of the Sun, which will be
   - Greater / Lesser — Than 90 Degrees, — When in Northern / When in Southern — Signs.

Required, The Angle at P. Which must be turned into Time, the which, being

Counted from Midnight
- Shews the *Beginning* of the *Morning Twilight*.
- And substracted from 12 Hours, shews the *End* of the *Evening Twilight*.

The Canon for Calculation.

By *Case* XI. of *O. A. S. T.*

	D.	M.
The *Sum* of the Three Sides of the Triangle is	199	18
The *Half Sum* is 99 d. 24 m. *Or.*	89	36
The *Difference* between the *Half Sum* 99 d. 24 m. And the *Side* opposite to P is	27	24

Then say,

(1.) As the Radius,
To the Sine of P N, (one of the Sides containing the enquired Angle P) 38 d. 30 m.
So is to the Sine of P ⊙. (the other Side containing the enquired Angle P) 88 d. 48 m.
To the Sine of 38 d. 29 m.

(2.) As

53
Fig.
V.

Fig.
LVI.
LVII.

25

Fig.
XLV.

(2.) As the Sine of 38 d. 29 m. last found. — Fig. XLV.
Is to the Sine of the *Half Sum* of the *Sides*, 99 d. 14 m. (or 89 d. 36 m.)
So is the Sine of the *Difference*, 27 d. 24 m.
To the Sine of 47 d. 41 m.
To which Sine add the *Radius*, and half thereof will be the Sine of 59 d. 19 m. Whose Complement 30 d. 41 m. is *Half* the *Angle* at P.
Thus the Angle at P, being 61 d. 22 m. that reduced into *Time*, makes 4 h. 6 m. *fere*.
At which Time the *Twilight Begins* in the *Morning*. And that substracted from 12 h. leaves 7 h. 54 m. for the *Time* that *Twilight* will *End* in the *Evening*.

PROB. XXXV.

The Elevation of the Pole, *and the* Altitude *of a* Star; *whose* Right Ascension *and* Declination *are both known: To find out the Moment of* Time, *and the* Azimuth *of the* Star *at that* Time.

I. *By the* Cœlestial Globe.

RECtifie the *Globe* to the *Latitude*, (suppose *London*) 51 d. 30 m. Bring the Place of the *Sun* (suppose 27 d. 54 m. of *Pisces*) to the *Meridian*, and the Hour-Index to 12, and the *Quadrant of Altitude* to the *Zenith*: Then turn the Body of the *Globe* about, till the *Star* (suppose *Caput Andromedæ*) meet with 5 d. 24 m. (the given *Altitude* thereof.) Then will the *Hour-Index* point at the Time required, *viz.* 3 h. 47 m. *Morn.*

II. *By* Trigonometrical Calculation.

The *Sun* at Noon is at the highest. Take away therefore the *Sun's Right Ascension*, from the *Right Ascension* of the *Star*; and the Remainder being turned into *Time*, shews how many *Hours Afternoon* the *Star* is at the highest. Therefore, — Fig. XLVI. XLVII.

When a Star is in the { Eastern / — / Western } Parts, as in { Fig. XLVI. / Fig. XLVII.

Fig. XLVI. XLVII.

In the Triangle P Z S there is

Given,
1. The Side P Z, the Distance of the *Pole* of the *Æquator*, and the *Horizon* of *London*, 38 d. 30 m.
2. The Side Z S, the Complement of the *Star's Altitude*, 84 d. 36 m.
3. The Side P S, the Complement of the *Star's Declination*, 62 d. 42 m.

Required,
1. The Angle P, and must be turned into Time: And {In Fig. XLVI. / In Fig. XLVII.} must be {*Taken* from / *Added* to} the *Time* of *Culmination*, that the *Time* of *Observation* may appear.
2. The Angle at Z, the *Azimuth* of the *Star*, at the Time of *Observation*.

The Canon for Calculation. By *Case* XI. of O. A. S. T.

1. For the Angle at P.

(1.) As the Radius,
Is to the Sine of P Z, 38 d. 30 m. (the Complement of the *Latitude*.)
So is the Sine of Z S (the Complement of the *Star's Altitude*, 84 d. 36 m.
To the Sine of 38 d. 18 m.

(2.) As the Sine of 38 d. 18 m.
Is to the Sine of the half Sum of the *Three Sides* of the *Triangle*, 92 d. 54 m. (or 87 d. 6 m.)
So is the Sine of the *Difference* (between the *Half Sum*, and the *Side* Z S, opposite to the Angle at P) 8 d. 18 m.
To the Sine of 61 d. 10 m. The Double whereof 122 d. 20 m. is the Quantity of the *Angle* at P. Which, turned into *Time*, gives 8 h. 10 m.

2. For the Angle at Z.

As the Co-sine of the *Star's Altitude* Z S, 84 d. 36 m.
Is the Sine of the Angle at P, 122 d. 20 m. *Or*. (comp. to 180 d.) 57 d. 40 m.
So is the Sine of the Side S P, (the Comp. of the *Star's Declination*) 62 d. 42 m.
To the Sine of the Angle at Z, 48 d. 57 m. And that is the *Star's Azimuth* from the *North-Eastward*.

Example

Example in *Caput Andromeda.*

Fig. XLVI. XLVII.

In the *Latitude* of 51 d. 30 m. the *Altitude* of the *Head* of *Andromeda*, was *Observed* in the *Eastern Part* to have 5 d. 24 m. of *Altitude* above the *Horizon*: The *Sun's Longitude* at that time being in 27 d. 54 m. of *Pisces* ♓. At which

	D.	M.
Time also the *Right Ascension* of the *Sun* is	358	04
The *Right Ascension* of *Cap. Andromeda*, found as before	357	45
To which add 360 d. it makes	717	45
From which substract the *Sun's Right Ascension* 358 d. 4 m. there remains	359	41

Which turned into *Time*, gives 23 h. and 57 m. which is as many *Hours* as the *Star* is coming to the *Meridian* after the *Sun*.

Thus the Angle at P, being found to be 122 d. 20 m. which (the *Star* being in the *Eastern* Part) must be substracted from the Time of *Culmination*, 23 h. 57 m. and then there will remain 15 h. 47 m. for the Time of *Observation* Afternoon, which is 3 h. 47 m. in the Morning.

PROB. XXXVI.

The Time *of Observation, and* Latitude *of the* Place, *being given: To find out the* Altitude *and* Azimuth *of a* Star *given.*

I. *By the* Cœlestial Globe.

RECtifie the *Globe* to the *Latitude*, the *Quadrant of Altitude* to the *Zenith*, the Place of the *Sun* to the *Meridian*, and the *Index* to 12. Then turn the Body of the *Globe* about, till the *Index* points at the *Time* given, and there keep it: Then the *Quadrant of Altitude* laid over the *Star* will give you upon it the *Star's Altitude*; and when it cuts the *Horizon*, the *Azimuth* thereof also.

II. *By* Trigonometrical Calculation.

This is but the Converse of the former *Problem*: And the Triangle made upon the *Globe* the same, *viz.* the Triangle Z P S, in Fig. XLVI, and XLVII. In which Triangle there is

Given,

Fig. XLVI. XLVII.

Given,
1. The *Side* Z P (the *Distance* of the *Poles* of the *Æquator*, and *Horizon* of *London*) 38 d. 30 m.
2. The *Side* P S, the Complement of the *Star's Declination (Cap. Andromeda)* 62 d. 42 m.
3. The *Angle* S P Z (the *Time* between the Time of *Observation*, and the Time of the *Stars* next *Culmination*, *viz.* 8 h. 10 m) which is 122 d. 20 m.

Required,
1. The *Side* Z S, (the *Complement* of the *Star's Observed Altitude*).
2. The *Angle* at Z (the *Azimuth* of the *Star*, from the *North* Part of the *Meridian*.)

The Canons for Calculation. By *Case* IX. and *Case* I. of O. A. S. T.

1. For the Angle at Z.

(1.) As the *Sine* of half the *Sum* of the *Sides* Z P and Z S, 50 d. 36 m.
Is to the *Sine* of their half *Difference*, 12 d. 6 m.
So is the Co-tangent of half the *Given Angle*, Z P S, 61 d. 10 m.
To the *Tangent* of 8 d. 30 m. the half *Difference* of the Two *Angles*, Z and S.

(2.) As the Co-sine of half the *Sum* of the *Given Sides*, 50 d. 36 m.
Is to the Co-sine of their *half Difference*, 12 d. 6 m.
So is the Co-tangent of half the given *Angle* Z, 61 d. 10 m.
To the Tangent of 40 d. 22 m. Which is the *half Sum* of the Two *Angles*.

Then 8 d. 30 m.		D. M.		D. M.	For the Angle	
	Added to	40 22	gives	48 52		Z.
	Subtracted from	40 22	gives	31 52		S.

So that the *Azimuth* of the *Star* is 48 d. 52 m. from the *North* Part of the *Meridian*.

2. For the Side Z S.

(3.) As the Sine of the *Angle* at Z, (last found) 48 d. 52 m.
Is to the Sine of the *Side* S P (the Complement of the Star's *Declination*) 62 d. 42 m.
So is the Sine of the *Angle* at P, 122 d. 20 m. (or 57 d. 40 m.)
To the Sine of 84 d. 36 m. for the *Side* Z S. Whose Complement, 5 d. 24 m. is the *Altitude* of the Star at the Time of Observation.

Astrono-

Aſtronomical Problems

Relating to

ASTROLOGY.

INTRODUCTION.

ASTrology conſiſteth principally of Two Parts, *viz.* the one *Mathematical*, as is the *Aſtronomical Part*; the other *Judiciary*, as is the *Aſtrological Part*.

The *Mathematical Part* teacheth how in a *Scheme* or *Figure* (as they call it) to repreſent the *Face* of the Heavens in *Plano*, for any Hour of the Day or Night, at all Times of the Year, and in all Parts of the World.

The *Aſtrological Part* teacheth how (from the Sight of the ſaid Poſition of the *Scheme* or *Figure* of the Heavens at the Time of its Erection) to give a determinate *Judgment* of what was demanded upon that *Erection* of the *Scheme* or *Figure*; as of *Annual Revolutions*, *Elections*, the *Nativity* of a Perſon, &c.

The principal Authors that have given their Opinions concerning the dividing of the Heavens into 12 *Manſions* or *Houſes*, are, 1. *Ptolomy*, 2. *Alcabitius*, 3. *Campanus*, and 4. *Regiomontanus*: Which laſt Way is now generally received and practiced among the *Aſtrologers* of theſe Times, and by them termed the *Rational Way of* Regiomontanus.

Now, becauſe (as I ſaid before) that the Erection of a *Figure* of the Heavens is the *Mathematical Part* of *Aſtrology*, I ſhall therefore ſhew how by the *Globes* to erect a *Figure* of the Heavens according to the *Rational Way of* Regiomontanus.

PROB.

PROB. I.

How to Erect a Figure *of the Heavens in the* Latitude *of* London, 51 *d.* 30 *m.* N. *for the* 10th *Day of* March, *at* 49 *m. after* 9 *in the* Forenoon; *at which Time the* Sun *entred into the first Scruple of* Aries, *in the Year* 1675.

THE Heavens are divided in XII *Houses* or *Mansions*, by 12 *Semicircles* of *Position*; for which Purpose, to some *Globes*, there is made one of *Brass*, which is fixed in the Intersections of the *Meridian* and *Horizon*, by the *Elevation* or *Depression* whereof the Heavens may be divided into Parts or Houses through each Degree of the *Eclipt*ick.

Of these XII *Houses*, or *Mansions* of Heaven, Four are called *Cardinal*, as, (1.) The *Horoscope*, or *Ascendent*, or *Cuspis* of the *First House*. (2.) The *Medium Cœli*, or Angle of the *South*, or *Cuspis* of the *Tenth House*. (3.) The *Descendent*, or *Angle* of the *West*, or the *Cuspis* of the *Seventh House*. (4.) The *Imum Cœli*, or the *Angle* of the *North*, or the *Cuspis* of the *Fourth House*.

Regiomontanus divides the Heavens into XII *Houses* according to his Way, by the *Circle* of *Positions* passing through every 30th Degree of the *Æquinoctial*, and cutting the *Ecliptick* at several Points, which are the *Cuspises* of the several *Houses*: —So that when the *Globe* is set to the *Latitude*, and the Hour-Wheel *Rectified* and brought to the Given Hour, you have the *Cuspises* of the Four *Cardinal Houses*: For,

The Degree of the *E-cliptick* cut by the		Gives the *Cuspis* of the		
	East Side of the *Horizon*,		First	House.
	South Part of the *Meridian*,		Tenth	
	West Side of the *Horizon*,		Seventh	
	North Part of the *Meridian*,		Fourth	

The *Cuspises* of the other 8 Houses are found by the Motion of the *Circle* of *Position*, as shall be shewed by and by.

The *Houses* are *denominated* by 1, 2, 3, 4, &c. to 12, from the *Ascendent* downwards to the *Imum Cœli*, up again to the *Descendent*, and again by *Medium Cœli*, down to the *Ascendent*. As in the following *Scheme*.

A

t this ſuffice for *Definition*, and now we will come to the ice by the *Globes*.

ſt, To the Day propoſed, the 10th of *March*, find (by the *Aſtronomical Problem*) the Sun's Place in the *Ecliptick* at 1, which you ſhall find to be in 0 d. 5 m. of *Aries*.

condly, Set the *Globe* to the *Latitude* 51 d. 30 m.

irdly, Bring the *Sun's Place* at *Noon* (0 d. 5 m. of *Aries*) to *Meridian*.

irthly, Turn the *Globe* about till the *Hour-Index* point to *Iour* given, *viz.* to 49 m. after 9 in the *Morning*.

ſtly, The *Globe* being fixed in this *Poſition*, you ſhall find he *Eaſt-Semicircle* of the *Horizon* doth cut the *Ecliptick* in 29 m. of *Cancer*, which is the *Sign* then *Aſcending*, and muſt aced upon the *Cuſpis* of the *Firſt Houſe*.

ien caſt your Eye upon the Interſection of the *South* Part of *Meridian* and the *Ecliptick*, and there you ſhall find the *E-ck* cut by the *Meridian* in 25 d. of *Aquarius*, and that Point e *Ecliptick* is then in the *Medium Cœli*, and muſt be ſet upon *uſpis* of the *Tenth Houſe*.

ſo, you find that the *Weſt Semicircle* of the *Horizon*, cuts the *tick* in 00 d. 29 m. of *Capricorn*; which Point is then *De-ing*, and muſt be placed upon the *Cuſpis* of the *Seventh Houſe*.

 Laſtly,

Lastly, You shall find that the *North* Part of the *Meridian* doth cut the *Ecliptick* in the 25th ⊙ d. of *Leo*, which Point is then upon the *Imum Cœli*, and must be placed upon the *Cuspis* of the *Fourth House*.

Thus have you found the Points of the *Ecliptick*, which do occupy the *Cuspises* of the Four *Cardinal Houses*: Now for the other *Eight Houses*.

Let the *Globe* still rest in its former *Position*, and then,

First, Bring the *Circle of Position* to its Place on the *East* Side of the *Horizon*; and being there fixed, raise it upwards towards the *Meridian*, till 30 d. of the *Æquinoctial* be intercepted between the *Horizon* and the *Circle of Position*; and then you shall find that the *Circle of Position* will intersect the *Ecliptick* in 20 d. of *Taurus*; which Degrees must be set upon the *Cuspis* of the *Twelfth House*.

Secondly, Move the *Circle of Position* yet higher towards the *Meridian*, till 30 d. more of the *Æquinoctial* be intercepted between it and the *Horizon*, (in all 60 d.) and when in doth so, you shall find the *Circle of Position* will cut the *Ecliptick* in 36 d. of *Pisces*; which Point must be set upon the *Cuspis* of the *Eleventh House*.

The *Meridian* gives the *Cuspis* of the *Tenth House* in 25 d. of *Aquarius*, as before.

Thirdly, Move the *Semicircle of Position* from the *East* Side of the *Horizon* to the *West* Side, and move it downwards from the *Meridian*, till 30 d. of the *Æquinoctial* be intercepted between the *Meridian* and *Circle of Position*, and then you shall find the *Circle of Position* will intersect the *Ecliptick* in 7 d. of *Aquarius*; which Point must be set upon the *Cuspis* of the *Ninth House*.

Fourthly, Move the *Circle of Position* yet lower by 30 d. *i. e.* 60 d. from the *Meridian* downwards, and then you shall find the *Position-Circle* to cut the *Ecliptick* in 21 d. of *Capricorn*; which Point must be set upon the *Cuspis* of the *Eighth House*.

The *Descendant* or *Cuspis* of the *Seventh House*, is the Intersection of the *West* Side of the *Horizon* and *Ecliptick*, which is in 00 d. 29 min. of *Capricorn*, as before.

And thus have you found the *Cuspises* of the *Four Houses* above the *Horizon*, beside the *Ascendant* and the *Medium Cœli*, viz. of the 12, 11, 9 and 8 Houses. Now the *Cuspises* of the

the Four other Houses under the Earth have the same Degrees of the opposite Signs upon them. For,

		House.			House.
20 d. ♉	Being upon the *Cuspis* of the	12	20 d. ♏	will be on the *Cuspis* of the	6
26 d. ♓		11	26 d. ♍		5
7 d. ♒		9	7 d. ♌		3
21 d. ♑		8	21 d. ♌		2

For the Six Signs
Aries, *Taurus*, *Gemini*, *Cancer*, *Leo*, *Virgo*,
are opposite to
Libra, *Scorpio*, *Sagitt.* *Capric.* *Aquar.* *Pisces*.

And this is the manner how (by the *Globe*) to erect a *Figure* according to the (reputed) *Rational Way* of *Regiomontanus*.

Now if you would insert the Places of the *Planets* into your *Figure*, (for it is them that the *Astrologer* principally giveth *Judgment* by) your best Way will be to have Recourse to some good *Ephemeris*, or *Calculate* them from *Astronomical Tables*).

Having in the foregoing *Precepts* shewed how to erect a *Scheme* (or *Figure* of the *Heavens*) by the *Globe*, I shall now (to retain the same Method throughout this whole *Treatise*) shew how to effect the same *Trigonometrically*; the which I shall do in the *Three* following PROBLEMS.

PROB. I.

The Time *of the* Day, *and* Latitude *of the* Place *given: How to Erect a* Scheme *or* Figure *of the* Heavens *for that* Time *and* Place.

By Trigonometrical Calculation.

YOU must first find (either by *Tables* already Calculated, or by the foregoing *Astronomical Problems*) the true *Place* of the *Sun* in *Longitude*; and also, his *Right Ascension*, both for the *Moment* of *Time* given.

Fig. XLVIII.

Then turn the *Time* given, (counted from the preceding Noon) into *Degrees* and *Minutes* of the *Æquator*; and add them (being so turned into *Degrees* and *Minutes*) to the *Right Ascension* of the *Sun* before-found; the *Agregate* (casting away 360 d. if

Fig. XLVIII. need be) is the *Right Ascension* of the *Mid-Heaven:* To which you must find what Degree of the *Eclipticks* doth answer; for that will be the *Sign* of the Xth *House*.

Then, To the *Right Ascension* of the *Mid-Heaven*, add 30 d. then 60 d. again 90 d. and 120 d. and 150 d. And thus you will find the *Oblique Ascension* of the XIth, XIIth, and Ist *House*; or of the IId and IIId *Horoscope*.

By the Erection of the *Circle* of *Position* above the *Horizon* at every 30 Degrees, there are made several *Spherical Triangles*, as H Æ 11, &c.

In which Triangle (besides the Right Angle at Æ,) there is

Given, { 1. The Side H Æ, (the Elevation of the *Æquinoctial* above the *Horizon* of *London*) 38 d. 30 m.
2. The Side Æ 11, 30 d. (and Æ 12, 60 d. &c.

To find, The Angle Æ 11 H, — Æ 12 H, &c. By *Case* XIII. of *R. A. S. T.*

As the Tangent of Æ H, 38 d. 30 m.
Is to the Radius:
So is the Sine of Æ 11, 30 d.
To the Tangent of 32 d. 9 m.
Whose Complement 57 d. 51 m. is the Quantity of the Angle H 11 Æ.

So will the Angle		Be found to be	D.	M.	to which the Angle		Is equal.
	H 11 Æ		57	51		Q 3 O	
	H 12 Æ		42	32		Q 2 O	
	H 1 Æ		7	10		Q 1 O	
	H 2 Æ		42	32		Q 12 O	
	H 3 Æ		57	51		Q 11 O	

These Angles being once found will be always in Use in the *Latitude* of 51 d. 30 m. For that they shew the Quantity of the Angle, that the *Ecliptick* makes with every *Circle* of *Position*.

Moreover,

In the Triangle 11 ♈ X, — 12 ♈ Y, &c. there is

Given, { 1. The Angle X 11 ♈, which is a contiguous Angle to the before-found Angle H 11 Æ.
2. The Side ♈ 11, the Complement of the *Oblique Ascension* of the XIth *House*, to a whole Circle, or 360 d.
3. The Angle 11 ♈ X, 23 d. 30 m.

To

To find the Side ♈ X, or the Distance of the *Sign* of the XIth House from the *Vernal Æquinox*. By *Case* IV. of *R. A. S. T.* *Fig.* XLVIII.

But it will be sufficient to count the *Signs* of the *Oriental Houses*: For those which are opposite to them, (as the IVth to the Xth, — the Vth to the IXth, &c.) are counted as many *Degrees* and *Minutes* as the others; but of an *Opposite Sign*.

Example, *Let it be required to* Erect *a* Scheme *of the* Heavens, (*in the* Latitude *of* London, 51 *d.* 30 *m.*) *for the 6th Day of* March, *the* 16 *h.* 5 *m. before* Noon.

The *Place* of the *Sun* then was in 27 d. 54 m. ♓.

The *Sun*'s *Right Ascension*, 358 d. 4 m.

The *Time* 16 h. 5 m. turned into Degrees, gives 241 d. 15 m. Which added to the *Right Ascension* of the *Sun*, make 599 d. 19 m. from which substract the Circle (360 d.) there remains 239 d. 19 m. for the *Right Ascension* of the *Mid-Heaven*. To which answer 1 d. 27 m. of ♐, which is the *Sign* of the Xth *House*. And from thence I collect the *Oblique Ascension* of the rest of the *Houses* as followeth, *viz.*

The *Oblique Ascension* of the	House to be		The *Sign* thereof is	D.	M.	
	XI	269 19		15	15	♐
	XII	299 19		1	09	♑
	I	329 19		29	27	♑
	II	359 19		28	46	♓
	III	29 29		11	45	♉

But because the IVth *House* is opposite to the Xth *House*, it will have as many Degrees and Minutes upon the *Cuspis* thereof, but of the opposite *Sign*, *viz.* ♊, and the Vth the same as the XIth, *viz.* 15 d. 15 m. of the opposite *Sign*, ♊. The VIth 1 d. 00 m. ♋. The VIIth 29 d. 27 m. of ♋. The VIIIth 28 d. 46 m. of ♍. The IXth 11 d. 45 m. of ♏. As may be seen in the *Erected* Fig. XLIX.

PROB. II.

How to Direct *any* Point *of an* Erected Scheme *to any other* Point.

1. THE *Point* that is *Directed* is called, otherwise, the *First Place*, or *Significator*: But that to which we direct it, the *Second Place*, or the *Promissor*: But the *Significators* are directed whither they tend: *Directed*, that is, *Going* according to the *Series* of the *Signs*; but *Retrograde* in *Antecedens*, or against the *Succession* of the *Signs*. *Fig.* XLIX.

2. Let

2. Let the *Position* made by any direct *Significator*, or by the *Promissor* to which the *Retrograde* is *Directed*, be called the *Horizon* of a *Star*: And it is either a right one, passing through both the *Poles*, as the *Meridian*; or an *Oblique* one, having One of the *Poles* elevated.

3. The *Oblique* one is, sometimes, the same as the *Horizon* of the *Place*; and consequently, the *Elevation* of the *Pole* is the same: But, when it is *different* (which most commonly happens) you must first enquire how much One of the *Poles* is *Elevated* above the *Horizon* of the *Star*: But if a Star be

Fig. I. II. III. IV.

Oriental,	With Declination,	*North*,	As in	Fig. I.
		South,		Fig. II.
		North,		Fig. III.
Occidental,		*South*,		Fig. IV.

In the Triangle P O S, or P H S, for the Situation of the *Star* above or below: There is

Given,
1. The Side P O, or P H, the *Elevation* of the *Pole* at *London*, 51 d. 30 m.
2. The Side P S, the Complement of the *Star's Declination*.
3. The Angle S P O, comprehended between them (whose *Measure* is the *Arch* of the *Æquator*, which is intercepted between the Sides P O (P H) P S: One of which being produced, shews the *Right Ascension* of the *Star*; the other, the *Right Ascension* of *Imum Cœli*.

Required, The Angle P O S, or P H S. By *Case* III. of *R. A. S. T.*

Again, In the Triangle R P O, or R P H, (which are in different *Hemispheres*, when O or H (before found) are greater than a *Quadrant* or 90 d.)

There is Given, besides the Right Angle at R,
1. P O or P H, the *Elevation* of the *Pole*.
2. The Angle O or H, before found, (which if it exceeds a *Quadrant*, is Complement to 180 d.

To find R P, The *Elevation* of the *Pole* above the *Horizon* of the *Star*. By *Case* III. of *R. A. S. T.*

4. Having found the *Elevation* (by the Third beforegoing) then look

when the *Horizon* hath any Star that is { Oriental, the Asc. / Occidental, the Obl. / Descension } both of the *Significator* and *Promissor*. Fig. I. II. III. IV.

5. Having found out the { *Significators Direct*, / *Promissor*, to whom the / *Retrograde* is *Directed*, } *Ascension* or *Descension* for the *Horizon* of the *Star*: Let the *Right* one, or the *Oblique* one, be substracted from the like *Ascension* or *Descension* of the other *Star* (adding 360 d. if Need be) the *Remainder* is the *Ark* of *Direction* sought for.

PROB. III.

The Ark *of* Direction *of any* Significator *being given*; *To find how far it will reach.*

THIS is but the Converse of the foregoing: Therefore look for the *Elevation* of the *Pole* above the *Horizon* of a *Direct Significator*, as before; and at that *Elevation* you will get (by the 3d.) the Oblique { *Ascension* / *Descension* } of the same *Significator* { *Oriental*, / *Occidental*, } to which { *Ascension* / *Descension* } add the given *Ark* of *Direction*, so will you get the *Oblique* { *Ascension*. / *Descension*. } To which find what Degree of the *Ecliptick* answers (being counted from the *Elevation* of the *Pole* above the *Horizon* of the *Significator*) which is the very *Place* to which the *Ark* is *Directed*, and to which the *Significator* will come.

But substract the *Ark* of *Direction* given, from the *Right Ascension* of the *Retrograde Significator*, and there will remain the Degree of the *Æquator*, which is found out by the *Retrograde Significator*, by *Direction*, tending to the *Promissor*: The *Circle* of *Position* passing through this *Degree* (which ought well to be observed) will be the *Horizon* of the *Promissor*: Above which, find the *Elevation* of the *Pole*, as before: For, if that Degree fall on the { *North*, Fig. I. / *South*, Fig. II. } the { Eastern / Western } Part of Heaven, the *Significator* inclining unto the { *North*, Fig. III / *South*, Fig. IV. } in the Triangle POS, or PHS.

For

Fig. L. I. II. III. IV.

For the Situation of the Degree observed, whether Above or Below.

From the given Angles, as above, (the Angle P being counted in the *Æquator*, not from the *Right Ascension* of the *Significator*, but from that Degree which hath, oftentimes, been taken notice of, as that at the *Medium* or *Imum Cœli*) I seek for the Angle O or H.

And moreover, in the other *Triangle*, R P, the very *Elevation* of the *Pole* above the *Horizon* of the *Promissor*: At this *Elevation* of the *Significator*, if you observe that Degree *Oriental* / *Occidental* } you will gain the *Oblique* { *Ascension.* / *Descension.* } From which take the given *Ark* of *Direction*, and the *Remainder* will be the *Oblique* { *Ascension,* / *Descension,* } to which the Degree of the *Ecliptick* answers, (but it must be counted from the *Elevation* of the *Pole* above the *Horizon* of the *Promissor*) and is the very *Place*, to which the *Significator* will come, by this *Ark* of *Direction* given.

The End of the Astronomical Problems.

A N-

Fig. L.
I. II.
III. IV.

ANCILLA MATHEMATICA.

VEL,

Trigonometria Practica.

SECTION V.

OF
SCIOGRAPHIA,
OR
DIALLING.

CHAP. I.

Of Dialling *in general.*

OF the *Mathematical Sciences*, I know none more Ingenious and Uſeful then This of the Deſcription of *Sun-Dials* upon all ſorts of *Plains* howſoever ſituate: Neither is there any thing that draws more Admiration from all knowing Men; then to ſee ſtreight Lines drawn upon a *Plain* at *Unequal Diſtances* to meaſure out exactly the *Equal Diviſions* of the *Time* of an *Artificial Day*: For, although the Sun appears in different places of Heaven, according to the different *Seaſons* of the *Year*; yet, the ſame *Streight Lines*, do ſtill determine the ſame *Hour*, at all thoſe different *Seaſons*. And as this *Science* is of ſuch

 Excellency

Excellency and *Use*, I shall in this *Section*, briefly shew (and that by an *Artifice* not usually practiced) how such *Hour-Lines* may be described upon all *Plains* in any part of the World: And that by *Trigonometrical Calculation*, as well as by projecting the *Circles* of the *Globe* upon a *Plain*.

Now, All *Plains* have a particular respect to some *Point*, in the *Latitude* of that Place in which they stand: Wherefore, some particular Place must be assigned for Exemplifying the same, in all the Varieties. We will therefore take *London*, the *Metropolis* of *England* (which is situate in 51 deg. 30 m. of *North Latitude*) for our *Examples* following.

CHAP. II.

Of the Diversity of Plains, *upon which* Hour-lines *may be described.*

ALL *Plains* upon which *Sun-Dials* may be made, in any *Latitude* or Place of the World, are situate either

Parallel
Perpendicular } To the *Horizon* of that
Oblique

Place (or *Country*) in which they are made. And, of these severally.

I. A *Plain* that lies *Parallel* (or *Level*) to the *Horizon*, Is called an *Horizontal Plain*, for the Latitude of that *Place* in which it is to stand.

II. Of *Plains* that are *Perpendicular*, or *Erect*, to the *Horizon*, there are several *Varieties*. For, of such *Plains*, if the *Face* thereof doth directly behold the

East
West
North } Point of the *Horizon* of the *Place*
South

wherein they stand, They are then called,

East
West
North } *Erect, Direct Plains.*
South

But

But if such *Erect Plains* do not behold the *Direct East*, *West*, *North* or *South* Points of the *Horizon*, they then will behold either

The { *South-East*, *South-West*, *North-East*, *North-West* } And then they are called

North or South } *Erect Plains*; *Declining* { East or West.

III. Of *Plains* that lie *Obliquely* to the *Horizon*: (such as the *Roofs* of *Houses*, the *Cooping* of *Walls*, &c.) are called *Reclining Plains*. And if such *Plains* do behold the direct

East, West, North, South } Points of the *Horizon*: They are then

called *Direct East*, *West*, *North* or *South* Plains: *Reclining* from the *Zenith* of the Place in which such Plain stands.

But if such *Reclining Plain* doth not respect the true *East*, *West*, *North*, or *South* Points, then they will lie open either to the

South-East, South-West, North-East, North-West } And then they are called

South or North } *Reclining Plains*: *Declining* { *East* or *West*.

And these are all the *Varieties* of *Plains*, upon which *Hour-Lines* may be described.

Only note,] That all *Reclining Plains* whatsoever; whether *Direct* or *Declining*, have *Two Faces*; the *Upper-Face*, which beholds the *Zenith* of the Place, is called the *Reclining Plain*: And the *Under-Face*, which beholds the *Horizon* of the Place, is called the *Incliner*: And one and the same *Dial* serves for both Places.

CHAP. III.

How to find the Situation *of any* Plain, *in respect of* Declination *and* Reclination, *in any Latitude.*

I. To find the *Reclination.*

Fig. LI. *Definition.* THE Quantity of the *Reclination* of a *Plain*, is the *Arch* of that *Vertical* or *Azimuth* Circle which is perpendicular to the *Reclining Plain*, comprehended between the Zenith of the Place and the Plain.

To find which, let A B C D represent such a *Reclining Plain*: Draw, first, thereon, by help of a Ruler and Quadrant, a Right Line G H, parallel to the *Horizon* of the place; which shall be the *Horizontal Line* of the Plain. Cross this Line G H, with another Right Line K S at Right-Angles to it; which Line K S shall be the *Vertical Line* of the Plain.

To this Line K S, apply a streight Ruler K L: And to that end of it which lyeth clear of the Plain, as at L; apply a *Quadrant* O E P, having a Thrid and Plummet hanging from the Centre. Then see what number of *Degrees* of the *Quadrant* are contained between O and E; for so much doth that *Plain Recline* from the *Zenith* of the Place; and is the *Reclination* of the Plain.

II. To find the *Declination.*

Definition. THE *Declination* of a *Plain*, Is an Arch of the *Horizon* comprehended between the *Pole* of the *Plain*, and the *Meridian* of the *Place.*—— Or, It is the distance of the *Plain* it self, from the *Prime Vertical Circle*, or *Azimuth* of *East* or *West.*

To find out the *Declination* of any *Plain*, there are required two *Observations* to be made by the *Sun*, both at the same *Time*, as near as may be.——The *First*, Of the *Horizontal Distance* of the *Sun*, from the *Pole* of the *Plain.*—And the *Second*, Of the *Sun's Altitude.*

I. To find the *Horizontal Distance.*

Apply one Edge of the *Quadrant* to the *Horizontal Line* of the *Plain*, so that the other may be *Perpendicular* to it; and let the *Limb* of the *Quadrant* be towards the *Sun*. The *Quadrant* thus applied

applied to the *Plain*, and held *Level* (as near as you can conjecture) hold up a *Thrid* and *Plummet* at full Liberty, near the *Limb* of the *Quadrant*; so that the *Shadow* of the *Thrid* may pass through the *Centre* and *Limb* of the *Quadrant*: And then observe the Degrees cut by the *Shadow* of the *Thrid* in the *Limb* of the *Quadrant*; and number them from that side of the *Quadrant*, that standeth *Square* or *Perpendicular* to the *Plain*: For those *Degrees* are the *Horizontal Distance* sought for. Fig. LI.

II. To find the Sun's *Altitude*.

Hold up a *Quadrant* with both your Hands, turning the *Left-side* of your *Body* to the *Sun*; then raise up, or depress the *Quadrant*, so held, till the *Sun* shining through the *Hole* in that *Sight* which is nearest the *Centre*, do cast its Beam of *Light* upon the *Hole* in the other *Sight* farthest from the *Centre*: And at such time mark exactly what *Degrees* of the *Quadrant* are cut by the *Thrid*, for those Degrees are the *Sun's Altitude* at that time.

The *Horizontal Distance*, and the *Sun's Altitude* thus *Observed*, at the same instant (as near may be) will help you to the *Plain's Declination*; by the *Rules* following. For,

I. By having the Sun's *Altitude*; you may find the Sun's *Azimuth*, as in PROBL. IX. of the Use of the *Cœlestial Globe* in *Astronomy*: And by CASE XI. of O. *A. S. T.* Then,

II. When you make your *Observation* of the *Horizontal Distance*; Mark whither the *Shadow* of the *Thrid* did fall between the *South*, and that *Side* of the *Quadrant* which was *Perpendicular* to the *Plain*.

First. If the *Shadow fall between* them: Then the Sun's *Azimuth* from the *South*, and the *Horizontal Distance* added together, do give the *Declination* of the *Plain*: And (in this Case) the *Declination* is unto the *same Coast* with the *Sun's Azimuth*; that is, *Eastward*, if the *Observation* were made in the *Forenoon*; or *Westward*, if in the *Afternoon*.

Secondly. If the *Shadow* fall *Not between* them:

Then, The *Difference* between the *Sun's Azimuth* and *Horizontal Distance* is the *Declination* of the *Plain*: And in this Case)——If the *Azimuth* be the *Greater* of the two, then the *Plain Declines* to the *same Coast* whereon the *Sun* is; But if the *H. Distance* be the *Greater*; then the *Plain Declines* the *contrary Coast*.

And

Fig. LI. And here *Note*] That the *Declination* thus found, is always accounted from the *South*; and that all *Declinations* are counted from either *North* or *South*, towards either *East* or *West*: And must never exceed 90 Degrees.

I. If therefore, The Degrees of *Declination* do exceed 90 deg. you must take the residue of that Number to 180 deg. and that shall be the *Plain's Declination* from the *North*.

II. If the *Degrees of Declination* exceed 180; then the *Excess* above 180, gives the *Plain's Declination* from the *North*, towards *that Coast*, which is *Contrary* to the *Coast* whereon the *Sun* was, at the time of *Observation*.

CHAP IV.

How Hour-Lines *may be described upon an* Horizontal Plain *in any* Latitude; viz. *of* London, 51 *d*, 30 *m*.

I. By the Globe.

ELevate the *Globe* to the *Latitude* of the place for which you would make your *Dial*, (suppose for *London*, in the *Latitude* of 51 deg. 30 min.) Then bring the *Vernal Equinoctial Colure* (which is the first point of *Aries* also) to the *Meridian*, and (if you will) the *Index* of the *Hour-Circle* to 12. This done,

1. Turn the *Globe* about *Westward*, till the *Hour-Index* points at 1 a Clock, or rather [till 15 degrees of the *Equinoctial* come to be just under the *Meridian*] and there keeping the *Globe*, look upon the *Horizon* how many degrees thereof are cut by the *Equinoctial Colure*; which you shall find to be 11 deg. 30 min. which set down in a little Table, as you see here is done; for this 11 deg. 50 min. is the distance that the *Hour-lines of* 11 and 1 a clock are distant from the *Meridian* upon the Dial Plain.

		d. *m.*
Latitude		51. 30
		d. *m.*
12		00.00
11	1	11.50
10	2	24.20
9	3	38. 3
8	4	53.35
7	5	71. 6
6		90. 0

2. Turn the *Globe* more *Westward*, till 30 degrees of the *Equinoctial* comes to the *Meridian*, and then see what degrees of the *Horizon* are cut by the *Equinoctial Colure*; which you will find to be 24 deg. 20 min. which note down in a Table as before, for that is the *Hour-distance* of 10 and 2 a clock from the *Meridian*.

3. Turn

3. Turn the *Globe* still more *Westward*, till 45 degrees of the *Equinoctial* come to the *Meridian*; and then shall the *Equinoctial Colure* cut 38 deg. 3 min. of the *Horizon* counted from the *Meridian*; which is the distance of 9 and 3 a clock.

Do thus with the other hours of 8 and 4, of 7 and 5, and so shall the *Colure* cut 90 degrees at 6 a clock, or when 90 degrees of the *Equinoctial* comes to the *Meridian*. And this being done, your Dial is so far made as the *Globe* can assist you.

II. *The Geometrical Construction of this* Dial, *in order to the Trigonometrical Calculation.*

1. With 60 Degrees of a *Scale of Chords*, describe a Circle representing your *Dial-plain*, and *Horizon* of the *Place*, (viz. *London.*) Cross it with 2 Diameters S N, representing the *Meridian* of the *Globe*, and *Hour-line* of XII; and the Line E W, for the *Hour-line* of VI: then will Z represent the *Zenith* of the *Place*, and be the Centre of the Dial. Fig. LII.

2. The *Latitude* of the Place being 51 d. 30 m. Set them from S to *a*, and from W to *b*: Then a Ruler laid from W to *a*, will cross the *Meridian* S N in P, the *Pole* of the *World*: And laid from E to *b*, it will cross the *Meridian* N S in Æ, the intersection of the *Meridian* and *Equinoctial*. And now you have three Points, *viz.* W, Æ, E; through which you may describe the *Æquinoctial* Circle W Æ E, whose *Centre* will be at *c*, always in some part of the Line N S, extended, if need be.

3. Divide the Semicirle W N E, into 12 equal parts, at the points ⊙ ⊙ ⊙, &c. And laying a Ruler to Z, and every of those points ⊙ ⊙ ⊙, it will cross the *Æquinoctial* Circle W Æ E, in the points ***, &c. dividing that Semicircle into 12 unequal parts.

4. A Ruler laid to P, the *Pole* of the World, and to the several points ***, &c. it will cut the Circle of your *Plain* in the points 1, 2, 3, &c. on the *East-side* of N, and 11, 10, 9, &c. on the *West-side* of N.

Lastly. If you lay a Ruler to the Centre Z, and the respective points 1, 2, 3, &c. and 11, 10, 9, &c. they shall be the true *Hour-lines* belonging to an *Horizontal Dial* for the *Latitude* of 51 de. 30 m. And their respective *distances* from N will be the same as in the *Table* they were found to be by the *Globe*.

III. By

Fig. LII.

III. *By Trigonometrical Calculation.*

In these Horizontal Dials, there is nothing to be found by *Calculation Trigonometrical*, but the *Hour-distances* upon the *Plain* from the *Meridian*; for which, This is

The Canon for Calculation.

As the Sine of 90 de.
Is to the Sine of the *Latitude* P N, 51 de. 30 m.
So is the *Tangent* of 15 d. (the *Æquinoctial Distance* of *One* hour; of 30 d. for *Two* hours; of 45 d. for *Three* hours, &c.)
To the *Tangent* of 11 d. 50 m. for 11 and 1—— of 24 d. 20 m. for 10 and 2—— for 38 d. 3 min. for 9 and 3, &c. as in the Table, before found by the Globe.

So have you all the Hour-lines between 6 in the morning and 6 at night; and for the Hour-lines of 4 and 5 in the morning, and of 7 and 8 at night, draw the same Lines before and after 6, through the Centre, as in the Figure, and they shall be the true Hour-lines: And so is your *Dial* finished.

The *Stile* must stand upright at 12 of the clock, not inclining on either side.

And in this manner may you describe Hour-lines upon an Horizontal Plain in any Latitude.

CHAP. V.

How to describe Hour-Lines *upon an* Erect direct South Plain, *in the* Latitude *of* London, 51 *deg.* 30 *m.*

AN *Erect Direct South Dial*, in any Latitude, is no other than an *Horizontal Dial* in that Latitude, which is equal to the Complement of that *Latitude* in which it is an *Erect Direct South Dial*: So that an *Erect Direct South Dial* in the Latitude of 51 d. 30 m. will be the same as an *Horizontal Dial* in the *Latitude* of 38 de. 30 m. which is the Complement of 51 de. 30 m.——So that the making of such a *Dial*, both by the *Globe* and by *Tri. Calculation* is the same with the other, only instead of 51 de. 30 m. *Latitude* you set your Globe to 38 d. 30 m. and so in the Calculation also: But in these *Dials* there needs no Hour-

Lines

Lines to be drawn through the Centre; for that the Sun never Shines upon them before 6 in the *Morning*, nor after 6 at *Night*. *Fig.* LIII.

The *Stile* of these *Dials* must stand upon the *Hour-Line* of 12, and must point downwards towards the *South Pole*. As in Figure LIII.

CHAP. VI.

To make an Erect Direct North Dial, *in the Latitude of* London 51 *de.* 30 *m.*

THE *North Erect Direct Dial*, is the same with the *South*, only the *Stile* must point upwards towards the *North-Pole*, and the hours about Midnight, as 9, 10, 11, 12, at Night; and 1, 2 and 3 in the Morning must be left out, and 4 and 5 in the Morning; and 7 and 8 at Night must be drawn through the Centre: So is your *North-Dial* also finished, as in Fig. LIV. *Fig.* LIV.

CHAP. VII.

To make an Erect Direct East *or* West Dial *in the Latitude of* London, 51 *de.* 30 *m.*

I. *By the* Globe.

THE *Globe* rectified to the *Latitude*, the *Index* to 12, the *Quadrant of Altitude* in the *Zenith*: If you turn the *Quadrant of Altitude* so about till the graduated edge thereof do behold the direct *East* or *West*-points of the *Horizon*, you shall find that it will lie in the very Plain of the *Meridian-Circle*, and so the *Pole* will have no elevation over it; for turning the *Globe* about, the *Equinoctial Colure* will not cut the *Quadrant of Altitude* in any particular degree, but it will cut all the degrees thereof at the same time; wherefore the Hour-lines of these Plains will make no *Angles* at the *Pole*, and therefore must be parallel one to the other, which the *Globe* evidently demonstrates, but will not conveniently give the parallel distance of each from other, they being *Fig.* LV.

Fig. LV. nearer or farther off each other according as the *Stile* is proportioned to the Plain, which I shall now come to shew.

For the Reasons aforesaid, there is no *Trigonometrical Calculation* required in the making of these *Dials*; and therefore, I shall proceed to

II. *The Geometrical Construction of these* Dials.

Let the Plain upon which you would make an *East or West Dial*, be A B C D.

1. Upon D (or any where towards the lower part of the Line B D, for an *East Dial*, or of A C for a *West*) with 60 degrees of your *Chord*, describe an Arch F G, upon which set the Complement of the *Latitude* of the place, *viz.* 38 deg. 30 min. from F to G, and draw the Line D G E for the *Equinoctial*.

2. Towards the upper part of this Line, as at P, assume any point, and through it draw the Line 6 P 6 perpendicular to the *Equinoctial*, for the Hour-line of Six.——Also, towards the lower part of the same Line, assume another point as L, and through it draw the Line 11 L 11 perpendicular also to the Æquinoctial for the Hour-line of Eleven.

3. With 60 degrees of your *Chords*, upon the point L, describe a small Arch of a Circle, as H K, and upon it (always) set 15 degrees (or one hours distance) from H to K, and draw the Line L K M, cutting the Hour-line of Six in M.

4. Upon M as a Centre, with 60 degrees of your *Chord*, describe an Arch of a Circle N O, which divide into five equal parts in the points ⊙ ⊙ ⊙ ⊙.

5. Lay a Ruler upon M, and each of these points ⊙ ⊙ ⊙ ⊙ and the Ruler will cut the *Equinoctial-line* E D in the points ****, through which points, if you draw Right Lines parallel to the Hour-line of 6, they shall be the Hour-lines of 7, 8, 9, and 10 of the Clock, the Hour-lines of 6 and 11 being drawn before.

6. For the Hour-lines of 4 and 5 in the Morning, before 6, they retain the same distance from 6, as do the hours of 7 and 8; and thus is your *Dial* finished.

The *Stile* must stand upon the Hour-line of 6; and be elevated so high as is the length of the Line M P, and may either be a Pin of Wyre, or a Plate of Brass or Iron.

The *West Dial* is the same with the *East*, only changing the names of the Hours.

For

277

Fig. LV.

Fig. LVI.

Fig. LV.

nea
tio

req
cee

al.

B
ye
m
F

pe
E
p
it
fe
a
d
L

fe
ii

a
t
l
e

For
4, 5, 6, 7, 8, 9, 10, 11 in the Morning, in the *East-Dial.*
Must be changed to
8, 7, 6, 5, 4, 3, 2, 1 in the Afternoon in the *West Dial*:
Which is all the difference.

Fig. LV.

CHAP. VIII.

To make an Erect Dial, *declining from the* South, Eastward, *or* West-ward; 30 *degrees in the* Latitude *of* 51 *deg.* 30 *min.*

I. *By the* Globe.

Fig. LVI.

THE *Globe* being Rectified to the *Latitude* of the place, the *Quadrant of Altitude* in the *Zenith*, the *Index* of the Hour-Circle at 12, and the *Equinoctial Colure* brought under the *Meridian*;

1. Count the Declination of the Plain upon the *Horizon*, from the *East or West*-points thereof (according as the Plain declines) towards the *South*: namely, 30 degrees; and to that point of the *Horizon* bring the *Quadrant of Altitude*, and there keep it.

2. Turn the *Globe* about till the *Index* of the Hour-wheel cuts 11 of the Clock, or rather (as I said before) till 15 degrees of the *Equinoctial* have passed the *Meridian*, and then shall you find the *Equinoctial Colure* to cut the *Quadrant of Altitude* at 9 deg. 50 min. if you count the degrees from the *Zenith point* downwards.

3. Turn the *Globe* farther about, till 30 degrees of the *Equinoctial* be past the *Meridian*, and then shall yon find the *Colure* to cut the *Quadrant of Altitude* at 18 deg. 14 min. counted from the *Zenith* downwards as before.

4. Do the like with all the rest of the Hours, and you shall find that at the several 15 degrees of the *Equinoctial*, the *Equinoctial Colure* will cut such degrees of the *Quadrant of Altitude* as are expressed in this Table, if you count them from the *Zenith* downwards, as is before directed.

Hours from Noon.		Hour-distances upon the Plain. d.	m.
	12	00	00
11	1	09	50
10	2	18	14
9	3	26	19
8	4	34	56
7	5	44	56
6	6	57	49
5	7	75	37

This done;

5. Bring the *Quadrant of Altitude* to the other side of the *Meridian*, and set it to 30 degrees, the Plains declination, counted from the *East* or *West*-points *Northward*, as you did

Fig.LVI before towards the *South*, which will be in the just opposite point of the *Horizon* to which it was before; and also, bring the *Equinoctial Colure* under the *Meridian*. Then,

6. Turn the *Globe* about (the contrary way to what you did before) till 15 degrees of the *Equinoctial* be past the *Meridian*, and then shall you find the *Equinoctial Colure* to cut at 12 deg. 23. min of the *Quadrant of Altitude* counted from the *Zenith*.

And so continuing turning the *Globe* about till 30, 45, and 60 degrees of the *Equinoctial* have passed the *Meridian*, you shall find the *Equinoctial Colure* to cut the *Quadrant of Altitude* at such degrees as are expressed in this Table.

Hours from Noon.		Hour-distances on the Plain.	
		d.	*m.*
12		00	00
1	11	12	23
2	10	29	19
3	9	52	42
4	8	80	07

The Hour-distances upon the Plain being thus attained, there are two other requisites in all upright declining *Dials* also to be found by the *Globe*, before the *Dial* can be finished. Namely,

1. The distance of the *Sub-stile* from the *Meridian*.
2. The height of the *Pole* above the *Plain*, or the height of the *Stile* above the *Sub-stile*.

To find both which,

Bring the *Equinoctial Colure* to the Plains declination 30 degrees counted upon the *Horizon* from the *South-Eastward*; and the *Quadrant of Altitude* to 30 degrees counted in the *Horizon* from the *East-Northward*: So shall the *Quadrant* cut the *Colure* at *Right Angles*. And

The number of degrees of the *Quadrant* contained between this *Intersection* and the *Zenith* (which here is 21 deg. 41 min.) is the distance of the *Sub-stile* from the *Meridian*. And the degrees of the *Colure* contained between this *Intersection* and the *Pole* (which here is 32 deg. 37 min.) is the height of the *Pole* above the Plain.

II. The *Geometrical projection* of this *Dial*, in order to the *Trigonometrical Calculation* of the *Hour distances* and other *Requisites* belonging to such an *Erect Declining Plains*.

1. Upon the Point Q, as a Centre, with 60 de. of a *Scale* of *Chords*, describe a Circle, representing your *Dial Plain*: And cross it with two Diameters Z Q N for the *Vertical*; and H Q O for the *Horizontal Line* of the *Plain*.

2. Set

2. Set 30 deg. the *Plains Declination*, from N to *c*, if the Plain Decline *Eastward*, as in this Example; or from N to *e*, if *Westward*: Then lay a Ruler from Z to *c*, and it will cut the *Horizontal Line* of the *Plain* in K, so have you three Points Z, K and N, by which to describe the Arch Z K N, representing the *Meridian* of the *Place*. And to find the *Pole* thereof, set 90 de: from *c*, to *d*, and then a Ruler laid from Z to *d*, will cut the Horizontal Line H O, in W, which is the Pole of the *Meridian Circle* Z K N: and the West-point of the Horiz.

Fig. LVI.

3. Set 51 deg. 30 m. the *Latitude* of the Place, from O to *a*, and from N to *b*: Then a Ruler laid from W to *a*, will cross the *Meridian* in P, the *Pole* of the *World*: And laid from W to *b*, it will cross the *Meridian* in Æ, the point through which the *Æquinoctial Circle* is to pass: And now you have two points W and Æ, through which the Æquinoctial Circle æ Æ æ may be described (by the XXI *Geometrical Problem*, Lib. 1.)

4. Through P, the *Pole* of the *World*, and Q, the *Pole* of the *Plain*, draw the right Line P Q, for the *Axis* of the *World*, and *Sub-stilar Line* of your *Dial*: And in this Line (extended) will the *Center* of the *Æquinoctial Circle* æ Æ æ be found.

5. From P, the *Pole* of the *World*, lay a Ruler to Æ, the intersection of the *Meridian* and *Æquinoctial*, and it will cut the *Plain* B: At this point B, begin to divide the *Semicircle* H N O of the Plain, into 12 equal Parts, at the points ⊙ ⊙ ⊙, &c.

6. From Q, (the *Pole* of the *Plain*) lay a Ruler to every of the points ⊙ ⊙ ⊙, &c. and it will cross the *Æquinoctial Circle* æ Æ æ, in the the Points ***, &c.

7. Lay a Ruler to P (the *Pole* of the *World*) and every of the points ***, &c. and it will cut the *Primitive Circle* representing the *Dial Plain*, in the points N, 9, 10, 11, &c. on the *West* side, and N, 1, 2, 3, &c, on the *East* side of the *Meridian*.

8. Lastly, Lines drawn from the Centre Q through these points, shall be the true *Hour-lines* of an *Erect Plain Declining* from the *South-Eastward* 30 de. in the *Latitude* 51 de. 30 m. And now,

Concerning the other *Requisites* belonging to this *Erect Declining Plain*.

These are all of them represented to the Eye in the *Scheme* of the *Projection* of the *Plain*: Wherein, by the intersection of the several *Circles* there is constituted a *Right-angle Spherical Triangle* Z T P, in which,

The

Fig. LVI.

The side { Z T is the *distance* of the *Sub-stile* from the *Meridian*.
Z P the Complement of the *Latitude* of the Place.
T P the *height* of the *Pole* above the *Plain*. }

The Angle { P Z T, the Complement of the *Plain's Declination*.
Z P T, the *Plain's difference* of *Longitude*.
Z T P, is a Right Angle. }

And all these *Sides* and *Angles* may be measured upon the Projection it self, by the *Precepts* deliver'd in the first *Book*, *Sect.* III. of *Spherical Trigonometry Geometrically performed* by *Projection*. And they will be found to be as here expressed, *viz.*

		de.	*m.*			*de.*	*m.*
Side	Z T	21	49	Angle	P Z T	60	00
	Z P	38	30		Z P T	36	25
	T P	32	36		Z T P	90	00

And now I shall shew how all these may be found
By Trigonometrical Calculation.

Before any *Erect Declining Dial* can be made, there are *Three things* which must be found, besides the *Hour-distances*, and those are these:

1. The height of the *Pole* (or *Stile*) above the *Plain*.
2. The *distance* of the *Substile* from the *Meridian*.
3. The *Plain's difference* of *Longitude*.

All which may be found in the *Triangle* Z T P: In which there is given (besides the Right Angle at T,)

(1.) The *Angle* T Z P, the *Complement* of the *Plain's Declination*, 60 deg.

(2.) The *Side* Z P, the Complement of the *Latitude* of the Place 38 de. 30 m.

By which may be found,

I. The *Side* T P: The *Height* of the *Pole*, or *Stile*, above the *Plain*:
By CASE I. of R. A. S. T.

As the Radius, S. 90 deg.
Is to the Side Z P, 38 d. 30 m. the Co-Latitude.
So is the Sine of T Z P: the Co-declination 60 de.
To the Sine of T P, 32 deg. 26 m.
Which is the *height* of the *Pole* (or *Stile*) above the *Plain*.

II. The

Fig. LVI.

The Side Z T, The *Distance* of the *Sub-stile* from the *Meridian*: By CASE II of R. S. T.

As the Radius Sine 90 de.
to the Tangent of Z P, the Co-Latitude 38 d. 30 m.
So is the Co-sine of T Z P, 30 deg.
o the Tangent of Z T, 21 de. 40 m.
Which is the *Distance* of the *Sub-stile* from the *Meridian*.

I. The *Angle* Z P T, The *Plains difference of Longitude*: By Case III. of *R. A. S. T.*

As the Radius,
to the Tangent of the Declination T Z P, 60 deg.
So is the Co-sine of Z P, the Co-Latit. 51 de. 30 m.
o the Co-Tangent of T P Z, 35 de. 25 m.
Which is the *Plain's difference of Longitude*.

These three Requisites being thus found by Trigonometrical alculation; The Plain's Difference of Longitude 36 de. 25 m. lling between 30 and 45 deg. (which are the Second and Third quinoctial Hour-distances) there will be contained therein two mpleat Hours, and 6 deg. 25 m. over: which shews that the b-stilar Line of the Dial will fall between the Hour-Lines of and X in the Morning (in this East Dial) but between the our of II and III in the Afternoon if the Plain had Declined estward:

Having proceeded thus far. Prepare a Table of Hours fit for Plain, such as is here done.

A Ta-

Fig. LVI.

A Table of the Hour Distances, for a South-Dial, Declining East or West, 30 deg. In the Latitude of 51 deg. 30 min.

	d.	m.
Stiles height	32.	36
Distance of the Sub-stile and Meridian.	21.	40
Difference of Long.	36.	25

Hours from the East	Hours from the West	*Equinoctial Hour Distances.* D.	M.	*True Ho. Distances upon the Plain fro. the Subst.* D.	M.
IV	VIII	83	35	78	12
V	VII	68	35	53	57
VI		53	35	36	08
VII	V	38	35	23	16
VIII	IV	23	35	13	14
IX	III	8	35	4	36
			Sub-	stile.	
X	II	6	25	3	28
XI	I	21	25	11	56
XII		36	25	21	40
I	XI	51	25	34	03
II	X	66	25	51	00
III	IX	31	25	74	21

Then against XII, set the *Plain's Difference of Longitude* 26 de. 25 min. (in the second Column) and from it substract 15 deg. and there will remain 21 d. 25 m. which set against XI. and I: And from 21 d. 25 m. substract 15 de. and there will remain 6 d. 25 m. which set against X and II. And (because it is less than 15 de.) write the Word Sub-stile over it, and substract it from 15 deg. and there will remain 8 deg. 35 m. which write over Sub-stile, and against the Hours of IX and III. Then to these 8 d. 35 m. add 15 deg. and and it makes 23 de. 25 m. which set against VIII and IV: And thus, by the continual Addition of 15 deg. you shall have such *Æquinoctial Distances* as in the Table: Which Table, thus prepared, the next thing will be

To find the true *Hour-distances* upon the *Plain*, from the *Sub-stile*. For which, This is
The Canon for Calculation.

As the *Radius*
To the Sine of the *Stiles height* 32 d. 36 m.
So is the Tangent of the *Æquinoctial Distance* 6 d 25 m.
To the Tangent of 3 deg. 28 min. Which is the *Distance* of the *Hour-lines* of X and II. upon the *Plain* from the Sub-stile.

And so will the Tangent of the next *Æquinoctial Distance* 21 d. 25 m. be, to the Tangent of 11 de. 56 min. for the *Distance* of the *Hour-lines* of XI and I, from the Sub-stile: And so for all the rest, as in the Table. And so you will find them to be in the Figure also.

Fig. LVII. And in the making of this *Dial*, you have made four; as in *Fig.* LVII.

For,

Fig. LVII.

For, if you hold the Paper upon which the *South-East declining Dial* is drawn, againſt the Light, then ſhall you diſcover the *Stile* to ſtand on the *Right-hand* of the *Plain*; whereas it now ſtands on the *Left-hand*; ſo the ſame *Hour-lines*, *Sub-ſtile*, *Stile* and all, being drawn on the back-ſide of the Paper, and thoſe that are the *Forenoon-hours* in the *Eaſt-decliner* numbred as the *Afternoon-hours* in the *Weſt-decliner*; that is, call 11, 1, and 10, 2, and 9, 3, &c. as in the Tables; ſo ſhall the *South-Dial* declining *Eaſt* 30 degrees, become a *South-Dial* declining *Weſtward* 30 degrees.

And if you turn the *South-Eaſt-Dyal* upſide-down, ſo that the *Stile* may point upwards towards the *North-Pole*, (and leave out the Hours about 12, as 9, 10, 11, and 1, 2, and 3, which in *North Dials* repreſent 9, 10, and 11 at Night, and 1, 2, and 3 in the Morning; all which time (in thoſe middle *Latitudes*) the Sun is under the *Horizon*) it will become a *North-Dial* declining *Eaſtward* 30 degrees.

Alſo if you turn the *South* declining *Weſt-Dial* upſide-down, and leave out the hours about *Midnight*, as 9, 10, 11, 12, 1, 2, and 3, it will then become a *North-Dial* declining *Weſtward* 30 degrees.

Now for ſuch *South* or *North Dials* as do decline far towards *Eaſt* or *Weſt*, as 60, 70, 80, or 85 degrees, there you ſhall find that the Hour-diſtances will fall ſo near together, that they will be of no competent diſtance one from another, except they be extended very far from the Centre; and therefore the old way hath been (in ſuch Caſes) to draw the *Dial* upon the Floor of a Room, extending the *Sub-ſtile*, *Stile* and *Hour-lines* till they appear of a competent diſtance from each other, and then according to the bigneſs of your *Dial-plain*, to cut off the *Hour-lines*, *Stile* and *Sub-ſtile*; and ſo transfer them from the Floor to the *Plain* upon which the *Dial* is to be made: but this way being to Mechanical for an Artiſt to exerciſe, I ſhall therefore here inſert a more artificial way of performing this work *Geometrically*, by which (although the *Dial* ſhould decline 80 or 88 degrees) upon a quarter of a ſheet of Paper you may draw your *Dial*, and have the *Stile* of a competent height, and all the Hour-lines at a convenient diſtance one from another. And ſo let this ſuffice to be ſaid in this place concerning *Upright declining Dials*; for I intend not here to teach the *Art of Dialling*, but ſhew the *Uſe* of the *Globes*: and from thence to Calculate the Requiſites from them Trigonometrically.

Fig. LVII.

C H A P IX.

Concerning such Erect South *or* North Dial Plains, *which decline many degrees towards the* East *and* West.

I. *By the* Globe.

Fig. LVIII.

THE Operation by the *Globe* is altogether the same, as in the last *Chapter*, and therefore need not be here again repeated. Wherefore, I will proceed to

II. The *Trigonometrical* Calculation.

This Work also is to be performed in like manner as the Former: And therefore I shall not again repeat it, but give an *Example* ready wrought.

Suppose an *Erect Plain* in the *Latitude* of *London*, 51 de. 30 m. should decline from the *South* towards the *East* 85. deg.

I. For finding the *Requisite*, if you work according to those *Canons* for *Calculation*, in the last *Chapter* delivered ; you will find

			de.	m.
The	*Height* of the *Pole* (or *Stile*) above the *Plain*	to be	3	06
	Distance of the *Sub-stile* and *Meridian*.		39.	22
	Plain's Difference of *Longitude*		86	05.

Hours *Morn.*	*Aft.*	*Æqui-noctial Distanc.* D.	M.	*Hour Distances on the Plain* D.	M.
XII		86	05	38	23
XI	I	71	05	8	58
X	II	56	05	4	36
IX	III	41	05	2	42
VIII	IV	26	05	1	31
VII	V	11	05	0	36
The Place		of	the	Sub-stile.	
VI		3	55	0	13
V	VII	18	55	1	04
IV	VIII	33	55	2	06

These *Requisites* thus found, you may proceed to the making of a *Table* for the *Hour-distances* in all respects, as in the last Chapter. By setting down the *Hours* proper for the *Dial* in order, as in this *Table*.

Then, against XII, set the *Difference of Longitude*, 86 d. 05 m. from which Substract 15 deg. and there will remain 71 d. 05 m. which set against XI and I. And from 71 d. 05 m. substract 15 de. the remainder will be 56 d. 05 m. which set against X and II. And so by the continual *Sub-*

straction

straction of 15 deg. you shall find 11 de. 05 m. to stand against the Hours of VII and V; under which write *The place of the Sub-stile*: And then, substract 11 d. 05 m. from 15 de. there will remain 3 de. 55 m. which must be set under the *Sub-stile*, against the *Hour* of VI: And then, by the continual *Addition* of 15 deg. thereto; you shall have such *Æquinoctial distances* as are expressed in the Table, against the *Hours* of V and VII, and IV and VIII. Then

Fig. LVIII.

II. For the *Five Hour distances upon the Plain*: If you work by the *Canon* delivered in this last Chapter, you will find them to be such as are exhibited in the Third *Column* of this *Table*.

By this *Table*, you see that the *Five Hour Distances* upon the *Plain* about the *Sub-stile* (and indeed all the rest, except the extreme Hour of XII) do fall so near together, that without a very large extention of them from the *Centre*, there will be no competent distance between *Hour-line* and *Hour-line*: Wherefore, laying aside your *Table*, proceed to make your *Dial Geometrically*, according to the following Precepts.

III. The *Geometrical Projection*, of *this* (or the like) *Dial*: when the *Pole* hath but small *Elevation* above the *Plain*.

1. Draw a Right Line A B, perpendicular by one side of your *Plain*, and towards the *Right Hand*, because the Plain declineth *Eastward*, And with 60 deg. of a *Scale* of *Chords*, describe an obscure Arch of a Circle C D E; and upon it set 38 de. 23 m. the *Sub-stiles distance* from the *Meridian*, from C to D, and draw the Line A D, for the *Sub-stile*.

Fig. LIX.

2. Take 3 deg. 6 m. the *Stiles height*; and set them from D to E, drawing the Line A E for the *Stile* of the *Dial*.

3. Now (because the *Stile* is but of small *Elevation*, *viz.* but 3 de. 6 m.) draw another Line G H, *parallel* to the Line of the *Stile* A E, at such convenient distance as in your judgment will best fit the *Dial Plain*, so that the Hours belonging to it, may come within the Limits of the Plain: So shall that Line, G H, so drawn, be the *Augmented Stile* of the Dial.

And, by the *Sub-stile*, and this *Augmented Stile*, the *Hour-Lines* may be described (at convenient distance) without any regard had to the *Centre* of the *Dial*. For,

4. Assume any two points in the *Sub-stilar* Line of the *Dial* A D, at some convenient distance from each other, as R and S: and through those two points, draw two infinite Right lines, both

Fig. LIX. of them at *Right Angles* to the *Sub-stilar Line* A D ; as the Lines Z R Z and X S X.

5. Set one foot of your Compasses in the Point R, and take the nearest distance to the new *Augmented Stile* G H ; and set that distance upon the *Sub-stilar* Line from R to K.——Also from the Point S, take the nearest distance to the *Augmented Stile* G H, and set that distance also upon the *Sub-stilar Line*, from S to L.

6. Upon these two Points K and L, (with 60 de. of a *Scale* of *Chords*) describe two *Semi-Circles* : and in either of them set off 86 d. 05 m. (the *Plains Difference of Longitude*) from R to M, and from S to M, both of them on the same side of the *Sub-stilar Line*, on which the first *Perpendicular* A B was drawn.

7. Divide either of the *Semicircle* into 12 equal parts (beginning at the points M) at the several points ⊙ ⊙ ⊙, *&c.*

8. Lay a Ruler to the point L, and to the several points ⊙ ⊙ ⊙, *&c.* the Ruler will cross the Line X S X, in the points * * *, *&c.*——Also, Lay a Ruler to K, and the respective Points ⊙ ⊙ ⊙, *&c.* the Ruler will cross the Line Z R Z, in the several Points * * *, *&c.*

Lastly, Lines drawn from the several Points * * * in the Line Z R Z, to the several points * * * in the other Line X S X each to its Correspondent, (which the *Sub-stilar Line* will direct you) those Lines so drawn shall be the true *Hour-Lines* proper for the *Plain* : And will be at a competent distance one from the other, as by the *Figure* they do appear ; the sight whereof will be more satisfactory than many Words.

Note, that in the making of this *Dial*, you have made four Dials, *viz.*

A { South Declining { East / West ; North Declining { East / West } 85 Degrees.

But you must change the *Names* of the *Hours*, and place the *Stile* on the contrary side of the Line A B for the *South Declining West*.——And by turning the Dial upside downwards for the North-decliners, so that the Stile may point upwards to the North Pole ; and the Hours about Midnight omitted : As in the last Chapter is directed.

CHAP.

Fig. LIX.

bow

the

ards

ma-

are

All

be-

ines)

Pro-

cepts

La-

Lati-

de of

In

m.

rom

the

Lati-

Lon-

deg.

eof:

deg.

de.

the

de of

Fig. LIX.

of

Z

the

dif

the

an

of

86

and

Lin

7

nin

⊙

* *

⊙

Po

Z

to

tho

Pla

as

fati

1

Bu

on

W

N

Po

C

CHAP. X.

Direct South *and* North Declining Plains, *and how* Hour-Lines *may be described upon them.*

UCH *Plains* as do directly behold the {*North* / *South*} point of the rizon, but do *Recline* (or fall backwards) from the *Zenith* towards e {*South* / *North*} are called {*North* / *South*} *Direct Plains Reclining* : So ma- r Dgrees as the *Reclination* is :. And of such *Plains* there are x *Varieties*; Three of *South*, and Three of *North Recliners* : All uich may be *Reduced* to *New Latitudes*, wherein they will be- me *Horizontal Plains* : And consequently *Dials* (or *Hour Lines*) ay be described upon them both by the *Globe*, *Spherical Pro- tion*, and *Trigonometrical Calculation*, according to the *Precepts* ivered in the *Fourth Chapter* hereof.

I. Of South Recliners.

Examples of all these *Varieties* of *Reclining Plains*, in the *Latitude* of *London*, 51 deg. 30 m. To find the *New Latitudes*.

I. Variety. Let there be a *Direct South Plain* in the *Latitude* of *London*, which *Reclines* from the *Zenith* thereof 20 deg. In what *Latitude* will that be an *Horizontal Plain* ?

The Plains *Reclination* 120 deg. being *less* then 38 de. 30 m. *Complement of the Latitude of London* ; Substract 20 deg. from d. 30 m. the *Remainder* (or *difference*) 18 de. 30 m. is the *-Latitude*. So that an *Horizontal Dial* made for that *Lati- e*, shall be a *South Recliner* 20 deg. in the *Latitude* of *Lon-*

II. Variety. If a *South Plain* in the Latitude of 51 deg. 30 m. should *Recline* 60 deg. from the *Zenith* thereof: In what Latitude will that be an *Horizontal Plain* ?

The *Reclination* of the Plain 60 deg. being *Greater* than 38 deg. m. the *Complement of the Latitude* of *London*, Substract 38 de. m. from 60 deg. and the *Remainder* 21 deg. 30 m. is the *-Latitude* : And an *Horizontal Dial* made for the Latitude of

21

Fig. LIX. 21 de. 30 m. shall serve for a *South Dial Reclining* 60 deg. in the Latitude of *London*.

III. Variety. If a *South Plain* in the Latitude of *London* should Recline from the Zenith thereof 38 d. 30 m. *Equal* to the *Complement of the Latitude* of *London*. Then

The *Difference* between the *Complement of the Latitude of London* and the *Reclination* being *nothing*; it shews, the *New-Latitude* to be no *Latitude*, that is neither *Pole* hath any *Elevation* above such a *Plain*: And therefore a *Dial* for such a *Plain* must be made in all respects as an *Erect Direct East* or *West Dial* is made, by the *Precepts* in *Chapter* VII. hereof: Only, the *Hour-Line* of VI there, must be the *Hour-Line* of XII in this: And as the *Stile there* was equal to the distance between *Six* and *Three* or *Nine* a Clock. So in this, it must be equal to the distance between XII and IX or III. And may be either a *Plate* of that *breadth*; or a *Wyre* or *Pin* of that *Length*.

II. Of North Recliners

I. Variety. If a *North Plain* in the *Latitude* of 51 de. 30 m. should *Recline* from the Zenith 20 deg. In what *Latitude* will that be an *Horizontal Plain*?

The *Reclination* 20 deg. being *less* than the *Complement of the Latitude* of *London*, 38 de. 30 m. Add the *Reclination* 20 de. and the *Co-Latitude* 38 deg. 30 m. together; their *Sum* 58 deg 30 m. is *The New Latitude*: And an *Horizontal Dial* for that Latitude shall be a *North Plain Reclining* 20 deg. in the Latitude of 51 de. 30 m.

II. Variety. If a *North-Plain* in the *Latitude* of *London*, 51 deg. 30 m. should Recline from the Zenith 75 deg. In what *Latitude* will such a Plain be *Horizontal*?

The *Reclination* 75 deg. being *greater* than 38 deg. 30 m. *Add* them together, and they make 113 deg. 30 min. which being above 90 deg. take the Complement thereof to 180 deg. which is 66 deg. 30 m. And that is the *New-Latitude*: So that an *Horizontal Dial* made for the Latitude of 66 de. 30 m. will be a *North Plain Reclining* 75 deg. in the *Latitude* of *London* 51 deg. 30 m.

III. Variety. If a *North Plain* in the Latitude of *London* 51 de. 30 m. should Recline from the *Zenith* thereof 51 de.

30

30 m. In what Latitude will such a Plain be Horizontal? *Fig. LIX.*

Here, the *Reclination* 51 de. 30 m. is *Equal* to the Latitude of *London*; And the *Sum* of the *Reclination* 51 de. 30 m. and the *Complement* of the *Latitude* 38 de. 30 m. Added together, their Sum is 90 deg. for *The New-Latitude*: And an *Horizontal Dial* made for that *Latitude* of 90 deg. (which is no other then a Circle divided into 24 *Equal Parts*, for the *Hours*; and a *Wyre* erected perpendicularly, of any *Length* for the *Stile*.)

These are all the *Varieties* of *South* and *North Reclining Plains*, Reduced to *New Latitudes*, wherein they will become *Horizontal Plains*: So that neither *Trigonometrical Calculation*, *Geometrical Projection*, or *Operation* by the *Globe*, need be here Repeated, they being all the same as in the IVth *Chapter* hereof.

Note also: That in making of any of these *North* or *South Reclining Dials*, you have made also A *Direct North* or *South Dial inclining from the Zenith towards the Horizon* so many Degrees as is the {*Re-* / *In-*} *clination*. So that when you have made a *South-Dial Reclining* from the *Zenith* 60 deg. (as is the *Second Variety* of *South Recliners* in this *Chapter*) you have made also a *North Dial inclining to the Horizon* 60 degrees, either by drawing of the Hour-lines and *Stile* through the *Centre*; or by turning the *Reclining Dial* about upon the *Hour-line* of VI. And then, as the *North Pole* is elevated upon the *South Recliner*; so much will the *South Pole* be elevated above the North Incliner, &c.

CHAP. XI.

Of Direct East *or* West Reclining Dials, *and how* Hour-lines *may be described upon them.*

As all *Direct North* and *South Reclining Dial Plains* were *Reduced* to *New-Latitudes* wherein they would be *Horizontal Plains*; and therefore made by the *Directions* given in the CHAP. IV. hereof: So all *Direct East* or *West Reclining Dial Plains*

Fig. LIX.

Plains in any one *Latitude*, may be *Reduced* to *Erect*, or *Upright*, *Declining Plains* in another *Latitude*: and therefore may be made by the *Precepts* delivered in the VIIIth. CHAP. hereof. Either, By the *Globe*; By *Spherical Projection*; or, By *Trigonometrical Calculation*: So that the Work of this CHAP. shall be only to shew

I. How to *Reduce* any *Direct East* or *West Reclining Dial Plain* in any Latitude, (suppose *London* 51 de. 30 m.) to a *New-Latitude*: wherein the *Reclining Plain* shall become an *Erect* (or Upright) *Plain*. And

II. What *Declination*, that *Upright Plain* shall have in that *New-Latitude*.

EXAMPLE.

Suppose then, that a *Direct East* or *West Plain*, in the *Latitude* of 51 de. 30 m. should *Recline* from the *Zenith* 40 deg. In what *Latitude* will that be an *Upright Plain?* And what *Declination* shall it have in that *Latitude?*

RULE.

The *Complement* of the *known Latitude*, is (always) The *New-Latitude*. And

The *Complement* of the *Reclination* is (always) The *Declination* in that *New-Latitude*.

So that if an *East* or *West Dial*, should *Recline* 40 deg. in the *Latitude* of 51 d. 30 m. That will be *An Upright Plain*, *Declining* 50 deg. in the *Latitude* of 38 de. 30 m.

For, 38 de. 30 m. being the *Complement* of the *Latitude* of *London* 51 d. 30 m. is the *New-Latitude*: —— And 50 deg. being the *Complement* of 40 de. the *Plains Reclination*, is the *Declination* in the *New-Latitude*.

So that, if (according to the *Precepts* delivered in the VIIIth. CHAP. hereof) you make an *Upright Dial* for the *Latitude* of 38 de. 30 min. to *Decline* 50 deg. Such a *Dial* will serve for an *East* or *West-Dial Reclining* 40 de. in the *Latitude* of (*London*) 51 de. 30 m.

Thus for the *Making* of the *Dial*.

But in the *placing* of the *Dial* (thus made) upon the *Reclining Plain*, this *Difference* is to be observed: For, Whereas in all *Upright Declining Plains*, the *Meridian*, or *Hour-line* of XII is always *Perpendicular* to the *Horizon* of the *Place* for which it is made.

But

But this *Declining Plain* when it is applyed to the *Reclining Plain*, the *Hour-line* of XII. must lie *Parallel* to the *Horizon* of the *Place*, as in the *Figure*. *Fig.* LX.

And here Note, that all *East Recliners* in the known *Latitude* (as here *London*) are *North-East-Decliners* in the *New-Latitude*: And all *West-Recliners*, are *North-West-Decliners*.

And Note farther; That upon all *East* and *West Reclining Plains* in *North-Latitudes*, that the *North Pole* is (always) *Elevated*: And upon the *East* and *West inclining Plains*, opposite to them, the *South Pole* is *Elevated*.

And Lastly, Note that when you have made an *East* or *West Reclining Dial*, you have made Four, *viz.* An *East* and *West Reclining*, and, An *East* and *West Inclining*.

CHAP. XII.

Of Dial Plains *that do both* Decline *and* Recline: *How* Hour-Lines, *&c. are to be described on them.*

THESE *Plains* may also be *Reduced* to *New-Latitudes* and *New Declinations*, where they may stand as *Upright Decliners*: And so *Dials* may be described on them, by the Directions in the VIIIth. CHAP. hereof: And to find the *New-Latitude*, and *New Declination* of any such Plain, The *Rules* following will direct.

EXAMPLE.

Suppose that in the *Latitude* of *London*, 51 d. 30 min: A *South Plain* should *Decline* towards the *East* or *West* 24 deg. 20 min. And also *Recline* from the Zenith 54 deg.

I. To find the *New-Latitude*

The *Canon* for Calculation.

As the Radius, Sine 90 de.	10.
Is to the Co-sine of the *Old Decl.* 24 de. 20 m.	9.959596
So is the Co-Tangent of the *Reclination* 54 d.	9.861261
To a Fourth Tangent, *viz.* 33 de. 30 m.	9.820857

This *Fourth Tangent* being thus found, the Rules following are to be observed.

I. In *South* Recliners.

RULE I. This *Fourth Tangent* must be compared with the *Old Latitude*, and the *Complement* of their *Difference* is the *NEW LATITUDE*.

So, In this *Example*, The *Fourth Arch* 33 de. 30 m. being substracted from the *Old Latitude* 51 de. 30 m. their difference is 18 deg. whose Complement 72 deg. is the *NEW LATITUDE*.

RULE II. If the *Fourth Tangent* fall out to be *Equal* to the *Old Latitude*, Then the *Difference* will be nothing; And so the *Plain* will be a *Polar declining Plain*: For the *Pole* will have no *Elevation* over it.

So, in the *Latitude* of 51 de. 30 m. If a *South Plain* should *Decline* towards the *East* or *West* 65 deg. 40 m. And *Recline* 18 de. 9 m. By the former *Canon*, the *Fourth Tangent* will be found 51 de. 30 m. *Equal* to the *Old Latitude*: So that the *Difference* is *Nothing*. And the *Hour-Lines* will be *Parallel* one to another as in the *Direct East*, *West*, and *Polar Dials*, in CHAP. VII. and CHAP. X.

RULE III. If the *Fourth Tangent* prove to be *Greater* than the *Old Latitude*, Then, The *North Pole* is *Elevated* in *South Decliners*: But if the *Fourth Tangent* be *Lesser* than the *Old Latitude*, Then, The *South Pole* is *Elevated* in *North Decliners*.

II. In North Recliners.

RULE I. The *Fourth Tangent* found as before, is to be compared with the *Complement* of the *Old Latitude*, and their *Difference* is the *NEW LATITUDE*.

RULE II. If the *Fourth Tangent* prove to be *Equal* to the *Complement* of the *Old Latitude*: that *Declining Reclining Plain* will be an *Æquinoctial Plain Declining*.

So, In the Latitude of 51 de. 30 m. If a Plain should *Decline* from the *North* towards the *East* or *West* 60 deg. And also *Recline* from the *Zenith* 32 deg. 11 m. The *Fourth Tangent* will be found to be 38 de. 30 m. *Equal* to the *Complement* of the *Old Latitude*: And will, therefore, be An *Æquinoctial Declining Plain*.

II. To find the *New Declination.*

The *Canon* for Calculation.

As the Radius, Sine 90 d. 10.
To the Co-sine of the Reclination 54 d. 9.769218
So is the Sine of the Old Declination, 24 d. 20 m. 9.614944

To the Sine of 41 de. 01 min. *9.384162

Which 41 de. 1 min. is the *NEW DECLINATION.*

And now, if (according to the *Precepts* delivered in the VIIIth. CHAP. hereof) you do make an Upright *South Dial Declining* 14 deg. 1 m. in the *Latitude* of 72 deg. That *Dial* shall serve for A *South Plain*, in the Latitude of 51 de. 30 m. *Declining* 24 deg. 20 m. and *Reclining* 54 deg.

And in such a *Dial* you will find the *Requisites* belonging thereunto, to be as the Table in the *Margin*. And thus much for the *Making* of the *Dial* according to its *New Latitude*, and *New Declination*.

New Latitude	72	00
New Declination	14	01
Stiles heigth	17	26
Dist. Sub. & Merid.	4	17
Difference of Long.	13	59

	Hours.		Eq. Di.		Tr. H D.	
41	VII	V	89	17	89	47
41	VIII	IV	74	17	46	48
41	IX	III	59	17	26	46
41	X	II	44	17	16	17
41	XI	I	29	17	9	32
41	XII		14	17	4	22
19	I	XI	00	43	0	12
19	II	X	15	43	4	49
19	III	IX	30	43	10	06
19	IV	VIII	45	43	17	05
19	V	VII	60	43	28	06
19	VI		75	43	49	39

But to apply this *Dial* to the *Reclining Plain* in the *Old Latitude*, you are not to place the *Hour-line* of XII *perpendicular* to the *Horizon*, as in *Upright Plains*, but it must make an *Angle* with the *Horizontal Line* of the *Reclining Plain*: And therefore the *quantity* of that *Angle* must be first found. Therefore,

III. To find the *Angle* made between the *Meridian* and the *Horizon.*

The *Canon* for Calculation.

As the Radius, Sine 90 d. 10.
Is to the Sine of the *Reclination* 54 d. 9.655347
So is the Tangent of the *Old Declination* 24 d. 20 m. 9.907957

To the Tangent of 20 deg. 6 m. *9.563304

Whose *Complement* 69 deg. 54 min. is the *Angle* that the *Meridian* (or *Hour-line* of XII.) must make with the *Horizontal Line* of the *Reclining Plain*. Thus for the *Quantity* of the *Angle*. But

IV. To know which way (or to what Coast) the *Meridian Line*, *Ascending* or *Descending Above* or under the *Horizontal Line* of the *Plain*, is to be drawn.

GENERAL RULES.

In { *North* Incliners. / *South* Recliners. }

- *Less* than Equinoctial, the Meridian must be drawn { Above / Below } { That *End* of the *Horizontal Line*, which lies *contrary* to the Coast of the Plain's Declination. }
- *More* than Equinoctial, the Meridian must be drawn { Below / Above } { That *End* of the *Horizontal Line*, which lies *the same way* with the Coast of the Plain's Declination. }

In-

In North Recliners / South Incliners:

- *Less* than a *Polar* the *Meridian* must be drawn — Above / Below — That *End* of the *Horizontal Line*, that looks *the same way* with the Coast of the Plain's Declination — And this *Meridian* thus drawn, in *North Recliners*, represents 12 at *Midnight*.
- *Equal* to a *Polar*, the Meridian must be drawn below the *Horizontal Line*, at that end which is contrary to the *Coast* of *Declination*: And the *Six a Clock Hour-line*, is (always) the *Substilar*.
- *More* than a *Polar*, the Meridian must be drawn — Below / Above — And from that *End* of the *Horizontal Line*, which lies *contrary* to the *Coast* of the *Plain's Declination* —— And, in *South Incliners*, is only serviceable to help to draw the rest of the *Dial*.

V. How the *Dial* (being made according to the *New Latitude* and *New Declination*) is to be transferred from the *Paper Draught*, upon the *Reclining or Inclining Plain*.

Having drawn an *Horizontal Line* upon your *Dial Plain*, in the most convenient part thereof, and made choice of a Point therein for the *Centre* of your *Dial*, apply the *Centre* of your *Paper Draught* to this *Centre*, moving the *Paper Draught* about till the *Meridian Line* thereof do make an *Angle* with the *Horizontal Line* drawn upon the *Plain*, equal to what you found it to be by the *Third Section* of this C H A P. And to its *Proper Coast*, as you found it by the *Fourth Section*: Then (if you have not erred in any of your former workings) will the *Stile* of your *Paper Draught* (or rather a pattern of it cut in *Past-board*) being placed upon the *Sub-stile* of the *Paper Draught*, have direct respect to the *Elevated Pole* —— And thus due respect being had to what is here deliver'd delivered, you may easily transfer the *Stile* and the rest of the *Hour-lines* to the *Plain*; putting thereon so many as the Plain is capable to receive at any time of the Year; and leaving out such as are superfluous.

CHAP. XIII.

Of the Inscription of other Great *and* Small Circles *of the* Sphere *upon all sorts of* Dial-Plains.

IN the foregoing Chapter is shewed how the *Meridians* (or *Hour-Circles*) of the *Sphere* may be described upon all sorts of *Plain Superficies* howsoever situate: And as those *Hour-Circles* were described, so may all other *Circles* of the *Sphere* be also delineated: And the manner how, I have largely treated of in my particular *Treaties of Dialling* already extant: So that in this place I shall be but brief therein, and yet sufficiently *Plain*. And whereas the *Hour* of the *Day* on all *Sun-Dials* is shewed by the whole *Axis* of the *Stile*; so all other *Circles* so described, are shewed by one *single point* taken in some convenient place of the said *Axis*, from which Point a Perpendicular let fall to the *Plain* upon the *Sub-stilar Line* shall be called the *Perpendicular Stile*: And the point upon the Plain, The *Foot* of the *Perpendicular Stile*; and the *Point* in the Axis (before assumed) the *Gnodus* or *Apex* of the *Stile*. —— Now, for the performances of the Work following (besides the *Trigonometrical Calculation*) the Practitioner must be provided with a *Scale* (or a *Sector* rather), which must have upon it *Scales* of *Sines*, *Tangents*, *Secants*, *Chords*, and *equal parts*: or if he have by him, *Tables* of *Natural Sines*, *Tangents*, and *Secants*, then a *Line of equal parts* only drawn from the Centre, will perform the work rather better than by the other *Scales*. Being thus provided; The first business must be

How to proportion the *Perpendicular Stile* to the *Plain*.

If your *Dial Plain* be small, consider whether it be *Direct*, or *Declining*; if *Direct* the *Sub-stile* may be placed in the *Middle*: If *Declining*, more to the side contrary to the *Coast* of *Declination*.

The Sub-stile well made choice of, and the *Perpendicular Stile*

Stile thereon *Erected*: Make the *Perpendicular Stile* the *Radius* (or Equal to the *Tangent* of 45 degrees) and make that part of the *Sub-stile* which lies beyond the foot of the *Stile* and towards the Centre, equal to the *Tangent Complement* of the heighth of the *Pole* (or *Stile*) above the *Plain*: and the other part of the *Sub-stile*, below the *Foot* of the *Stile*, Equal to the Tangent Complement of the *Meridian Altitude* of the *Sun*, when he is in that *Tropick* which is to be most remote from the *Centre* of the *Dial*.

CHAP. XIV.

Of the Inscription of the Signs *or* Parallels *of the* Sun's Course.

A *Sign* is the Twelfth part of the the *Ecliptick*, and therefore contains 30 degrees.

A *Parallel*, is the *Sun's Diurnal Motion* Day by Day; and because there are 47 deg. between the two *Tropicks*, there may be so many *Parallels*, that is, *Circles*, which the Sun describeth every 24 Hours: and although there be 47 of these, yet in the *Latitude* of 51 deg. 30 min. we account but *nine*, viz. those which are the *Day* from *Sun* to *Sun*, when it is 8, 9, 10, 11, 12, 13, 14, 15, or 16 just hours long. The *Description* of these *Parallels* and *Signs* is made the same way; only due respect must be had to the quantity of the *Sun's Declination*: For (in all *Direct Horizontals*) the *Perpendicular Stile* being made *Radius*, the *Tangent Complement* of the Sun's height, in any *Sign* or *Parallel*, at any hour of the *Day*, set off, from the *Foot* of the *Stile*, and extended to that *Hour-line*, gives a *Mark* upon the *Hour-line*, by which the *Parallel* of that *Day* shall pass: So that this *Work*, repeated so often as the number of *Parallels* to be inscribed, and the *Hour-lines* require; shall give respective *Points* enough, in each *Hour*, to draw each *Parallel* by.

Example. In the Latitude of 51 de. 32 m. the Sun being in the beginning (or entring) of *Pisces*; The *Sun's height* above the *Horizon* at every *Hour* may be found (by CASE IX of O. *A. S. T.*) to be as followeth, viz.

At

At			Deg. M.	The Complement	
	12		27. 01		62. 59
	1	11	25. 37		64. 23
	2	10	21. 49		68. 11
	or				
	3	9	15. 57		74. 03
	4	8	8. 32		81. 28

Now, the *Perpendicular Stile* being *Radius*, the Tangents of the Complements of the respective *Altitudes*, as 62 deg. 59 m. the Complement of 27 de. 01 m. set from the foot of the *Per. Stile*, on the *Hour-line* of 12 (or *Sub-stile*) shall give a point thereon, by which the *Parallel* of *Pisces* must pass: And so, the Tangent of 64 de. 23 m. set from the foot of the *Per. Stile*, upon the *Hour-lines* of 11 and 1 a Clock, shall give you two other points by which the said *Parallel* shall pass: And so for all the rest of the *Hour-lines*, through which points found upon all the *Hour-lines*, a *Line* drawn by an even hand, shall be the *Parallel* required; for along that Line will the *Shadow* of the *Top* of the *Per. Stile* (as it creepeth along) pass, when the Sun is in the beginning of *Pisces*, *viz.* about the 9th of *February*.

And therefore, generally in *Verticals*, as also in all *Recliners*; that is to say, upon all *Plains* whatsoever: Draw an *Horizontal Dial* proper to the *Plain*, and inscribe the *Signs* or *Parallels* upon it, by setting off from the *Foot* of the *per. Stile*, the *Tangents Complements* of the *Sun's height* at every *hour* in the beginning of every *Sign* above that Plain (taken as an *Horizontal*, the Foot of the *per. Stile* being ever *Radius*) and at the end of these *Tangents* so set off upon every respective *Hour-line*, will be a Point; By which Points, *Lines* drawn with an even Hand, shall trace out upon the *Dial Plain*, the *Parallels* required.

Example. Suppose a Plain Decline 30 deg. and Recline 55 deg. the height of the *Pole* above the Plain 19 deg. 25 min. And the *Sun's height* at the beginning of *Taurus* to be at the several *Hours*, as in this Table.

At			De. M.	Complements	De. M.
	12		38. 05		07. 55
	11	1	73. 30		16. 30
	10	2	60. 03		29. 57
	or				
	9	3	46. 01		43. 59
	8	4	31. 53		58. 07
	7	5	17. 47		72. 13

Then,

Then, The Tangents of the Complements of these *Hour-distances* (as 7 deg. 55 m. for 12 : 16 deg. 30 m. for the Hours of 11 and 1) set off from the Foot of the *per. Stile* (the said *Stile* being the *Radius* to those *Tangents*) to the obscure *Horizontal Hours* of 12, 11, 10 : and 1, 2, 3, *&c.* give the true *distances* between the *Foot* of the *Stile*, and those *auxiliary Hours*, for the *Parallel* of *Taurus* ; and so points for the describing of other *Parallels* of *Declination* : Having first (by *Trigonometrical Calculation*) found the *Horizontal Distances*, and the *Sun's Altitude* at his entrance into those Parallels of *Signs* or *Declination*, in such *Latitude* as you have need of. All which are taught how to do in the foregoing parts of this Book.

CHAP. XV.

Of the Inscription of the Vertical Circles *(commonly called* Azimuths) *upon all* Dial Plains.

THESE are great *Circles* of the *Sphere*, whose *Poles* lie in the Horizon, and intersect one another in the *Zenith* and *Nadir* Points of the *Place* wherein the *Dial* is to stand.

The whole *Horizon* being divided into 32 equal parts ; these *Circles* passing through those *Divisions*, are called *Points* of the *Compass*, and denominated accordingly, as *South*, *S by E*, *S S E*, &c. But the better way of accounting them is by 10, 20, 30, *&c. Degrees* from the *Meridian* on either side thereof.

First, in all *Horizontal Dials* ; the *Perpendicular Stile* being chosen, making the Foot thereof the *Centre* ; at any convenient distance, describe a *Circle* ; and account from the *Meridian* both ways, *Arches* equal to 10, 20, 30, *&c. Degrees* : From which Divisions, right Lines drawn to the *Foot* of the *Stile* aforesaid, shall represent those *Azimuths* upon that *Dial*.

Secondly, Upon a *Prime Vertical* (or *South*) *Dial* : Through the Foot of the *Per-stile*, draw a Right-Line Parallel to the *Horizon* ; and making the said *Stile Radius* ; upon the *Parallel Line*, set off, both ways from the *Meridian Tangents* of 10, 20, 30, *&c. Degrees* ; through which Divisions, Right-lines drawn, all at *Right Angles* with the *Parallel Line*, shall be the *Azimuths*.

Thirdly, Upon any *Declining Vertical*, the ſame being done; ſhall give the *Azimuths* of 10, 20, 30, *&c.* degrees from the *Meridian* of the *Plain*; or from the *Meridian* of the *Place*, juſt allowance being made for the *Difference* of *Meridians*.

Fourthly, In *South Declining Reclining* Plains, the *Per. Stile* being choſen, and made the *Radius*, the *Tangent Complement* of the *Reclination*, applyed from the *Foot* of the *Per. Stile* to the *Meridian* of the *Place*, ſhall determine the *Zenith* of the *Place*: through which, and the *Foot* of the *Stile*, (that is the *Zenith* of the *Plain*) a right Line drawn, ſhall be a *Perpendicular* to the *Horizontal Line*, and ſhall concur with the *Æquator* in the *Hour-line* of 6; and therefore, if from the *Foot* of the *Stile* upon the ſaid *Perpendicular*, towards the *North* (for the former application was made towards the *South*) be ſet off the *Tangent* of the Reclination, a *Line* drawn from the end thereof, at *Right Angles* with it, ſhall be the *Horizontal Line*: Upon which, the *Tangents* of 10, 20, 30, *&c.* (the *Secant* of the *Reclination* being now made *Radius*) ſet from the ſaid *Right Angle*; *Lines* drawn from them to the *Zenith* of the *Place*, ſhall be the *Azimuths*.

Fifthly, The *Diſtance* between the *Meridians* being known, upon the *Horizontal Line*; the *Azimuths* which were accounted from the *Meridian* of the *Plain*, may be fitted for the account from the *Meridian* of the *Place*, with eaſe.——For *Example*, let that diſtance be the *Tangent* of 20 deg. Then that *Azimuth* which is 10, from the one; is 10 from the other alſo: And that which is 30 on the ſame ſide of the Sub-ſtile, is 10 on the other ſide of the *Meridian* of the *Place*: And the like method ſerves for any diſtance.

CHAP. XVI.

Of the Inſcription *of* Almicanters *or Circles of the* Sun's Altitude *upon* Dial Plains.

THESE are *leſſer Circles* of the *Sphere*; and may be called the *Parallels* of *Declination* from the *Horizon*; they having, in all reſpects, the ſame relation and habitude to the *Azimuths*, as the *Signs* and *Parallels* of *Declination* have to the *Meridians*; although theſe be counted by 15 *deg.* and thoſe uſually by 10. And

And therefore, as in the description of the *Signs* and *Parallels*; so in these,

Let an *Horizontal Dial*, proper to the *Plain*, be first (obscurely) described; and then, as it was there shewed, that the points through which the *Signs* or *Parallels* must pass, upon every *Hour-line*, might be had by applying the *Tangents* of the *Complements* of the *Sun's height* of those *Hours* in those *Parallels*, from the *Foot* of the *Per. Stile*, to the respective *Hour-lines*: —— So here, making use of that *Azimuth* which is perpendicular to the Plain, (which in all *Plains* is that which passeth through the *Foot* of the *Per. Stile*) the rest of the *Azimuths* being also inscribed, the *Tangents Complements* of the *Sun's height* above the *Plain*, when he is in any *Azimuth*, applyed from the *Foot* of the *Per. Stile* to the said *Azimuth*, gives a *Point*, through which that *Circle* or *Almicanter*, upon that *Azimuth* must pass.

Now to know what *Altitude* the Sun will have, when he will be upon any *Azimuth*, in any *Parallel* of *Declination* (or degree of the *Ecliptick*) is taught in the *Section* of *Astronomical Problems*; Or, by the Resolving of an *Oblique Angled Spherical Triangle*, wherein is always Given, *Two Sides*, and the *Angle* opposite to one of them to find the third *Side*, (By CASE V. of O. A. S. T.) Which *third Side* so found, is the *Complement* of the *Altitude* which is in this case required; and must accordingly be set from the *Foot* of the *Per. Stile* unto the *Azimuths*, &c.

CHAP XVII.

How to inscribe the Jewish, Babylonish *and* Italian Hours, *upon all* Dial-Plains.

FOR the Inscription of these *Hours* upon *Dial-plains*, there needs no *Trigonometrical Calculation*: For the two *Tropicks*, the *Æquator*, and other *Parallels* of *Declination* being already described (or such of them as shall be needful) together with the *common Hour-lines* proper for the *Plain*, Points through which these *Hour-lines* may pass, may be found by these following Directions.

The *Babylonish Hours* are accounted *Equal Hours* from *Sun-rising*, and may be inscribed upon any *Plain*, by help of those two *Pa-*

rallels of *Declination* which shew the *Longest* and *Shortest Day*, consisting of *whole* (or entire) Hours: As with us 16 and 8; and the *Æquator*. For,

A Line drawn through { 9, 7, 5 } In the Parallel of { 8 of Declination, Æquinoctial, 16 of Declination } Shall be the *Hour* of *One* from the *Sun's rising*. And likewise in the same order; through 6, 8 and 10, shall give the *Second Hour* from *Sun-rising*; and in the like *Order*, all the rest.

In Winter, when the *Parallel* of 8 *Hours* shall fail, the other two Points will serve to draw it by, because those *Hours* are *Streight Lines*. But after the first *Six Hours* are inscribed, the *Æquinoctial* also failing, some other *Diurnal Arch*, (as of 9 or 10 Hours) must be described to supply that want.

The *Italian Hours* are accounted by 1, 2, 3, &c. from *Sun-setting*: And for the Inscription of these, the same *Parallels* of 8 and 16 *Hours*, with the *Æquator*, will serve: For a Line drawn through them, in the Hours of 9, 7 and 5, *Afternoon* (observing the same as before) shall be the *hour* of *One*; the like through 7, 5 and 3, shall be the *hour* 23: The *Night hours* of 9, 10, &c. are the *Morning hours* produced.

The *Jewish hours* are reckoned like the *Babylonish*, from *Sun-rising*, but unequally; their *Sixth hour* being (always) *Noon*; and every *hour*, one *Twelfth* part of the *Artificial Day*, of what length soever that be.

For the *Inscription* of them: The *Vulgar Hours* proper for the *Plain* being first drawn, and the *Diurnal Arches* of 15, 12, and 9 *hours*, divide the degrees in each by 12; and that Quotient by 15; or else (which is all one) *divide* the said *Arches* by 180, the three *Quotients* shall give the just *Times*, in *hours*, and usual parts of *hours* from 12 a Clock, upon the two *Parallels* and the *Æquator*: through which, *Lines* drawn by a *Ruler*, shall be the *Jewish hours* required.

Example. In Latitude 51 deg. 32 min. the Diurnal Arch of 15 hours is in Degrees 225, which divided by 180, the Quotient is 1 ¼ h. and so much the *Jewish hours* of 5 and 7 are distant from *Noon*; one hour and a quarter being a *twelfth part* of the *Diurnal Arch* of 15 hours: And this *hour* and *quarter* being *doubled*, gives the place for 4 and 8: *Tripled*, the place of 3 and 9, &c. from *Noon*, upon that *Parallel* of 15 hours.

In like manner, the *Diurnal Arch* of 9 hours is 135 deg. which divided by 180, the Quotient is $\frac{135}{180}$, that is 3 quarters of an hour: Which shews the place of the *Jewish hours* of 7 and 5, to be three quarters after, or before, *Noon* : and that doubled is *One hour and a half*, which gives the place of 8 and 4 ; all one with our 1½ and and 10½ ; and so *Tripling* and *Quadrupling* and *Quintupling* 3 quarters, you have the places of the *Jewish hours* upon this *Parallel* of 9 hours length of the Day.

And these parts *Doubled* and *Tripled*, as is said, will always (in this *Parallel* and the former) fall upon *even hours*, *halves* and *quarters* of *hours* : And that is the only reason why these two *Parallels* of 15 and 9, are preferred ; there being no necessity of using them, more than the *Tropicks* or other *Parallels*, only this conveniency of even parts.

Lastly. In the *Diurnal Arch* of 12, that is, the *Æquator*, the *Common* and the *Jewish hours* concur ; that is, the *Jewish hours* of 5 and 7, with our *hours* of 11 and 1 : Their 4 and 8, with our 10 and 2, &c. So that a *Line* drawn from 1¼ in the *Parallel* of 15, to 1 in the *Æquator*, and from thence to ¾ in the *Parallel* of 9, is the 7th *Jewish hour*. And so are all the rest to be inscribed.

Trigo-

Trigonometria Practica.

SECTION VI.

OF NAVIGATION.

I Intend not here to treat of *Navigation* in the general; it being an Art that requires (for the true Underſtanding, either the Theory, or Practice of it) an inſpection into divers other *Sciences Mathematical*; of which, that of *TRIGONOMETRIA*, (or the *Doctrine* of *Triangles*) is the *Principal*; for that the ſolution of all ſuch *Problems* which are of daily Uſe at Sea, are performed thereby; and thoſe are ſuch as concern *Longitude*, *Latitude*, *Rumb* (or courſe) and *Diſtance*, &c. And therefore, I ſhall *Define*, Firſt, what is meant by *Longitude*, *Latitude*, *Rumb*, *Diſtance*, &c. And Secondly, any two of them being known; how to find the other two; and that by *Trigonometrical Calculation*: with ſome other *Problems* pertinent to that Art. And I ſhall perform them, (1) By Plain Sailing. (2) By Mercators Sailing; And (3) By the Middle Latitude.

DEFINITIONS.

I. *Longitude*, Is the *Diſtance* of a *Place* from ſome known *Meridian* to that *Place*; and is always counted upon the *Æquinoctial* Circle, from that known Meridian towards the *Eaſt* or *Weſt*.

II. La-

II. *Latitude*, Is the *Distance* of any *Place* from the *Æquinoctial* Circle; counted upon that *Meridian* Circle which passeth over that *Place*, towards either of the *Poles*; either *North* or *South*: and accordingly the *Latitude* is *Denominated* either *North* or *South Latitude*.

III. *Rumb*, (or *Course*) Is that *Angle* which a *Ship* in her *Sailing* makes with the *Meridian* of the Place from whence the Ship came, and the Place where the Ship then is: But the *Complement* of the *Rumb*, is that *Angle* which the *Rumb* makes with that *Parallel* of *Latitude* in which the Ship is; and is the *Complement* of the *Rumb* to 90 deg.—The *Rumb* is made known to the *Mariners* at all times, by help of his *Compass*.

IV. *Distance*, Is the *Number* of *Leagues*, *Miles*, *Centesms*, &c. that a Ship hath *Sailed* upon any *Rumb* or *Course*.—And this is known to the *Mariner* by the vering (or running out) of the *Log-Line* in any known quantity of *Time*.

V. If your *Rumb* (or *Course*) be directly *East* or *West*, you alter not your *Latitude* at all:—If your *Course* be *Northward*, you continually *Raise* the *North Pole*; and you increase your *Latitude Northward*: Or the *South Pole*, if you *Course* be from the *Æquinoctial* Southward.

So that

The *Raising* of the *Pole* is, When you Sail from a *Lesser* Latitude to a *Greater*: And the *Depressing* of the *Pole* is, when you Sail from a *Greater* to a *Lesser* Latitude.

These things known, before I proceed to the Solution of the several *Questions* or *Problems* in *Navigation*, unto which the *Doctrine* of Triangles is subservient, it will be necessary to say something concerning the *Situation* of *Places* upon the *Earth* or *Sea* in respect of *Longitude* and *Latitude*: And of their *Distances* one from another in *Leagues*, *Miles*, or *Minutes*: And then, shew how to lay down (upon a *Blank Chart*) any two (or more) *Places*, according to their respective *Latitudes* and *Difference* of *Longitudes*. All which shall be comprehended under these *Heads* following.

I. *Of the* Situation *of* Places, *in respect of* Longitude *and* Latitude.

IN this *Problem* there are variety of *Cases*, according as the *places* may be situate one from the other, in respect of the *First*, or *General Meridian*, as to their *Difference of Longitude*; and in respect of the *Æquinoctial*, as to their difference of *Latitude*: And for the better understanding of these *Varieties*, I shall exhibit all of them in one *Scheme*: In which,

Fig. LXI.

N Æ S is the *First Meridian*, or beginning of *Longitude*.
W Æ E the *Æquinoctial Circle*.
Æ E the *East part* thereof.
Æ W the *West part* thereof.
N the *North Pole*.
S the *South Pole*.
D B and T A, two *Parallels* of *North Latitude*.
O R and Z X, two *Parallels* of *South Latitude*.
N I L M S a *Meridian* of *West Longitude*.
N H C S a Meridian of *East Longitude*.

CASE I. If the *Places* lies in the *Æquinoctial Circle*, as the *Places* L, G, H, and K; and so have no Latitude—Or in the same Parallel of *Latitude*, as the *Places* D, V, X, and B, in the *Parallel* of 40 de. of *North Latitude*; and the Places O, M, C and R in the *Parallel* of 35. 85 deg. of *South Latitude*: Then, (1.) If the Two Places proposed, do lie both on the *East-side* of the *First Meridian*, as H, K—X, B—C R: Or, both on the *West-side*, as D, V— L, G—O, M: Then, The *Lesser Longitude substracted* from the *Greater*, gives the *Difference of the Longitudes* of those *Places*. But, (2.) If of the two *Places* proposed, one do lie on the *East-side*, and the other on the *West-side* of the *First Meridian*: As V and X— L and H, M and C: Then, the two *Longitudes added* together, gives the *Difference* of *Longitude* between those two Places. But, Note,

If the *Sum* of the two *Longitudes* do exceed 180 deg. *substract* the *Sum* of them from 360 deg. and the *Remainder* is the *Difference* of *Longitude* of those two *Places*.

CASE

CASE II. If One *Place* lie in the *Æquinoctial Circle*, as L, and the other under the same *Meridian* N L S, as the *Places* V, in the *Parallel* of 40 deg of *North*, and M, in the *Parallel* of 35. 85 deg. † of *South Latitude*: Then, the *Latitudes* themselves are esteemed (or are to be taken) for the *Difference* of *Latitude*.

Fig. LXI.

† *In all this* Section, *the* Degrees *are not divided into* 60 *min. but into* Centesms, *or* hundred Parts.

CASE III. If Two Places proposed differ both in *Latitude* and *Longitude*, do lie both of them on the *North*, as V in the *Parallel* of 40 deg. and I in the *Parallel* of 18. 25 deg. the *lesser Latitude* taken from the *greater*, leaves 21. 75. deg. for the *Difference* of *Latitude* of those Two *Places*: The like, if both the *Places* had lain on the *South-side* of the *Æquinoctial*, as at C and Y.—But if the Two *Places* proposed had been V, in 40 deg. of *North Latitude*, and C, in 35. 85 deg. of *South Latitude*; then, the *Sum* of them two *Latitudes* 75. 85 deg. would be the *Difference* of *Latitudes*, which can never exceed 90 deg.

How to Reduce Degrees *and* Minutes, (*or* Degrees *and* Centesms, *or hundred parts of* Degrees) *into* Miles, *or* Minutes.

THIS is done by multiplying the *Degrees* by 60. and adding the odd *Minutes*, if any be, and the Product will be *Miles*. Also *Degrees*, and *Centesms* of *Degrees*, multiplied by 60, the Product will be *Degrees*, and *Centesms*. So,

26 de. 33 m. multiplied by 60: The Product of 26 by 60, is 1560, to which add the 33 Minutes, and it makes 1593 Miles.

Or, 72. 75 deg. multiplied by 60, will produce 4365 *Miles*: And 26. 47 deg. by 60, will produce 1588. 20 *Miles*; that is 1588 *Miles*, and $\frac{2}{10}$ or $\frac{20}{100}$ of a *Mile*; and so of any other.

III. *How to lay down* (*upon a* Blank Chart) *any two* Places, *which differ both in* Longitude *and* Latitude.

LET the two *Places* proposed be the *Harbour* of *Jamaica*; which lies in 18. 25 deg. of North *Latitude*, and 78. 35 deg. of West *Longitude*——And the *Cape of Good Hope*; which lies in 35. 85 deg. of *South Latitude*, and 27. 50 deg. of *East Longitude*.

Fig. LXII.

The Geometrical Conſtruction of the *Figure.*

Firſt, Draw a Right-Line Æ Æ for the *Æquinoctial Circle* of the *Terreſtrial Globe:* And another Right Line N S, for the *Principal*, or *Firſt Meridian*, from whence the *Longitude* of *Places* begin to be accounted either *Eaſt* or *Weſtward* from it: which two Lines croſs each other at Right Angles in the Point *æ.*

Secondly, The *Longitude* of *Jamaica* being 78. 35 deg. Reduce them into *Miles* (by multiplying them by 60) and they make 4701 *Miles*, which taken from any *Scale* of *Equal Parts* ſet upon the *Æquinoctial Circle* from *æ* to *a*, towards the *Weſt*, becauſe the *Longitude* was *Weſtward*: And through that point *a*, draw a Right Line *n a s*, parallel to N *æ* S, the Firſt *Meridian*, ſo ſhall that Line *n a s*, be the *Meridian* under which the Harbour of *Jamaica* lieth.——Alſo the *Longitude* of the *Cape of Good Hope*, being 27. 50 deg. Reduce them into Miles (by Multiplying them by 60) and they make 1650 *Miles*, which ſet from *Æ* to *b*, and through *b*, draw a Right Line *n b s* parallel to N *æ* S, ſo ſhall this Line *n b s*, be the *Meridian* under which the *Cape of Good Hope* lyeth.

Thirdly, The *Latitude* of *Jamaica Harbour* being 18. 25 degrees, reduce them into *Miles*, and they make 1095, which ſet upon the Line *n a s* from *a* to H, towards *n* (becauſe the *Latitude* was *North*) ſo ſhall H be the Point repreſenting the *Harbour* of *Jamaica*. And, If you draw a Line through this point H, parallel to the *Æquinoctial*, as the Line H *b*, that Line ſhall repreſent the *Parallel* of the *Latitude* of the *Harbour* of *Jamaica, viz.* of 18. 25 deg.——Alſo, The *Latitude* of the *Cape of Good Hope* being 35. 85 deg. Reduce them into *Miles*, and they make 2151, which ſet upon the Line *n b s*, from *b* downwards towards *s*, (becauſe the *Latitude* is *South*) to C, ſo ſhall the point C, be the Point repreſenting the *Cape of Good Hope*: Through which, if you draw a Right Line C A, parallel to the *Æquinoctial*, it ſhall repreſent the *Parallel* of the *Latitude* of the *Cape of Good Hope, viz.* of 35. 85 deg. of *South* Latitude.

Laſtly, Draw the Lines H C, C A and A H; ſo ſhall you have conſtituted a *Right-lined Triangle*, Right-*angled* at A; from whence ſhall be deduced and reſolved all ſuch *Problems* and *Navigation*, as concern *Longitude*, *Latitude*, *Rumb* (or Courſe) and *Diſtance.*

PROB.

Fig. LXII.

PROB. I.

TWO *Places*, H in 18. 25 de. of *North Latitude*, and 78. 35 d. of *West Longitude* : And C, in 35. 85 deg. of *South Latitude*; and in 27. 50 de. of *East Longitude* being *Given* : To *Find*,

I. The *Rumb* leading from one *Place* to the other, the *Angle* H C A.

II. Their *Distance* : upon the *Rumb*; H C.

The *Difference* of the *Longitudes*, and *Latitudes* of the two *Places* being *Found*, and Reduced into *Miles*, as is shewed in the two foregoing *Problems* : Then, in the Triangle H A C Right-angled at Æ, you have given, (1.) The Perpendicular H A, the *Difference* of *Latitude* 3246 Miles : And, (2.) The Base A C, the *Difference* of *Longitude* 6351 Miles : By which you may find the *Rumb* H C A (by C A S E I) and the *Distance* H C (By C A S E VII) of R. A. P. T.

1. For the *Rumb* H C A.

As A C, the *Dif.* of *Longitude* 6351 M.	3.802842
Is to the *Dif.* of *Latitude* H A 3246 M.	13.511348
So is the *Radius*, Tang. 45 deg	10.
To the Tang. of 27. 07 deg.	9.708506

Which is the Angle H C A, whose *Complement* H C n——A H C, is the *Rumb*, 62. 93 deg. that is, *North-Easterly* 62. 93 deg. from C to H, and *South-Westerly* 62. 93 deg. from H to C.

2. For the *Distance* H C.

As the Sine of A H C the *Rumb* 62. 93 deg.	9.949610
Is to A C, the *Dif.* of *Longitude*, 6351 M.	13.802842
So is the *Rad.* Sine 90 deg.	10.
To H C their *Distance*, 7132. 4 M.	3.853232

PROB. II.

THE *Rumb* C H A, 62. 93 deg. And the *Difference* of *Longitude* A C 6351 M. *Given*, to *Find*,

I. The *Difference* of *Latitude* H A, And

II. The

Fig. LXII.

II. The *Distance* H C.

By CASE II. and CASE V: Of R. A. P. T.

1. For the *Difference of Latitude* H A.

As the Sine of A H C, 62. 93 deg.	9.949610
Is to the Sine of H C A 27. 07 de.	9.658086
So is AC, the *Diff. of Longitude* 6351.	3.802842
	13.460928
To the *Diff. of Latitude* H A, 3246 M.	3.511318

2. For the *Distance* H C.

As the Sine of the *Rumb* A H C 62. 93 de.	9.949610
Is to A C, the *Diff.* of *Long.* A C, 6351.	13.802842
So is Rad. S. 90 d.	10.
To the *Distance* H C 7132. 4 M.	3.853232

PROB. III.

THE *Distance* H C 7132. 4 M. And the *Difference of Longitude* A C, 6351 M. *Given*; to *Find*,

I. The *Rumb* A H C, And

II. The *Difference of Latitude* A H.

By CASE III and CASE VI, of R. A. P. T.

1. For the *Rumb* A H C.

As the *Distance* H C, 7132. 4.	3.853174
Is to the *Dif. of Longitude* A C, 6351 M.	13.802842
So is *Rad.* S. 90 de	10.
To the Sine of the *Rumb* 62. 93 de.	9.949668

2. For the *Difference* of *Latitude* H A.

Having found the Rumb, by the Last, Say.

As the Sine of the *Rumb* A H C, 62. 93 de.	9.949610
Is to the Dif. of Longitude A C, 6351 M.	3.802842
So is the Sine of H C A, 27. 07 deg.	9.658086
	13.460928
To the Dif. of Latitude H A. 3246 M.	

PROB.

PROB. IV.

THE *Distance* C H, 7132. 4 M. And the *Difference of Latitude* H A 3246 M. *Given*: To find,
I. The *Rumb* H C A,
II. The *Difference of Longitude* A C.
By CASE III. and CASE VI. of R. A. P. T.

1. For the *Rumb*.

As the Distance H C 7132. 4 M.	3.853232
Is to the Dif. Latit. H A 3246	13.511348
So is the Radius Sine 90 de.	10.
To the Sine of 27. 07 de. H C A,	9.658116

Whose Complement is A H C the Rumb 62. 93 de.

2. For the *Difference of Longitude* A C.

As the Radius Sine 90 de.	10.
Is to the *Distance* H C 7132. 4 M.	3.853232
So is the Sine of H C A 27. 07 de.	9.658086
To the *Diff. of Longitude* A C 6351 M.	13.511318

PROB. V.

A *Ship* at H, the Harbour of *Jamaica*, in 18. 25 de. of *North Latitude*; and *West-Longitude* 78. 35. de. being bound for the *Lyzard*, which lyes in 50. 34 de. of *North Latitude*; and 5. 96 de. of *West Longitude*: And she sails upon these several *Courses, viz.* (1.) *North-East* 16. 25 deg. 42. 16 *Miles* from H to K.——(2.) *North-East* 60. 00 deg. 25. 36 *Miles* from K to L——(3.) Directly *North*, 32. 00 Miles from L to M——(4.) *North-East*, 60. 12 deg. 28. 16 *Miles*, from M to N.——(5.) *North-East*, 46. 50 deg. 40. 25 *Miles*, from N to O.——(6.) Directly *East* 28. 00 *Miles*, from O to P——And now I would know,

I. The *Difference* of *Latitude*, P R, and *Difference* of *Longitude*, H R, at the end of each of these *Courses*.

II. How the Ship being at P, *bears* from the *Harbour* at H, from whence she came.

III. In what *Longitude* and *Latitude* she then is.

IV. The

Fig. LXII.

IV. The nearest *Distance* from P to H.

I. *The* Geometrical Construction *and (gradual)* Trigonometrical Calculation, *of the whole* Travers.

DRAW a Right Line N S for the *Meridian*, and another E W at Right-angles thereto, for the *Parallel* of the *Latitude* of *Jamaica*, crossing each other at H, the point of the *Harbour*, where the *Voyage* begins. Then,

I. The First *Course* being *North-East* 16. 25 de. make the Angle *a* H K equal thereto: And the *distance* Sailed being 42. 16 Miles, set them from H to K; and through K, draw *a* K parallel to W E: So shall you have constituted a *Triangle* H *a* K, right-angled at *a*, in which there is given, (1.) the *Rumb* (or *Course*) *a* H K 16. 25 deg. (2.) the *Distance* Sailed upon that *Rumb*, 42. 16 Miles: by which you may find, (1). the *Difference* of *Longitude a* K, 11. 80 M. And (2.) the *Difference* of *Latitude* H *a* 40. 47 M. By CASE IV. of R. A. P. T

1. For the Difference of *Longitude* a K.

As the Radius Sine 90 deg.	10.
Is to the *Distance* sailed H K, 42. 16 M.	1.624900
So is the Sine of the *Rumb* a H K. 16. 25 d.	9.446893
To *a* K 11. 80 M. the Difference of *Long.*	✠1.071793

2. For the Difference of *Latitude* H *a*.

As the Radius, Sine 90 deg.	10.
Is to the Co-sine of the *Rumb* a K H, 73. 75 de.	9.982294
So is the *Distance* sailed H K, 42. 16 M.	1.624900
To H a 40. 47 M, the Dif. of *Latitude.*	✠1.607194

And in such difference of *Longitude* and *Latitude* will the Ship be, when arrived to K.

II, The Second *Course* being *North-East* 60. 00 de. through the point K, draw a Right Line *b e* parallel to N S; And on the point K protract an Angle of 60. 00 deg. as *b* K L, and set the distance sailed 25. 36 M. from K to L; and through L, draw L *b* parallel to W E: Then in the *Triangle b* K L, there is given the Angle b K L the *Rumb* 60. 00 d. and the Side K L 25. 36 M

36 M. the distance upon the *Rumb*, whereby may be found *b* L the Differ. of *Long.* and K *b* the Difference of *Latitude*, as in the former *Course*.

Fig. LXII.

1. For the Difference of *Longitude* b L.

As the Radius	10.
To the *Distance* Sailed K L 25. 36 M.	1.404149
So is the Sine of the *Rumb* b K L 60. 00 d.	9.937530
To the Difference of *Longitude* b L 21. 96 M.	✠1.341679

2. For the Difference of *Latitude* K b.

As the Radius	10.
To the *Distance* Sailed K L, 25. 36 M.	1.404149
So is the Co-sine of the *Rumb* b L K 30 00 d.	9.698970
To the Difference of *Latitude* K b 12. 68 M.	✠1.103119

III. The Third *Course*, being *directly North*, and the *Distance* Sailed 32.00 M. Through the Point L, draw a Right *Line* L c parallel to N S. And because the *Course* was directly *North*, set 32. 00. M. (the *Distance* sailed) from L to M; and to that Point is the Ship now arrived; which being in the same *Longitude* as when she was at L, the difference of *Longitude* is nothing. But the difference of *Latitude* is the same with the *Distance* Sailed; *viz.* 32. 00 Miles, or *Minutes*. So that for this *Course* there is no need of any *Trigonometrical Calculation*.

IV. The Fourth *Course* being *North-East* 60. 12 deg. continue the Line L M to *c*. And on the Point M, protract the Angle of the *Rumb*, *c* M N 60. 12 m. and set the *distance* sailed 28. 46 M. from M to N: And through N, draw N *c*, parallel to W E. Then, in the Triangle M *c* N, there is given the *Angle* *c* M N 60. 12 de. the *Course* or *Rumb*: And the Side M N 28. 46 M. the *Distance* Sailed; by which you may find the Side *c* N, the Difference of *Longitude*; and M *c* the Difference of *Latitude*: In the same manner as in the two first *Courses*.

1. For the Difference of *Longitude* *c* M.

As the Radius	10.
Is to the *Distance* Sailed, M N 26. 46 M.	1.422589
So is the Sine of the *Rumb* *c* N M, 60 12 d.	9.938054
To the Difference of *Longitude* c N 22. 94 M.	✠1.360643

2. For

Fig. LXII.

2. For the Difference of *Latitude* M c.

As the Radius	10.
Is to the *Distance* Sailed M N, 26. 46 M.	1.422589
So is the Co-sine of the *Rumb* c N M 29. 88 M.	9.697391
To the Difference of *Latitude* M c, 13. 18 M.	+1.119980

V. The Fifth *Course* being *North-East* 46. 50 d. through the Point N, draw a Right Line *d* N *f*, parallel to the *Meridian* N S: And upon the Point N, protract the *Angle* of the *Rumb*, *d* N O, 46. 50 deg. And set the *Distance* sailed 40. 25 M. from N to O; and through O, draw *d* O parallel to W E. Then, in the Triangle N *d* O, you have given the *Angle* *d* N O the *Rumb*, 46. 50 deg. And the Side N O, the *Distance* Sailed 40. 25 M. by which may be found, the difference of *Longitude* *d* O, and the difference of *Latitude* N *d*, as before.

1. For the Difference of *Longitude*, *d* O.

As the Radius Sine 90 de.	10.
To the *Distance* sailed, N O 40. 25 M.	1.604766
So is the Sine of the *Rumb* d N O, 46. 50 d.	9.860562
To the Difference of *Longitude* d O, 29. 19 M.	+1.465328

2. For the Difference of *Latitude* N d.

As the Radius	10.
To the *Distance* Sailed, N O, 40. 25 M.	1.604766
So is the Co-Sine of the *Rumb* d O N, 43. 50 d.	9.837812
To the Difference of *Latitude* N d 27. 70 M.	+1.442578

VI. The Sixth *Course* being directly *East*: And the *Distance* Sailed 28. 00 Miles, continue the Line or *Parallel* *d* O, and set the *Distance* sailed upon it, *viz.* 28. 00 M. from O to P, and to that Point is the Ship now arrived: which being in the same *Parallel* of *Latitude* as when she was at O, the Difference of *Latitude* is nothing: But the Difference of *Longitude* is equal to the *Distance* Sailed; namely 28. 00 M. So that for this *Course*, there needs no *Trigonometrical Calculation*.

Fig. LXIII. Having thus finished all the *Courses*, you find the *Difference* of *Longitude* and *Difference* of *Latitude* to be such as are set down in the *Scheme* of the *Travers*. And thus is the *First part* of the *Problem* resolved. Now for the Second.

VII. From

Fig. LXIII.

VII. From the Point P (in your *Travers*, *Fig.* LXIII) draw a Right Line to H, the *Harbour* from whence you began, and from the same Point P, let fall the *Perpendicular* P R, upon the Line W E: So shall you have constituted a *Right-angled Plain Triangle* H R P, Right-angled at R: By the Solution whereof, the other parts of the *Problem* may be *Trigonometrically* resolved.

VIII. Collect the several *Differences* of *Longitude* and *Latitude*, into one Sum as is done in the following Table.

In the *Course* from		Difference of Longitude is	Miles	Difference of Latitude is	Miles
H to K	a K	15. 80		H a	40. 47
K to L	b L	21. 96		K b	12. 68
L to M	M	——		L M	32. 00
M to N	c N	22. 94		M c	13. 18
N to O	d O	29. 19		N d	27. 70
O to P	O P	28. 00		O	——
	H R.	113. 89		P E.	126. 03

By the *Table* you find the *Sum* of all the *Differences* of *Longitude* to be 113. 89 *Miles*, or *Minutes*: Which reduced into *Degrees* and *Minutes* (by dividing them by 60) they make 1. 91 deg. Which substracted from the *Longitude* of *Jamaica*, 78. 35 deg. there remains 76. 34 deg. for the *Longitude* in which the *Ship* is, being arrived to P.

Also, you find by the *Table*, the *Sum* of all the *Differences* of *Latitude* to be 126. 33 *Miles*: which reduced into *Degrees*, makes 2. 10. deg. Which being added to 18. 25 deg. the *Latitude* of *Jamaica*, it makes 20. 35 deg. for the *Latitude* in which the *Ship* is at P. And thus is the *Second* part of the *Problem* resolved.

Now for the *Third*.

IX. The next thing to be found is the *Rumb*, *Course*, or *Bearing* of the *Ship*, she being at P, to H, the *Harbour* from whence she came.——In the Triangle P R H, there is given, (1) The

Fig. LXIII. Side H R, the *Difference* of *Longitude* 113. 89 min. (2) The *Difference* of *Latitude* P R, 126. 03 min. To find the Angle H P R, the *Rumb*: by CASE I. of R. A. P. T.

As the Side P R, Dif. *Latit*. 126. 03 M.	2.100474
Is to the Side H R, Dif. *Lon*. 113. 89 M.	12.056485
So is Radius, Tang. 45. 00 de.	10.
To the Tangent of the *Rumb* H.P R 42. 11 d.	9.956011

So that the *Course* from P to H is *South-West*, 42. 11 deg. and from H to P, *North-West* as much.

X. The last thing to be found is, the *shortest Distance* (or *Distance* upon the *Rumb*) between H and P: By CASE V. of R. A. P. T.

As the Sine of the *Rumb*, H P R, 42. 11 d.	9.826435
Is to the Difference of *Longitude* H R, 113. 89 M.	12.056485
So is the Radius, Sine 90 d.	10.
To the *Distance* upon the *Rumb*, P H, 169. 83 M.	2.230050

PROB. VI.

IF from the *Latitude* of 18. 25 deg. at H, I sail upon some *Rumb* between the *North* and the *East*, as H P 169. 83 *Miles*, and then find that I have departed from the *Meridian* from whence I came 113. 89 min. I would then know,

I. Upon what *Rumb* I steered; And
II. How much I have *altered* my *Latitude*.

THIS *Problem* may be resolved in the *Triangle* H R P: In which there is *Given*, (1) The Side H P, the *Distance Sailed* upon the unknown *Rumb*, 169. 83 M. (2) The Side H R, the *Departure* from the *Meridian* of H, from whence you failed, 113. 89 M. by which you may find the Angle P H R, the *Rumb* failed upon: By CASE III, of R. A. P. T.

As the *Distance* failed H P, 169. 83 M.	2.229014
Is to the Radius, Sine 90 de.	10.
So is the *Departure* H R, 113. 89 M.	12.053191
To the Sine of P H R, 42. 11 deg.	9.824177

Whose Complement 47. 89 deg. is the Angle P H R, that is *North East* 47. 89 de. the *Rumb*.

Then

Fig. LXIII.

Then for the Difference of *Latitude* P R.

As the Co-sine of the *Rumb* H P R, 42. 11 d.	9.82643
Is to the *Departure* H R, 113. 89 M.	2.056485
So is the Sine of *Rumb* P H R, 47. 89 d.	9.870321
	11.926806
To the Side P R 126. 03 M. the *Diff.* of *Latit.*	2.100371

Which reduced into deg. makes 2. 10 deg. which is the difference of *Latitudes*, which being added to 18. 25 deg. The Latitude of H, from whence you came, makes 20. 35 de. For the *Latitude* in which you are at P.

The Eight foregoing PROBLEMS, were all resolved by *Right-angled Plain Triangles*: These which follow, fall under the CASES of *Oblique-angled Plain Triangles*.

PROB. VII.

Fig. LXIV.

A *Ship* at C, in 22 deg. of *North Latitude*, Sails *North-East* 22. 50 deg. 64. 50 *Miles* to D: And from D, she Sails between the *South* and the *East* 124. 00 *Miles*; and then finds that she is in the same *Latitude* from whence she first came: I would now know.

I. What was the *Rumb*, (or *Course*) from D to F.
II. The Difference of *Longitude* betwen C and F.
III. What *Latitude* the Ship was in, when she was at D.

The Geometrical Construction of the *Figure*.

First, Draw a Right Line N S for the *Meridian*; and at Right Angles thereto another W E, for the *Parallel* of *Latitude* 22. 00 deg. *North*.

Secondly, Upon E, Protract the Angle of the *Rumb North-East* 22. 50 deg. *a* C D and set thereon 64. 50 M. the *distance sailed* from E to D; and through the point D, draw an obscure Line D *a*, parallel to the Line W E.

Thirdly, Take the *Distance Sailed* from D to F 124.00 Miles in your Compasses, and setting one foot in D, with the other describe the Arch *b c*, crossing the *Parallel* of *Latitude* W E, in the point F.

 Fourthly,

Fig. LXIV. Fourthly, Joyn C D and F D, so shall you have constituted an *Oblique-angled Triangle* D C F, In which there is *Given* (1) The *Side* C D, 64. 50 M. (2.) The *Side* D F 124. 00 M. and (3.) The *Angle* D C F, 67.50 de. the Complement of the *Rumb* sailed upon from C to D; to find (1.) The Angles at D and F, the *Rumbs*: (2.) The Difference of *Longitude* between C and F: And (3.) The Difference of *Latitude*, C a.

By Trigonometrical Calculation.

1. For the *Rumb* (or *Course*) D F C, By CASE I. of O. A. P. T.

As the Dist. sailed from D to F 124. 00 M.	2.093421
Is to the Co-sine of the *Rumb*, from C to D, 22. 50 d.	9.965615
So is the *Distance* sailed, from C to D 64. 50 M.	1.809559
	11.775174
To the Sine of the Angle D F C 28. 72 d.	9.681753

Whose Complement 61. 28 de. is the *Rumb* from F to D, that is *North-West* 61. 28 de: And *South-East* 61. 28 de. from D to F.

2. For the *Difference* of *Longitude* C F.

The Angles at C and F being known, the Angle at D is also known, and will be found to be 96. 22 de. Then,

As the Sine of D F C, 28. 72 deg.	9.681720
Is to the *Distance* sailed from C to D 64. 50 M.	1.809559
So is the Sine of the Angle C D F, 83. 78 de.	9.997411
	11.806970
To the *Diff.* of *Longitude* C F, 133. 43 M.	1.125250

3. For the *Diff.* of *Latitude*, D G, equal to C a.
By CASE IV. of R. A. P. T.

As the Radius Sine 90 de.	10.
To the *Dist.* sailed from C to D, 64. 50 M.	1.809559
So is the Co-sine of the *Rumb* from C to D 67. 50 d.	9.965615
To the Dif. of *Latitude* D G = a C, 59. 59 L.	+1.775174

PROB.

PROB. VIII.

TWO *Ships* set Sail, from the *Port* P, One of them Sails *North-West* 70. 00 deg. 106. 00 *Miles* to O; The other Sails *South-West* 74. 00 deg. 236. 60 Miles, to M. I demand, *Fig.* LXV.

I. How the two *Ships* at O and M, do *Bear* from each other: And

II. Their *Distance* M O.

The Geometrical Construction.

Having drawn your *Meridian* N S, and *Parallel* W E, crossing each other at Right Angles in the point P, which let be the *Port* from whence the two Ships sailed. Then,

First, Upon P, describe an Arch of a Circle, as *n o m* : And because the *Course* from P to O, was *North-West* 70. 00 deg. set 20 deg. (the Complement thereof) from *o* to *m*; and through *m*, draw a Line, as P *m* O, setting thereon 106. 00 *Miles*, from P to O.

Secondly, Because the *Course* from P to M, was *South-west* 74. 00 deg. set 16. 00 deg. (the Complement thereof) from *o* to *n*, And through *n*, draw P *n*, setting thereon the Distance sailed, *viz.* 236. 60 Miles from P to M.

Thirdly, Joyn O and M. So shall you have constituted an *Oblique-Angled Triangle* P O M: In which there is *Given*, (1.) The Angle O P M, 36 00 deg. the *Sum* of the Complements of the *Bearings* of the two Ships. (2.) The two Sides P O 106. 00 Miles and P M 236. 60 Miles. The several *Distances* which the two Ships sailed: And by help of these, you may Find. (1) The *Bearing* of the two *Ships* one from the other: And (2.) Their *Distance*.

By Trigonometrical Calculation.

I. For the *Bearings*, O M P and M O P;
By CASE II. of O. A. P. T.

The *Sum* of the *Sides*, P O and O M, is, 342.06. M.
Their *Difference*. 130.06. M.
The *half Sum* of the Un-known *Angles* M and O, is 72.00. De.
Being thus prepared, say,

As

Fig. LXV.

As the *Sum* of the *Sides* 342. 06	2.534102
Is to the *Difference* of the *Sides* 130. 06	2.113144
So is the Tangent of the ½ Sum of the *Angles* 72 d.	10.448224
	12.561368
To the Tangent of 46. 80 degrees.	10.027266

Which 46. 80 deg. *Added* to 72. 00 deg. (the *half Sum* of the Angles O and M) gives 118. 80 d. for the Angle M O P; And 46. 80 d. subſtracted from 72 00 d. leaves 25. 20 deg. for the leſſer Angle O M P.

Now for the *Bearing*, ſubſtract the *Bearing* of the Ship from P to O, *viz.* 72. 00 deg. from 118. 80 deg. (the quantity of the M O P) there will remain 48. 80 deg. for the Angle M O Q, which is *South-West* 48. 80 d. and ſo doth the Ship at O *Bear* from the Ship at M. And the Ship M, from O, *North-East* 48. 00 deg.

2. For the diſtance of the Ships M and O.

As the Sine of M O P, 118. 80 d. (61. 20)	9.942656
Is to the *Distance* ſailed from P to M. 236. 60 d.	2.374014
So is the Sine of O P M, 36. 00 deg.	9.769218
	12.143232
To the *Distance* O M 158. 70 M.	2.200576

PROB. IX.

Fig. LXVI.

THERE are Two Iſlands, as K and M, which lie directly *North* and *South* of each other, and are diſtant 37. 75 Miles; And there is a third Iſland at L, which is diſtant from that at M, 59. 64 Miles; and from that at K, 80. 92 Miles: Now I would know, How the *Iſland* at L, bears from the other two Iſlands M and K.

The Geometrical Conſtruction

Having drawn N S for the *Meridian*, and W E for a *Parallel* of *Latitude*, croſſing each other in the point K, which let be the place of one of the *Iſlands*; Then

The *Iſland* M, lying directly *North* therefrom, at the Diſtance of 37. 75 M. Set 37. 75 from K to M.—— And becauſe the *Iſland* at L is diſtant from the *Iſland* at K, 80. 92 M, with that diſtance, ſet one foot of the Compaſſes in K, and with the other deſcribe the Arch *a b*.——Alſo, the diſtance of the *Iſland* at L, being

being distant from M, 59. 64 Mil. with that distance set one foot in M, and with the other describe the arch *c d*, crossing the former arch *a b*, in L: ——Lastly, Joyn L M, and L K, so shall you have constituted an *Oblique-Angled Triangle* K L M; in which all the three *Sides* are *Given* And the *Angles* at K and M are Required:

Fig. LXVI.

The *Trigonometrical* Calculation.

As the *Greatest Side* K L; 80. 92 M.	1.908056
Is to the *Sum* of K M and L M, 97. 39 M.	1.988425
So is the *Difference* of those *Sides*, 21. 89 M.	1.340246
	3.328671
To the Part of the *Greater Side* L O, 26. 34.	1.420615

This L O, 26. 34, substracted from K L 80. 92 leaves 54. 58 for the other part of the longest side K O; and the half thereof, 27. 29 is the part K P or P O. And now, the *Oblique-Angled Triangle* K L M, is Reduced into two *Right-angled Triangles* M P K, and M P L; in both which the *Hypotenuses* and *Bases* are given, by which the other *Angles* may be found, By CASE III. of R. A. P. T. According to which, you shall find the *Angle* M K P. to be 43. 71 de. and the *Angle* M L K 25. 94 deg.—So that L Bears from K *North-East* 43. 71 deg.——And M from *North-East* 69. 65 deg.

PROB. X.

There are two *Head-Lands* at B and C, which are distant 356.00 *Miles*; And there is a *Port* at A, from whence two *Ships* sail to those two *Head-Lands*, the Ship from A to C, sails *South-East* 34. 90 deg. and that from A to B, *South-West* 72. 10 de. And both the Ships together have sailed 424. 12 Miles. I demand the *Distance* that each *Ship sailed*; Or, how far are either of the *Head-Lands* from the *Port* at A?

Fig. LXVII.

The Geometrical Construction.

ONE Ship sailing *South-East* 34. 90 deg. and the other *South-West* 72. 10 deg. the distance of those *Rumbs* is 107. 00 de. Wherefore,

First,

Fig. LXVII. First, Draw two Lines at pleasure, as A B and A C, making the *Angle* at A, equal to 107. 00 de.

Secondly, Take 356. 00 Miles, in your Compasses, and enter them between the two Lines at pleasure, as at B and C: constituting the *Triangle* A B C.

Thirdly, Continue the Side B A to D, making A D equal to A C, and joyn C D, constituting another Oblique-angled Triangle D C A, in which there is given. (1) The Angle at C A B, 73.00 deg. (the Complement of B A C to 180 deg.) And (2) The Angles A C D and A D C, each of them equal to half the Angle B A C, *viz.* to 53.50 deg. Then in the Oblique-Angled Triangle D C D, say,

As the given Side B C, 356.00 M.	2.551450
Is to the Sine of B D C, 53.50 de.	9.905178
So is the Side B D, 424.12 M.	2.627488
To the Sine of 73.27 deg.	12.532666
	9.981216

Which is the quantity of the Angle D C B: From which substract the Angle A C D, 53.50 de. there will remain 19.77 d. for the Angle A C B. Then in the Triangle B A C,

As the Sine B A C, 107.00 d. (73.00)	9.980619
Is to the Side B C, 356.00.	2.551450
So is the Sine of A C B, 19.77 de.	9.529231
	12.080681
To B A, 125.90 M.	2.100062

Which substracted from B D, 424.12, there remains 298 M, for the Side A C. the Distance that the other Ship sailed. So that the Port A, is distant from the Head-Land C 298.22 M. And from that at B, 125.90.

PROB. XI.

Fig. LXVIII. Two Ships set sail from two Head-lands at B and C, distant from each other 356. 00 Miles; and meeting together at A, they observe that the two *Head lands* bear so from their *Ship*, that they make an *Angle* of 107. 00 deg. Also, the way that both the Ships have made from B and C to A, doth amount to 424. 12 Miles: I demand how much of this each Ship sailed.

The

Fig. LXVIII.

The Geometrical Construction.

First, DRAW a Right-line at pleasure, as B A D, and upon any point thereof, as A, protract an Angle of 107. 00 deg.

Secondly, The Distance of the two *Head-lands* being 356. 00 Miles enter them between the two sides, at B and C, and joyn B C, making the Triangle B A C.

Thirdly, Make A D equal to A C, and joyn C D, so have you constituted another Triangle A C D.

Now in the Triangle A B C, is Given. (1) The Side B C, 356 00. (2) The Angle B A C 107. 00 deg: And in the Triangle A C D, the Angle B A C being 107. 00 de. the Angle D A C must be 73. 00 de. And the Sides A C and A D, being equal (by construction) the Angles A D C and A C D, must be also equal, viz. (each of them) 53. 50 deg. And being thus far prepared, you may proceed to find the Angle A C B.

By Trigonometrical Calculation.

As the Side B C, 356. 00	2.551449
To the Sine of the Angle A D C, 53. 50 d.	9.905178
So is the whole Side B A D, 424. 12 M.	2.627006
	12.532178
To a fourth Sine, 73 06 d.	9.980729

This 73. 06 d. should be the quantity of the Angle D C B, but (by the construction) the Angle you see is Obtuse; and therefore the Complement of 73. 06 deg. to 180, *viz.* 106. 94 deg. is the quantity of the whole Angle D C B. From which if you substract the Angle A C D, 53. 50 deg. there will remain 53. 44 deg. for the Angle A C B.

Then

As the Sine of B A C, 107. (73. 00d.)	9.980619
Is to B C 356. 00.	2.551449
So is Sine A C B, 53. 44 deg.	9.904842
	12.456291
To the Side B A, 299. 00 ———	2.475672

And so much did the Ship from B, sail to A, and that substracted from 424 12 there remains 175. 12, and so much did the Ship from C to A Sail.

II. *Of* MERCATOR's Sailing, *by the True* Sea-Chart.

ALL the foregoing *Problems* are performed by that kind of *Navigation*, commonly called *Plain Sailing*, or *Sailing* by the *Plain Sea-Chart*; in which *Chart*, the *Degrees* of *Longitude* and *Latitude*, in all Places, are supposed to be equal; which is *Erroneous*, though most practised: But there are two other ways of *Sailing*, both more exact than the former, the one called *Mercator's*, the other, *Sailing* by the *Middle Latitude*.

That of *Mercator's* requires that the *Degrees* of *Latitude* in that *Chart* be *inlarged* as they go farther from the *Æquinoctial* towards either of the *Poles*, which is done by *Reducing* them into *Meridional Parts*: but the *Degrees* of *Longitudes* into *Miles* and *Centesms*, as in *Plain Sailing*: And for the *Reducing* of *Degrees* and *Minutes* of *Latitude* into *Meridional Parts*, I have here inserted a *Table* for the ready performance thereof, to every *Degree* and *Quarter*, or 25 *Centesms* of a *Degree*.

A

75
011
082
153
226
300
374
449
526
603
682
762
843
925
3009
3094
3181
3269
3359
3451
3545
3641
3739
3839
3940
4047
4155
4266
4380
4498
4619
4744

II.

S

A

Plain
titu
thou
both
other
Th
Char
either
nal P
as in
of L
the
25. C

A Table of Meridional Parts.

Degrees	Centesines. 0	25	50	75	Degrees	Centesines. 0	25	50	75
0	0	15	30	45	31	1960	1976	1993	2011
1	60	75	90	105	32	2028	2046	2064	2082
2	120	135	150	165	33	2100	2117	2135	2153
3	180	195	210	225	34	2171	2190	2208	2226
4	240	255	270	285	35	2244	2263	2281	2300
5	300	315	330	346	36	2318	2337	2355	2374
6	361	376	391	406	37	2393	2411	2430	2449
7	421	436	451	466	38	2468	2487	2507	2526
8	482	497	512	527	39	2549	2564	2584	2603
9	542	557	573	588	40	2623	2642	2662	2682
10	603	618	634	649	41	2702	2722	2742	2762
11	664	679	695	710	42	2782	2802	2822	2843
12	725	741	756	771	43	2863	2884	2904	2925
13	787	802	818	833	44	2946	2967	2988	3009
14	848	864	879	895	45	3030	3051	3073	3094
15	910	926	942	957	46	3116	3137	3159	3181
16	973	988	1004	1020	47	3203	3225	3247	3269
17	1035	1051	1067	1082	48	3292	3314	3337	3359
18	1098	1114	1130	1146	49	3382	3405	3428	3451
19	1161	1177	1193	1209	50	3475	3498	3522	3545
20	1225	1241	1257	1273	51	3567	3593	3617	3641
21	1289	1305	1321	1338	52	3665	3690	3714	3739
22	1354	1370	1386	1402	53	3764	3789	3814	3839
23	1419	1435	1451	1468	54	3865	3890	3916	3940
24	1484	1509	1517	1533	55	3968	3994	4021	4047
25	1550	1567	1583	1600	56	4074	4101	4128	4155
26	1616	1633	1650	1667	57	4183	4211	4238	4266
27	1684	1700	1717	1734	58	4295	4323	4352	4380
28	1751	1768	1785	1802	59	4409	4439	4468	4498
29	1819	1837	1854	1871	60	4528	4558	4588	4619
30	1888	1906	1923	1941	61	4650	4681	4712	4744

A Table of Meridional Parts.

Degrees	Centesmes. 0	25	50	75
62	4715	4807	4840	4872
63	4905	4939	4972	5006
64	5040	5074	5109	5144
65	5179	5215	5251	5287
66	5324	5361	5399	5436
67	5475	5513	5552	5592
68	5631	5672	5712	5754
69	5795	5837	5880	5923
70	5967	6011	6056	6101
71	6147	6193	6240	6288
72	6336	6381	6432	6484
73	6535	6587	6640	6693
74	6747	6802	6857	6914
75	6972	7030	7089	7150
76	7211	7274	7338	7403
77	7469	7536	7605	7675
78	7746	7819	7894	7970
79	8048	8127	8209	8292
80	8377	8465	8555	8647
81	8742	8839	8939	9042
82	9148	9258	9371	9488
83	9609	9735	9865	10000
84	10141	10288	10441	10601
85	10770	10946	11133	11330
86	11539	11761	11399	12255
87	12521	12821	14015	13513
88	13920	14381	14914	15545
89	16318	17316	18729	21170
90				

The

The Use of the Table of *Meridional Parts.*

What *Meridional Parts* do answer to 26 deg. of *Latitude.*

Latitude	26. 00 Deg —	1616 Merid. Parts.
	24. 25 Deg —	1509 M. P.
	47. 50 Deg —	3247 M. P.
	63. 00 Deg —	4905 M. P. &c.

2. What *Degrees* and *Centesms* do answer to.

679	Merid. Parts.	Answer	11. 25 Deg.
3522	M. P.		50. 50 Deg.
9488	M. P.		83. 75 Deg.

For the Resolving of the following *Problems*, which (in kind) will be the same with those in *Plain Sailing* as to the *Trigonometrical Work.* But in the performance of them, I will set the places proposed down upon a *Sea-Chart*, made according to the *Projection* of *Mercator*, where the Degrees of *Latitude* are *enlarged* according as they tend nearer and nearer to the *Pole* : And on the out-side of such a *Chart*, I have described a *Plain Sea-Chart* also, where the *Meridians* and *Parallels* of *Latitude* are every where of an equal distance; and so wrought the *Questions* according to both *Charts*, by which the *difference* will more plainly appear: And the *Chart* which I have here made to work the following *Examples upon*, begins (at the bottom of it) at about 49.50 deg. of *Latitude*, and extends upwards to 55. 50 deg. of *Latitude* ; and at the *Top* and *Bottom* to *Six* Degrees *Difference* of *Longitude* : But the *Parallels* of *Latitude* in the *Plain Chart*, (which are distinguished by *Pricked Lines*) extend to above 59 deg. of *Latitude* within the same Bounds.

Fig. LXIX.

PROB. I.

The *Latitudes* of two *Places*, and their *difference of Longitude* being known ; To find, (1) The *Rumb* leading from one to the other : And (2) Their *Distance* upon that *Rumb* : according to *Mercators Chart.*

By *Trigonometrical Calculation*

LET one Place be at A in the *Latitude* of 50 deg. and the other at C, in the *Latitude of* 55 deg. but differing in *Longitude* Eastward from B, 6. 50 deg. which two places thus set down in the

Fig. LXIX. the *Chart*, draw the Line A C, which is the *Rumb* (or *Course*) from A to C: And thus have you upon your *Chart* constituted a *Right-angled Plain Triangle* A B C, in which you have given, (1) B C, the *Difference of Longitude* 6.50 deg. which you must *Reduce* into *Miles* (or *Minutes*) of *Longitude* by multiplying them by 60, (as in *Plain Sailing*) and they make 330. 00 Miles.—— (2) A B, the *Difference of Latitude* 5. 00 deg. which must be Reduced into *Meridional Parts* thus:

The *Meridional Parts* for 50. 00 d. are	3475
For 55. 00	3968
Their *Difference*——	493

Which are the *Meridional Parts* answering to the 5. 00 deg. *Difference* of *Latitude* of the two *Places* A and C: And by these you may find;

(1.) The *Rumb* B A C.

(2) The *Meridional Distance* upon the *Rumb* A C.

By Trigonometrical Calculation.

(1.) For the *Rumb* B A C

As the Merid. Dif. of *Latitude* A B, 493 M. P.	2.692847
Is to the Dif. of *Longitude* B C 330 Miles	12.518514
So is the Radius, Tangent 45 deg.	10.
To the Tangent of B A C, 33. 80 degrees.	9.825667

Which is the *Rumb*, leading from A to C: whose Complement 56. 20 deg. is the Angle B C A, or the *Complement* of the *Rumb*:

(2.) For the *Meridional Distance* on the *Rumb* A C.

As the Co-sine of the *Rumb*: B C A, 56. 20 d.	9.919592
Is to the *Merid. Differ.* of *Latitude* A B 493 M. P.	12.692847
So is Radius Sine 90 deg.	10.
To the *Merid. Distance* A C, 593.27 M. P.	2.773255

Now to find the *Meridional Degrees* answerable to these *Meridional Parts*, you must

First, Substract the M P O of A B, from the M P of A C, their *Differenee* is 100. 27 M. P. the half whereof is 50.13 M. P.

Secondly, *Add* this *half Difference* 50. 13, to the *Meridional Parts* of the *Greater Latitude* 55. 00 deg. *viz.* 3968, and it makes 4018. 13, which are the *Merid. Parts* answering to 55.50 deg. of *Latitude.*—— Also, *substract* this *half Difference* 50. 13, from the *Merid. Parts* of the *Lesser Latitude* 50. 00 deg. *viz.* 3475, and the Remainder

Remainder will be 3424. 87, which are the *Merid. Parts* answering to 49. 38 deg. of *Latitude*. *Fig.* LXIX.

Thirdly, The *Difference* of these two *Latitudes* last found, *viz.* 55. 50 de. and 49. 38 deg. is 6. 12 deg. And that is the *true Distance* upon the *Rumb* between A and C, in *Degrees*.

And according to this Method, may all the other *Problems* (before wrought by the *Plain Sea-Chart*) be performed by *Tri. Calculation*; by the same *Canons*: Only remember, to *Reduce* the *Difference of Latitudes* and *Distance* on the *Rumb*, into *Meridional Parts*, but the *Difference* of *Longitude* into Miles, as in the other.

Now that you may see the *Difference* between these two ways; See the *Figure* of the *Chart*, wherein the two *Places* are laid down by the *Plain Chart*; as in the Triangle AOS, which Triangle being resolved (as is before shewed) you will find

(1) The *Rumb* OAS, to be 47. 73 deg.
Differing from the truth 13. 93 deg.

(2) The *Distance upon the Rumb* AS, to be 7. 43 de.
Differing from the other 1. 31.

And let thus much suffice for *Sailing* according to *Mercator*.

There is a third way of *Sailing*, which differs not much from this way of *Mercator's*, but may be performed without *Reduction* or use of *Meridional Parts*; which is called *Sailing* by the *Middle Latitude*: of which a little.

III. *Of* Sailing *by the* Middle Latitude.

I Shall exemplefie this way of Sailing by *Four* of the most usual *Problems* in *Navigation*: And they shall be the same as in the foregoing: whereby the Difference will the better appear.

PROB. I.

The *Longitude* and *Latitude* of two Places, A and C, Given to find,

(1) The *Rumb* BAC.
(2) The *Distance* upon the *Rumb*: AC.

LET

LET one Place be in 50. 00 de. the other in 55. 00 de. of *North Latitude*: And 6. 50 deg. *difference in Longitude*: Then the *Middle Latitude* between 50. 00 deg. and 55.00 deg. is 52. 50 de. And its Complement 37. 50 deg. Then

(1) For the *Rumb*: The Proportion is,

As the *Difference of Lat.* 300. 00 Miles.	2.477121
Is to the *Dif.* of *Longit.* 330. 00 Miles.	2.518524
So is the Co-sine of *Middle Latitude*, 37. 50 de.	9.784447
	12.302971
To the *Tangent* of 33. 81 deg. Which is the *Rumb* B A C.	9.825850

(2) For the *Distance* upon the *Rumb*. The Proportion is,

As the Co-sine of the *Rumb* 56.19 de.	9.919542
To Radius Sine 90 de.	10.
So is the *Difference* of *Latitude*, 300 M.	12.477121
To the *Distance* 361.06 M.	2.557579

PROB. II.

THE *Latitudes of two Places*, A and C, and the *Rumb*, B A C, (or *Course*) between them *Given*: to *Find*,

(1) Their *Distance* A C.

(2) Their *Difference of Longitude*, B C.

For the *Distance* it may be found as in the former *Problem*. But

(2) For their *Difference of Longitude*, This is The Proportion.

As the Co-sine of the *Middle Latitude*, 37.50 de.	9.784447
Is to the *Difference* of *Latitude* 300 M.	2.477121
So is the Tang. of the *Rumb*, 33.81 de.	9.825876
	12.302997
To the *Difference of Longitude* B C, 330.01 M.	2.518550

PROB.

PROB. III.

THE *Distance* A C, 361.06 M. And both the *Latitudes* Given, To find the *Rumb*.

The Proportion.

As the Distance A P, 361.06	2.557579
To the *Difference of Latitude* 300 M.	12.477121
So is *Radius* 90 deg.	10.
To the Sine of 56.19 deg.	9.919542

Whose Complement 33.81 deg. is the *Rumb* B A C. from the *Meridian*.

PROB. IV.

HAving one *Latitude*, the *Rumb*, and *Distance* upon the *Rumb*, to find the *Difference* of *Latitude*.

The Proportion.

As the Radius Sine 90 deg.	10.
To Co-sine of the *Rumb*, 56.19 deg.	9.919542
So is the *Distance* upon the *Rumb*, 361.06	2.557579
To the *Difference of Latitude*, 300 M.	12.477121

Which Reduced into Degrees, is 5.00 deg.

PROB. V.

HAving the *Rumb* and both *Latitudes* given, to find the *Difference of Longitude*.

The Proportion.

As the Co-sine of the *Mid. Latitude*, 37.50 de.	9.784447
To the *Difference* of *Latitude*, 300.00 M.	2.477121
So is the Tangent of the *Rumb*, 33.81 deg.	9.825876
	12.302997
To the Tangent of 300.03 M.	2.518550

Which is 5 Deg.

The Differences in these Three Ways of Sailing :- In this Example.

	Plain Sail.	Mercator.	Mid. Latit.
Differ. of Long.	330.00 M.	330.00 M.	330.00 M.
Differ. of Latit.	300.00 M.	300.00 M.	300.00 M.
Rumb.	47.73 De.	33.80 Deg.	33.81 Deg.
Compl. Rumb.	42.27 De.	56.20 Deg.	56.19 Deg.
Distance.	445.97 M. Or 6.50 Deg.	593.27 M.P. Or 6.12 Deg.	361.06 M. Or 6.01 Deg.

JANUA

JANUA MATHEMATICA

VEL,

Trigonometria Practica.

SECTION VII.

OF

Theorical ASTRONOMY.

IN the Fourth *Section* of this Third PART, you have the *Doctrine of Triangles* applyed to *Practice* in that part of *Astronomy* which concerns the *Diurnal Motion* of the *Sun* and *Fixed Stars*: Now, in this Section, I shall discourse something of the other Stars or *Planets*, which have a *Secondary Motion* of their own; shewing also, how subservient *Trigonometry* is for the *Calculating* of their true *Places* at any time, as also in computing of their *Magnitudes*, *Distances*, &c. And in order thereunto I shall first give you a brief description of the Planetary *System*.

CHAP. I.

Of the Planetary System.

THIS *System* may be sufficiently represented as in the *Figure* thereof: Wherein,

First, The *Sun* being in the *Centre*, hath only a Rotation from *West* to *East*, upon its own *Axis*, in the space of 26 days, or

 much

much thereabouts: His *Centre* being an immoveable *Point*; to which the *Revolutions* of the other *Planets* are referred.

Secondly, The *Planets* are moved about the *Sun*, from *West* to *East*, in several *Orbs*, which return into themselves, every *Planet* in *Time* proportionable to the *Magnitude* of his *Orb* and *Distance* from the *Sun*; the Motions of the *Primary Planets*, ♄, ♃, ♂, ⊕, ♀, and ☿, being *uniform*, perpetually *constant* and *regular*.

The Figure of the System.

Fig. LXX. The Circle ☿ BCD ☿, denotes the *Way* and *Revolution* of the Planet *Mercury*, about the *Sun*, in 88 Days.——The Circle ♀ E F G ♀, the *Revolution* of *Venus* in 225 Days.——The Circle ⊕ H I K ⊕, the *Revolution* of the *Earth*, with the *Moon* in one Year—The Circle ♂ L M N ♂, the *Revolution* of *Mars*——The Circle ♃ O P Q ♃, the *Revolution* of *Jupiter*, with his four *Companions* (or *Satellities*) in twelve Years——The Circle ♄ R S T ♄, the *Revolution* of *Saturn*, with his *Ring* and *Moon*, in Thirty Years.——The *Moon* also moves round the *Earth*, every Month——*Jupiters* four *Companions*, move about him according to their *Distances* from him.——The First (and nearest to him) in *One Day* and 18 *Hours*——The Second, in *Three Days* and 13 *Hours*——The Third, in *Seven Days* and 4 *Hours*.——The Fourth, in 26 *Days* and 5 *Hours*——*Saturns Moon* moveth about him in *Sixteen Days*: And all of them from *West* to *East*, according to their *Planets Revolution* about the *Sun*.

These Planets, whose *Revolutions* respect the *Sun* only, as *Saturn Jupiter*, *Mars*, the *Earth*, *Venus* and *Mercury*, are called *Primary Planets*: The other that move about *Saturn*, *Jupiter* and the *Earth*, *Secondary Planets*.

The *Secondary Planets*, are all of them much *Less* in *Magnitude* than their *Primary*; and all the *Planets* together, much *Less* than the *Sun*, from whom they receive their *Light*, *Motion*, &c.

CHAP.

Fig. LXX.

CHAP. II.

Of the Theory of the Sun *and the other* Primary Planets.

THAT the Motion of the *Primary Planets* about the Sun were *Elliptical*, was first discovered by the learned *Kepler*, which he deduced from the acurate *Observations* of the Noble *Dane Tychobra*. And therefore, in such an *Elliptical Figure*, may be described all such *Points*, *Lines*, *Arches*, &c. as are requisite to be known, and to inform the *Fancy* in the *Trigonometrical Calculation* of the *Planets Places*. But before I proceed so far, I must shew,

How to describe an *Ellipsis*.

Concerning the *Definition* of this *Figure*, only that it is one of the *Sections* of Cone, and the many ways that it may be Artificially *described*, I shall say nothing, referring the Reader to *Midorges*, and others that have largely written of the *Sections* of a *Cone*. But to *Describe* such a *Figure Mechanically*. Thus:

About any Two *Right-Lines* Given, for the Two *Diameters* of the *Ellipsis*, to describe such an *Ellipsis*.

Fig. LXXI.

Let the Line P Y be the *Longer*, and R M the *Shorter Diameter* given: Let them cross one another at *Right-angles* in B. This done, Take half of the *Longest Diameter* B P or B Y, in your Compasses, and setting one foot in R or M, the other will reach upon the *Longest Diameter* to the *Points* Z and S, which will be the *Centres* upon which the *Ellipsis* P R Y M must be described. Wherefore,

In the two Points Z and S, Fix two *Pins*, *Nails*, or the like; and about them put a *String* so long, that being doubled it may reach from the *Pin* at S, to the end of the Diameter at P, or from the *Pin* at Z to the end Y; so then the whole length of the String, will be twice as long as the Distance S P or Z Y; which *String* at that length, joyn at both ends: Then, putting this *String* over the two *Pins*, at Z and S, with a *Third Pin* (which may be a Black-Lead Pencil, or such like) move the *String* about upon the two Fixed *Pins* Z and S; and it will by its *Motion* describe an *Ellipsis*, such as is the *Figure* P C R E Y F M D P.

CHAP.

Fig. LXXI.

CHAP. III.

Of the Motion *of the* Planets *in the* Ellipsis.

IN this *Elliptical Figure*, the Line P Y is called the *Transverse*, (or *Longer*,) And R M, the *Conjugate* (or *Shorter*) *Diameter* of the *Ellipsis*.—S the *Lower*, and Z the *Upper Focus* of the *Ellipsis*; upon which two Points the *Elliptical Figure* was described— And the *Elliptical Figure* it self P C R E Y F M D P be the *Orbit* of the *Earth*, or any other of the *Primary Planets*:— S the Place of the *Sun*; (the common *Node*, and *Centre* of the *Planitary Orbs*, to which the *true Motions* of the *Planets* are referred,) in the *Lower Focus* of the *Ellipsis*.— Now, a Planet moving in the *Ellipsis*; when it shall be in P, it is in *Aphelion*, or at its *Greatest Distance* from the Sun— When in Y *Peribelion*, or at its *Nearest Distance* to the *Sun*— About the upper *Foci* of the *Ellipsis* Z, the *mean Motion* of the *Planet* is regulated: B is the *common Centre* of the *Figure*: B Z or B S, the *Excentricity*: And S Z the double *Excentricity*.

The *Figure* thus described, the next thing to be known is, after what manner the true Place of a Planet may be determined therein.

First, The mean *Anomaly* of a *Planet*, is its *equal Distance* from P, the *Aphelion Point* of that *Planet*: And this *Anomaly* is to be accounted from P, to P again. And here it is to be observed, that a *Planet* moving in the *Elliptical Arch* P R Y M, upon the *Focus* Z, is as long time tracing the lesser *Arch* C P D, as he is the greater C Y M; the Reason is, for that (the *mean Motion* being made upon Z) the *Planet* describeth *equal Angles* in *equal* Intervals of *Time*: And it is also evident, That if the *Motion* be *equal* upon one *Focus* Z, of the *Ellipsis*, it must be *unequal* upon the other S, in which the Sun is seated.

For instance: Suppose the Planet to be in C, which is 90 deg. from the *Aphelion Point* P, then will the *Angle* of its *mean Anomaly* be P Z C; but the *Co-equated Anomaly* will be the *Angle* P B C: And the *Angle* at the *Sun* P S C, which is the *Angle* to be found.

CHAP.

CHAP. IV.

to find the Angle *that a* Planet (*in any part of the* Ellipsis) *makes with the* Sun; *and also, the* Planets *true* Place *in the* Ellipsis.

Fig. LXXII.

...t S be the Sun, X the Centre of the Planets *Mean Motion*; the Points H, A, T, several Places of the *Planet* in its *Orb*, the *Aphelion* point P, then will the several *Angles* P X H, A, P X T, represent the *Anomaly* of the *Planet*: And the ...s P S N, and P S M; the *Angles* at the *Sun*. Now these *An*...ay be found as followeth.

...ppose the *Mean Anomaly* of the *Earth* being at H, to be 30 deg. represented by the Angle P X H, whose Complement B X H, ...o deg. By help whereof, with the *Common Radius*, and *Ex*...*city*, the true *Place* of the *Planet* may be found by *Trigono*...*al Calculation*. For,

...ft, In the Triangle H X B, there is Given, (1) The Angle B, 150.00 d. the Complement of the *Planets Mean Anomaly* ...o deg. (2) The Side B H, the *Common Radius* 100000. And ...he Side B X, 1685, equal to half X S, the *Excentricity*. By ...t you may find the Angle B H X, By CASE I. of O. A. Then,

...s the *Common Radius*, B H, 100000 Parts	5.000000
X B (half the *Excentricity*) 1685 P.	3.226599
the Sine of H X B, the Complement of the ...*maly* H Z P, 30 d. to 180 deg.	9.698970
	12.925569
the Sine of B H X 0.483 deg.	7.925569

...t therefore the Angle X B H is 29.518 deg. the double ...hereof, 59. 036 deg. is the *Anomaly of Variation*.

Then say, Secondly,

...s the *Radius*, Sine 90 d.	10
...) the Sine of the *Greatest Variation* (which is) 0.24 deg.	6.594172
...he *Variation* last found, 59.034 d.	9.933210
the Sine of 0. 019 deg.	16.527382

...hich is the *Variation* required.

Now,

Fig. LXXII. Now, forasmuch as the *Planets* are moved in *Ellipses*, and not in perfect *Circles*, the next thing, therefore, to be enquired is, the *Planets* Place in the *Ellipsis*: And to that purpose, about Z, let there be described a little Circle K O N, upon which, from K to N, I number the *Anomaly* of the *Epicicle* 59.034 deg. before found, and then will the Angle P B Z, be equal to the *corrected Anomaly*; and so is N the place of the *Planet* in the *Ellipsis*; and the Angle Z B N, the *Æquotion* of the *Epicicle*; or (which is the same) the *Difference* between the *Place* of a *Planet* in the *Circle*, and in the *Ellipsis*: And may be found *Trigonometrically*, thus. For,

Thirdly, In the Triangle B Z N, you have given, (1) The Side B Z, the *common Radius* 100000 P. (2.) The Side Z N, 00008 P, the *Semidiameter* of the *Epicicle*. (3) The Angle B Z N 120.98 Deg. by which you may find, the Angles B N Z and Z B N, (by CASE II. of O. A. R. T.) Thus,

The Angle N Z B being 120.98 deg. the Complement to 180 deg. is 59.02 deg. the half whereof is 29.51 deg. The Side B Z is 100000 P. the Side Z N 00008 P, their *Sum* 100008 P, their *Difference* 99992. Then,

(3) As the *Sum* of the Sides B Z and Z N, 100008	5.000035
Is to their *Difference* 99992	4.999965
So is the Tang. of half Z B N and Z N B 29.505 d.	9.752818
	14.752783
To the Tang. of the *half Difference* 29.502 d.	9.752748

This *half Difference*, *added* to the *half Sum*, gives 59.007 deg. for the *Angle* Z N B: And *subtracted* from it, leaves 0.003 deg. for the *Angle* N B Z.

Then again say,

As the Sine of B N Z, 59.007	9.993121
Is to B Z 100000 P.	5.000000
So is the Sine of B Z N, 59.02 d.	9.933140
	15.933143
To the Side B N 100004 P.	5.000022

From the Angle B P Z 29.502
Subtract N B Z 005
There remains P B N 29.497 Whose Complement to 180 deg. is 150.503 deg. and is the Angle S B N.

Fourthly,

Fourthly, In the Triangle B S N, there is given, (1) B N 100004. (2) B S 1685. (3) The Angle contained by them, S B N 150.50 deg. *Fig.* LXXII.

Then,

As the Sum of B N and B S, 101689.	5.007278
Is to their Difference 98319	4.992637
So is the Tangent of half B S N and B N S 14.77d.	9.421029
	14.413666
To the Tangent of 14.30 d. the half difference.	9.406388

Which added and substracted from the half Sun, gives 29.07 deg. for B S N the *Angle at the Sun.*

And 0.47 deg. for the *Opt. Æquation* B N S.

Then from the Mean *Anomaly* P X H	30.00 deg.
Substract the Angle at the *Sun* B S N,	29.07
There remains the *Prostapherisis.*	0.93

Again,

As the Sine of B S N, 29.04 de.	9.686117
Is to B N, 100004.	5.000021
So is the Sine of B S N, 150.50 deg. (or 29.50.)	9.692338
	14.692359
To S N, 101564.	5.006232

Note that when a *Planet* is in any of the *Points* P, A, or Y, there is no *Variation*, but being out of any of these *Points*, it hath *Variation* more or less, and is at its *greatest* when the *Elongation* is 45 deg. from any of those *Points.*——And note also, That this *Variation augments* the *Places* of all the *Planets* (*Venus* and the *Moon* excepted) in the *First* and *Third Quadrants* of the *Orbit*; But in the *Second* and *Fourth*; *it diminishes* the Planets place; according to the quantity of the *Angle* of *Variation.* And as this *Example* was wrought, the *Earth* or other *Planet* being supposed in the *First Quarter* of the *Ellipsis*, the same is to be understood and performed, if the *Planet* were in the *Second*, *Third* or *Fourth Quadrant* thereof, as is plain by the *Figure.*

Fig. LXXII.

CHAP. V.

Of the Two-fold Inequality *of the other* Primary Planets.

THE *Sun* seems to be moved under the Plain of the *Ecliptick*; But it is the *Earth* that performeth this *Motion* about the *Sun*, who is seated in the *Focis* of the *Earths Ellipsis*: So that, as the *Earth* is moved under the *Ecliptick Orb*, so much the *Sun* appears to move on the contrary part thereof: And from hence may be concluded, that the *Sun* (or *Earth*) can be subject to no other *Inequality*, than what is produced by their own *Simple Motion* in the *Ellipsis*: And this is true also, in all the *Primary Planets* as in respect to the *Sun*; who is seen, from them, to change his *Place*, according to the quantity of their Motions.

Now, For as much as the *Earth* is far distant from the *common Centre* of their *Orbs* (which is the *Sun*) therefore it is, that (by their different *Motions*, and various positions of their *Orbs*) they seem to us to be subject to *a second Inequality*; which is not essential in their *Motions*, but accidental only; being caused through the *great distance* of the Point to which their *Places* are referred: So that the *Planets* seem to us to have a different species of *Motion* from the *Natural Motion* in their respective *Orbs*: they appearing at one time *Direct*, at other times *Retrograde*, and sometimes not to move at all, but appear as *Stationary*: To be nearer to the *Earth* at one time than at another; and appearing to the Eye of a *Greater* and *Lesser Magnitude*: The cause of all which various *Passions* will manifestly appear from the *Calculation* of their *Places*, which shall be the Work of the following *Chapter*.

CHAP. VI.

To Calculate the true Place of a Planet Trigonometrically.

RETaining the same Method deliver'd in the last Chapter, I shall give an Example in the finding out of the place of the Planet *Mars*. The Sun being in 5.85. deg. of *Scorpio*, and the mean Anomaly of *Mars* from his Aphelion point 55.47 deg.

The

The rest, as in the following Table. And from these things given, the true place of the Planet *Mars* may be found. *Fig.* LXXII.

A Table of the Semidiameters *of the* Orbits *and* Epicicles; *the* Excentricities; *the Greatest* Variations *and* Inclinations *of the* Planets, *in order to the* Trigonometrical Calculation *of their* Places.

	♄	♃	♂	⊖	♀	☿
Semidiam. of the Circle P A Y.	952500	521300	152040	100000	72405	38192
Semidiam. of Epicicle Z N.	788	299	327	8	1	429
Excentricity B X = B S.	54800	24960	14115	1788	530	8100
The Greatest Variation of the Ano	0.09 d	0.05 d	0.25 d	0.02 d	0.02 d	1.30 d
Angle of the greatest Inclin. H B V.	2.51 d	1.36 d	1.85 d	0.00 d	3.38 d	6.90 d

The Trigonometrical Calculation.

IN the Figure the Semicircle P H Y, is half the Orbit of *Mars*, P his *Aphelion* point, H his place in his Orbit, B the Centre, B X the Excentricity: H B the Semidiameter of the Orbit. *Fig.* LXXII.

I. Then, In the Triangle X H B, there is given, (1) The Side H B, the Semidiameter of the Orbit, 15204. (2) The Side X B, the Excentricity 1411. (3) The Angle H X B the Complement of the mean Anomaly P H: 55.47 to 180 deg. viz. 124.53 deg.

By which you may find the Angles B H X, and X B H. (By CASE II. of O. A. P. T.

As the Sum of the Sides H B and X B 16615.	4.2205003
Is to their Difference 13793.	4.1396587
So is the Tang. of half X H B and X B H 27.73 d.	9.7207213
	13.8603800
To the Tangent of 23.50 d.	9.6398797

Which substracted from 27.73 (the half Sum) leaves 4.15 de. for the Angle at H: And added to 27.73 d. gives 51.31 deg. for the Angle X B H. And that doubled, makes 102.62 d. Which is the Anomaly of Variation.

Which found; say,

Fig. LXXII. As Radius — 10.

To the Sine of the greatest Variation of *Mars*, 0.25 d. 7.639816
So is the Sine of Variat. last found 102.62 (77.38) 9.989379
To the Sine of H B Z, 0.244 deg. the Variat. 7.629195

To the Angle X B H — 51.32 deg.
Add the Angle H B Z — 0.244 the Variation.
The Sum is X B Z — 51.564 deg.
The double whereof is — 103.128 And is the *Motion* of the *Epicicle* a Z N.

Describe the *Epicicle*, and upon it set the *Motion* of the *Epicicle* 103.128 deg. from a to N, so will the Angle a O N be equal to the corrected *Anomaly* P B Z; so that N is the Place of the Planet *Mars* in the *Ellipsis*: And the Angle Z B N the *Equation* of the *Epicicle*; which is the difference between the place of a Planet in the *Circle*, and in his *Ellipsis*; which may be thus found. For,

In the Triangle B Z N there is given, (1) The Side Z B, the Semidiameter of the Circle 100000, (2) The Side Z N, the Semidiameter of the Epicicle 33. (3) The Angle B Z N, the Complement of the Motion of the Epicicle N Z a (103.64) *viz.* 77.36 de.

And by these may be found, the Angle N B Z.

Thus:

As the Sum of the Sides Z B and Z N, 100008 — 5.00008
Is to their Difference 99992 — 4.99996
So is the Tang. of half Z N B and Z B N, 51.32 de. — 10.09659
15.09655
To the Tangent of 51.31 d. — 10.09647

Which added to 51.32, gives 102.63 de. for the Angle Z N B; and substracted therefrom, leaves 0.01 deg. for the Angle N B Z.

Then for the Side N B.

As the Sine of Z N B 102.63 de. (77.37) — 9.989362
Is to the Sine of N Z B 77.36 deg. — 9.989345
So is the Side Z B, 100000 — 5.000000
14.989345
To the Side B N, 99996 — 4.999983

From

Fig. LXXII.

From the Angle P B Z — 51.32 deg.
Substract N B Z — 0.244

There remains the Angle X B N — 51.076 whose Complement to 180 deg. is — 128.924 for the Angle N B S.

And then,

In the Triangle B N S, there is Given, (1) The Side B N 99996. (2) The Side B S, 1411, And (3) The included Angle N B S 128.92 deg. From whence may be found the Angle at the Sun B S N; and the Side S N.

Thus,

As the Sum of the Sides N B and S B, 101407.	5.006082
Is to the Difference of those Sides 98585.	4.993811
So Tang. of half the Angles B S N and B N S, 25.54 d.	9.679276
	14.673087
To the Tangent of 24.92 deg.	9.667005

Which added to 25.54 de. gives, 51.08 deg for the Angle B S N; and substracted, leaves 0.62 de. for the Angle B N S.

Then, For the Side N S,

As the Sine of S N B, 0.62 deg.	8.034260
Is to the Side B S, 1411.	3.149219
So is Sine N B S 128, 92 de. (51.08)	9.891115
	13.040334
To the Side S N 101405.	5.006074

Which is the Distance of ♂ from ☉.

Place of the Aphelion.	150.35 deg.
The Angle at the Sun.	51.08.
Heliocentrick Place of *Mars*.	201.43
The North Node Sub.	47.55
The Argument of Latitude.	153.88

CHAP.

CHAP. VII.

To find the Place of a Planet in the Ecliptick, &c.

Fig. LXXIII. FOR that a Planets Place in Longitude in the Ecliptick, differeth something from the Longitude in its own *Orb*, I shall therefore shew (by Trigonometry,) How to find the *Reduction*, and the *Ecliptick Place*, with the *Curtation* and *Parallax* of the *Orb*, and consequently, the *Geocentrick Place* and *Latitude*.

In order whereunto, in the Scheme; Let S represent the *Sun*, H the Place of *Mars* in his *Orbit*; V his place in the *Ecliptick*; E the *Earth*: A the *North Node* of *Mars*; B the *South Node*; (the two Intersections of the *Planets Orb* with the *Ecliptick*; A D B, a *Semicircle* of the *Ecliptick*: D R, the limit of the Planets *Greatest* North *Latitude*. These things premised, I proceed to

The Trigonometrical Calculation.

1. In the Spherical Triangle B H V, Right-angled at V, there is given. (1) The Side B H 26.65 de. (the Complement of the *Argument of Latitude*.)——(2) The Angle H B V, the *Greatest Inclination* of *Mars*, 1.85 de. By which you may find the Side B V: (By CASE I. of R. A. S. T. Thus

As Radius, 90 de.	10.
To Co-sine H B V, 88. 15 de.	9.999773
So is the Tangent of B H, 26.63 de.	9.700577
To the Tangent of B V 26.64 de.	9.700350

The *Difference* between B H and B V, is 0.01 de. is to be substracted from 201.43 d. the place of *Mars*, because the Arch B V is lesser then the Arch B H.

Heliocen. Place of *Mars*, in his Ellipsis		201.43 deg.
Reduction	Subst.	.01
Heliocen: Place Corrected		201.42
Place of the Sun		215.85
Difference		14.43

Which is the Anomaly of Commutation.

2. In the same Triangle, there is given as before, whereby the *Inclination* of the *Orbit* from the *Plain* of the *Ecliptick* H V, may be found (by CASE II. of R. A. S. T. Thus

As

As Radius, 90.00 de.	10.	*Fig.*
Is to the Sine of B H, 26.65 de.	9.651800	LXXIII.
So is the Sine of B H V, 1.85 de.	8.508974	
To the Sine of H V, 0.83 de.	8.160774	

Unto which, the Angle H S V is equal.

3. For the *Curtated Distance* S.V. In the Right-angled Plain Triangle H S V, there is given, (1) The Angle V S H, 0.83 d. (2) The Side S H 15204, the *distance* of *Mars* in his *Orbit* from the *Sun*: By which you may find the Side S V, (By CASE IV. of R. A. S. T.) Thus

As Radius, Sine 90.00 de H V S,	10.
Is to the Side HS, 15204.	4.181958
So is the Sine of V H S, 89.17 d.	9.999954
To the Side S V, 15202.	14.181912

The *Ecliptic Place* of the Planet *Mars*, with his *Inclination* and *Curtation* thus attained: The next thing to be enquired after is, the *Parallax* of the *Earths Orb*; and his *Geocentrick Place*, in *Longitude* and *Latitude*.

4. In order whereunto, in Figure LXXIII. the Circle OXEZ, representing the *Earths Orbit*; Number the quantity of the *Anomaly* of *Commitation* 14.43 de. from X (the opposite Place of *Mars* from the *Sun*) to E, drawing the Line E V; which will constitute an Oblique-angled Plain Triangle S V E: in which there is Given, (1) The Side SV 15202. (2) The Side S E, 10000, (3) The included Angle V S E 165.57 de. (the Complement of the *Angle of Commitation* to 180 deg.) By which you may find (1.) the Angles S V E and V E S. And (2) the Side V E. *Fig.* LXXIII.

As the Sum of V S and S E 25202.	4.401435
Is to their Difference 5202.	3.716170
So is the Tang. of half the Angles at E and V, 8.22 d.	9.159743
	12.875913
To the Tangent of 1.71 deg. their difference.	8.474478

Which added to the half Sum of the Angles at V and E (8.22 d.) gives 9.03 de. for the Angle V E S: And substracted therefrom, leaves 6.51 de. for the Angle S V E: Which is the *Parallax of the Orb*.

Then

Then for the Side V E,

Fig. LXXIII.

As the Sine of V E S 9.93 de.	9.236650
Is to the Sine of V S E 163.56 (16.44)	4.181900
So is the Side V S, 15202	9.451803
	13.633703
To the side V E 24959	4.397053

Then,

To the *Heliocentric* Place of *Mars.*	201.43
Add the *Parallax* of the *Orb* S V E	6.51
The Sum is	207.94

Which is the *Geocentric Long.* of *Mars.*

That is 6 s. 27 deg : Or 27 deg. of ♎.

CHAP VIII.

Of the Proportions of the Semidiameters *of the* Sun, Earth, Moon, *and the other* Planets.

I. Of the *Sun, Earth,* &c. *Moon* ;

BY the best Telescope-Observations.

To the mean distance of the *Earth* from the *Sun*	10.00000
The Semidameter of the *Sun* is of those Parts	46300
The Semidiameter of the *Earth*	727
To the mean distance of the *Moon* from the *Earth*	1.00000
The Semidiameter of the *Earth* is	1650
The Semidiameter of the *Moon*	446

And from hence, at all times, the Distance of the *Luminaries* being first found, their apparent Semidiameters may be obtained : And for the Semidiameter of the *Earths Shadow*, in *Lunar Eclipses* ; I have here inserted the Diagram of *Hyparcus*.

In which Diagram.

Fig. LXXIV.

A denotes the *Centre* of the *Sun.*
A D his *Semidiameter.*
B the *Centre* of the *Earth.*

B D

Fig. LXXV.

B E her *Semidiameter*.
A E D, or A B D, the apparent *Semidiameter* of the *Sun*.
A E H, or B D E, the *Horizontal Parallax*.
C G F equal to H E D, the *Semi-Angle* of the *Cone* of the *Earths Shadow*.
B C and B F, being equal to the *Distance* of the *Moon* from the *Earth*.
B T E, her *Horizontal Parallax*.
C B F, the apparent *Semidiameter* of the *Earths Shadow*.

From hence;

1. The Semidiameter of the *Sun*, the *Horizontal Parallax* being substracted, is equal to the *Semi-angle* of the *Cone* of the *Earths Shadow*.
   So A E D — A E H = H E D
2. The *Horizontal Parallax* of the *Moon*, the *Semi-angle* of the *Cone* of the *Earths Shadow* being substracted, is equal to the apparent *Semidiameter* of the *Shadow*.
   So B F E — C G E = G B F.
3. The Sum of the *Horizontal Parallaxes* of the *Sun* and *Moon*, is equal to the Sum of the apparent *Semidiameter* of the *Sun* and *Shadow* of the *Earth*.
   So, B D F + B F D = A B D + C B F.

Therefore, From the Sum of the *Horizontal Parallax* of the *Sun* and *Moon*, substract the apparent *Semidiameter* of the *Sun*; and there will remain the apparent *Semidiameter* of the *Earths Shadow*.
So, B F D + B D F — A B D = C B F.

II. For the Proportional Magnitudes of these Three Bodies.

The *Semidiameter* of the	*Sun* 46300 Log.	4.665581
	Earth 727	2.861534
	The Log. Difference	1.804047
	Multiply by	3
The Logar. of 258309		5.412141

So that the *Body* of the *Sun* exceeds the *Body* of the *Earth* 258309 times.

Fig. LXXV.

The *Semidiameter* of the { *Earth* 1650 Log.		3.217484
{ *Moon* 446 Log.		2.549335
The Logar. Difference		0.568149
Multiply by		3
To the Logar. of 50.63		1.704447

So that the *Earth* is greater than the *Moon* 50 times, and $\frac{63}{100}$ parts.

III. Of the *Semidiameters*, and *Proportions* of the other *Primary Planets* to the *Earth*.

From the accurate Observations of *Hugenius*, *Gassendus*, and *Horrox* are determined.

The apparent Semidiameters of these 5 Planets in their mean distance from the Earth,		D.	D.	Which in such Parts as the Sediameter of the Earth is 727, gives the Semidiameter of		
	♄	10	30		♄	4855
	♃	24	0		♃	6054
	♂	4	0		♂	296
	♀	10	30		♀	509
	☿	5	0		☿	242.

From hence may be gathered, That the *Body* of *Saturn* is *Greater* than the *Earth* 298 times.

The *Body* of *Jupiter* 577 times.

But the *Earth* is *Greater* than the other Three : For it

Exceeds { ♂ 15 / ♀ 3 / ☿ 27 } Times.

GEO-

of
nd
ti-

th
th
e-
he

th *Fig.*
at LXXVI.

is

he

Fig. LXXV.

T

To

III.

Ho

Th
me
in
fro

GEOMETRICAL ASTRONOMY,

SHEWING

A Plain and Easie Way to Project Theories of the Planets (in Plano:) And by them to find their true Places, both in Longitude and Latitude, at any time.

And their Distances from the Sun or Earth.

CHAP. I.

How to lay down (in Plano*) the* Theory *of the* Earth, *or other* Planet.

FOR the performance hereof you must have a *Scale* of *Chords*, and a *Scale* of 1000 *equal Parts* to the same *Radius*: (Or, rather, a *Sector*, with such *Scales* upon it.) Then,

I. How to describe and divide the *Ecliptick*.

With 60 deg. of your *Scale of Chords*, describe a *Circle*; which divide into 12 equal Parts, for the 12 *Signs*, noting them with their respective *Characters*, ♈, ♉, ♊, *&c.* This *Circle* thus described and charactered, represents the *Ecliptick Circle* in the Heavens; whose *Centre is* ☉, the fixed place of the *Sun*.

In this *Ecliptick Circle*, every of the *Planets* hath a *Point*, which is called the *Aphelion* of that *Planet*: For when the *Planet* is in that Point, it is at its greatest distance from the *Sun*. Now, the *Aphelial Points* of each *Planet*, are these which are exhibited in this little Table.

Fig. LXXVI.

Fig. LXXVI.

The Aphelial Point of			Deg.	
	♄ Saturn	Is in	27.50	♐
	♃ Jupiter		7.82	♎
	♂ Mars		0.35	♍
	♀ Venus		2.82	♒
	☿ Mercury		14.95	♐
	⊖ The Earth		7.00	♑

II. To draw the *Aphelial Line.*

From the *Aphelial Point* of the *Planet* (as in our *Example* MARS,) 0.35 de. ♍ draw a Right-line to the Centre ⊙, as the Line *Ap.* ♂ ⊙; and that shall be the *Aphelial* Line of MARS.

For the *Aphelial Line* of the *Earth*: The EARTHS *Aphelial Point* being in 7.00 deg. of ♑ *Capricorn*, from that Point draw a Right-line to ⊙ the Centre, as the Line ⊙ *Ap.* ⊖, and that shall be the Earth's *Aphelial Line*, in the *Theories* of all the *Planets.*

III. To find the *Excentric Points* of any of the 3 Superior Planets ♄, ♃, ♂; they will be found by the numbers in this little Table. For,

If from the *Aphelial Point* of			
	♄	to the Centre ⊙ be 1000,	946
	♃	the *Excentric* Point shall be	954
	♂	from the *Aphelial Point*	915

Wherefore, take the length of the *Aphelial Line Ap.* ♂. ⊙ in your Compasses, and open the *Sector* in 10, and 10 of the equal Parts. Then, if from the *Sector* you take 915, (the *Excentric* of *Mars*) it will reach from the *Aphelial Point Ap.* ♂, upon the *Aphelial Line* of *Mars*, to X, so shall X be the *Excentric* of MARS; upon which Point (at the distance of *Ap.* ♂ X) if you describe a Circle, as *Ap.* ♂, ☊, ♂, that Circle shall be the *Orbit* of the *Planet* MARS, in which it moveth.

For the *Aphelial* and *Excentric Points* of the EARTH in the Theories of the three Superiour Planets, they will be at the Numbers exhibited in this little Table. For,

The

The distance of the *Aphelial Point* of the ⊖ from the *Centre*, shall be in the *Theory* of			And the distance of the *Excentric* of the ⊖ from the *Aphelion* of the ⊖, shall be in the *Theory* of		
	♄	101		♄	099
	♃	186		♃	183
	♂	611		♂	600

Wherefore, Out of your *Sector*, (it being still opened to the former distance) take the distance of the *Aphelial Point* of the *Earth* (for MARS 611) and set it (upon the *Earths Aphelial Line*) from the Centre ⊙ to *Ap.* ⊖, so shall the point *Ap.* ⊖ be the *Aphelial Point* of the *Earth* —— And the distance of the *Excentric* of the *Earth*, from the *Aphelial Point* of the *Earth* being 600, take 600 out of the *Sector*, and set it from the *Earths Aphelial Point Ap.* ⊖ downwards towards the Centre ⊙, and that shall be the Centre whereon to describe the *Orbit* of the *Earth* on the *Theory* of MARS. Which Point falling so near the Centre ⊙, you may describe the *Earths Orbit* upon the Centre ⊙ it self, the difference being only $\frac{11}{1000}$ Parts: And in the *Theories* of the other two *Superior Planets*, the difference wil be much less; for in the *Theory* of *Saturn* it will be $\frac{2}{1000}$ Parts: And in the *Theory* of *Jupiter*, but $\frac{3}{1000}$ Parts: So that in small *Theories* they need not be much regarded.

4. For the *Aphelial* and *Excentric Points* of the *Earth*, and the two *Inferior Planets* ♀ and ☿, the *Tables* following will exhibit.

1. For the *Aphelial Points.*

If the distance of the Aphelial Point of the Earth to ⊙ be 1000, the Excentric Point of		Shall be from the Centre ⊙	
	⊖		1.000
	♀		.716
	☿		.461

2. For the *Excentric Points.*

And from the *Aphelial Point* of		to the Excentrick Point of		shall be	
	⊖		⊖		.982
	♀		♀		.711
	☿		☿		.381

5. For

Fig. LXXVI.

5. For the Nodes of the Planets.

The Ascendant Nodes of the several Planets (for the Earth hath none:) This Table will tell you to what Points in the Ecliptic they are to be drawn. For

The Node of		Is in	d.	
	♄		22.45	♋
	♃		5.50	♋
	♂		17.55	♉
	♀		14.00	♊
	☿		14.15	♉

Wherefore, the Node of the Planet ♂ being in 17.55 de. of ♉, lay a Ruler from the Centre ☉ to 17.55 de. of *Taurus*, and there draw an obscure Line: This Line is the common Section of plain Planets Excentric, with the Plain of the *Ecliptic*; and where this Line crosseth the Orbit of *Mars*, there write ☊ *Dragons Head*; at which Point, when any Planet is in its Orbit, it goes into *North Latitude*: And at the opposite Point, through the Centre, is the place of ☋ the *Dragons Tail*, where it goes into *South Latitude*.

Of the Inclination of the *Planets*.

The Greatest Inclination of the Planets are such as are expressed in this Table.

For			Degrees.
	♄	2.53	
	♃	1.32	
	♂	1.84	
	♀	3.37	
	☿	6.90	

The *Aphelian Lines* and *Orbits* of the *Earth*, and any of the *Planets* being drawn within the *Eclіptick Circle*, the next thing to be done is, to shew how to place the *Earth* and any *Planet* in their respective *Orbits*, for any *Time* proposed: And also, How to *Reduce* their Places so found in their *Orbits* at that time, to their respective Places in the *Ecliptick Circle*.

In order whereunto it will be necessary to shew,

I. How *Time* is to be *Accounted*.

II. To

II. To *Collect* the *Anomalies* of the *Earth* and *Planet* for any *Time* proposed. *Fig.* LXXVI.

III. How *Degrees of Anomaly* in the *Orbits*, are to be *reduced* to *Degrees* in the *Ecliptick.*

CHAP. II.

How Time *is to be Accounted.*

TIME is to be thus Accounted: Any *Day* begins upon its own *Noon*: So that 12 at *Noon* of the first Day of *January*, is the *Common Term* of the *Old* and *New Years*; it being the *End* of the *Former*, and the *Beginning* of the *Latter.*

Before the *Longitude* or *Place* of a *Planet* in the *Ecliptic* can be found; the *equal Motion* of the *Anomaly*, for the Time proposed must be known; and that *Motion* for any proposed *Time* may be collected out of these TABLES.

Epoches

Æpoches *of the* ANOMALIES *for the* YEARS.

Epoches for	Earth	♂	♃	♂	♀	☿
		100	100	100	100	100
Years.	De. Pts.	De. Pts.	De. Pts.	De. Pts.	De. Pts.	De. Pts.
1692	194. 34	346. 17	246. 08	126. 89	247. 04	167. 86
1700	194. 26	83. 88	128. 88	218. 08	248. 42	245. 58
08	194. 18	181. 59	11. 68	309. 26	249. 80	323. 30
16	194. 11	279. 30	254. 48	40. 45	251. 17	41. 01
24	194. 03	17. 01	137. 28	131. 63	252. 55	118. 73
32	193. 95	114. 72	20. 08	222. 81	253. 93	196. 45
40	193. 87	212. 43	262. 88	314. 00	255. 31	274. 17
48	193. 80	310. 14	145. 68	45. 18	256. 68	351. 89
56	193. 72	47. 86	28. 48	136. 40	258. 06	69. 61
64	193. 64	145. 57	271. 28	227. 58	259. 43	147. 33
72	193. 57	243. 28	154. 08	318. 76	260. 81	225. 05
80	193. 49	341. 08	36. 88	49. 94	262. 19	302. 77
88	193. 41	341. 00	279. 68	141. 12	263. 57	20. 49
96	193. 34	78. 71	162. 48	232. 30	264. 94	98. 21
1800	193. 30	225. 27	283. 88	277. 90	85. 63	317. 35
1	359. 74	12. 21	30. 33	191. 27	224. 27	53. 69
2	359. 49	24. 41	60. 66	22. 53	89. 54	197. 38
3	359. 23	36. 62	90. 99	213. 80	314. 32	161. 08
4	359. 96	48. 86	121. 40	45. 59	180. 69	218. 86
5	359. 71	61. 06	151. 73	236. 86	45. 46	272. 55
6	359. 45	73. 27	182. 06	68. 13	270. 23	326. 24
7	359. 19	85. 47	212. 39	259. 39	135. 00	19. 93

Motion

Motion of the Anomaly, in Months Of the Common YEAR.

	Earth	♄	♃	♂	♀	☿
	100 De. Pts.	100 De. Pts.	100 De. Pts.	100 De. Pts.	100 De. Pts.	100 De. Pts.
Janua.	30. 55	1. 04	2. 58	16. 24	49. 67	126. 86
Febru.	58. 15	1. 97	4. 90	30. 72	94. 52	241. 45
March	88. 17	3. 01	7. 48	47. 16	144. 19	8. 31
April	118. 27	4. 01	9. 97	62. 88	192. 25	131. 08
May	148. 03	5. 05	12. 55	79. 13	241. 92	257. 94
June	178. 58	6. 05	15. 04	94. 85	289. 98	20. 71
July	208. 95	7. 09	17. 62	111. 09	339. 65	147. 57
August	239. 50	8. 13	20. 19	127. 34	29. 31	274. 43
Septem	269. 07	9. 13	22. 68	143. 06	77. 38	37. 20
Octob.	299. 62	10. 17	25. 26	159. 30	127. 04	164. 06
Novem	329. 19	11. 17	27. 75	175. 02	175. 11	286. 83
Decem.	359. 74	12. 21	30. 33	191. 27	224. 77	53. 69

Months in the Leap-Year.

	Earth.	♄	♃	♂	♀	☿
Janua.	30. 55	1. 04	2. 58	16. 24	49. 67	126. 86
Febru.	59. 14	2. 01	4. 99	31. 44	96. 13	245. 54
March	89. 69	3. 04	7. 56	47. 69	145. 79	12. 40
April	119. 26	4. 05	9. 95	63. 41	193. 86	135. 17
May	149. 81	5. 08	12. 63	79. 65	243. 52	202. 03
June	179. 38	6. 09	15. 12	95. 37	291. 58	24. 80
July	209. 93	7. 12	17. 70	111. 62	341. 25	151. 66
August	240. 49	8. 16	20. 27	127. 86	30. 92	278. 52
Septem.	270. 05	9. 16	22. 77	143. 58	78. 98	41. 20
Octob.	300. 61	10. 20	25. 34	159. 83	128. 64	168. 15
Novem	330. 18	11. 20	27. 84	175. 55	176. 71	290. 92
Decem.	360. 73	12. 24	30. 41	191. 79	226. 37	57. 78

DAYS.

Days	☉ D. P.	♄ D. P.	♃ D. P.	♂ D. P.	♀ D. P.	☿ D. P.
1	0. 99	0. 03	0. 08	0. 52	1. 60	4. 09
2	1. 97	0. 07	0. 17	1. 05	3. 20	8. 18
3	2. 96	0. 10	0. 25	1. 57	4. 81	12. 28
4	3. 94	0. 13	0. 33	2. 10	6. 41	16. 37
5	4. 93	0. 17	0. 24	2. 62	8. 01	20. 46
6	5. 91	0. 20	0. 50	3. 14	9. 61	24. 55
7	6. 90	0. 23	0. 58	3. 67	11. 21	28. 65
8	7. 88	0. 27	0. 66	4. 19	12. 82	32. 74
9	8. 87	0. 30	0. 75	4. 72	14. 42	36. 83
10	9. 86	0. 33	0. 89	5. 24	16. 02	40. 92
11	10. 84	0. 37	0. 91	5. 76	17. 62	45. 00
12	11. 83	0. 40	1. 00	6. 29	19. 23	49. 11
13	12. 81	0. 43	1. 08	6. 81	20. 83	53. 20
14	13. 80	0. 47	1. 16	7. 34	22. 43	57. 29
15	14. 78	0. 50	1. 25	7. 86	24. 03	61. 38
16	15. 77	0. 53	1. 33	8. 38	25. 63	65. 48
17	16. 76	0. 57	1. 41	8. 91	27. 24	69. 57
18	17. 74	0. 60	1. 50	9. 43	28. 84	73. 66
19	18. 73	0. 63	1. 58	9. 96	30. 44	77. 75
20	19. 71	0. 67	1. 66	10. 48	32. 04	81. 85
21	20. 70	0. 70	1. 75	11. 00	33. 64	85. 94
22	21. 68	0. 73	1. 83	11. 53	35. 25	90. 03
23	22. 67	0. 77	1. 91	12. 05	36. 85	94. 17
24	33. 65	0. 80	1. 99	12. 58	38. 45	98. 22
25	24. 64	0. 83	2. 08	13. 10	40. 05	102. 31
26	25. 63	0. 87	2. 16	13. 62	41. 66	106. 40
27	26. 61	0. 90	2. 24	14. 15	43. 26	110. 49
28	27. 60	0. 93	2. 33	14. 67	44. 86	114. 58
29	28. 58	0. 97	2. 41	15. 20	46. 46	118. 68
30	29. 57	1. 00	2. 49	15. 72	48. 06	122. 77
31	30. 55	1. 04	2. 58	16. 24	49. 67	126. 86

The manner of *Collecting* the *equal Anomalies* from these TABLES, is thus:

First, Exscribe the *Epocha* which belongs to that Year which most nearly precedeth the *Year* wherein you seek the Place of any *Planet*.

Secondly, Under that *Epocha* (or number) write the Motions belonging to so many *Years*, *Months* and *Days*, as are compleatly expired since the *Year* of the *Epocha*; all which Numbers must be taken out of their proper *Tables*, and set orderly one under another; which the disjunction of the Numbers will give Direction enough how to do.

Thirdly, All these Numbers must be added into *One*; and their *Sum* shall give the *Anomaly* for the *Time* proposed —— And if the *Sum* rise to be above a whole *Circle*, or 3.60 Degrees; you must then cast away the number of 360 as oft as you may, and the remaining Number must be taken for the *Anomaly*.

Fourthly, These things are to be done both in the *Earth* and the *Planet* severally; as is done in the Example following.

CHAP. III.

How to Collect the Anomalies *of* Saturn *and the* Earth, *for the 16th Day of* May, *at Noon*, *in the Year of* Christ 1712, *being* Bissextile, *or* Leap-Year.

	Anomaly. ⊖	Anomaly. ♄
	Deg.	Deg.
Æpocha for the Year 1700	194. 26	83. 88
Motion in Years Compleat { 4	359. 96	48. 86
Motion in Years Compleat { 7	359. 19	85. 47
April Compleat B.	119. 26	4. 05
Days Compleat 15	14. 78	. 50
The Sum of the Anomalies	1947. 45	222. 76
The 2 Circles Subst.	720.	
Equal Anomalies for the Time.	327. 45	222. 76
Which reduced to the EclipticK, are	♐ 15. 50	♌ 15. 00

First, Out of the Table of *Years*, exscribe the *Æpocha* for 1700: And under it 4 *Years* and 7 *Years*, (which together make 11 compleat *Years*:) Under them exscribe *April Compleat* (putting B to it, for that it is a *Leap-Year*:) And under that exscribe *Days Compleat* 15. Thus is the first part of your Work performed.

Secondly, Repairing to your *Tables*, you shall find the *Anomaly* of the *Earth* which stands against 1700, to be 194.26 de. and the *Anomaly* of ♄ to be 83.88 deg. both which set in their proper *Columns* under ⊖ and ♄, against 1700. Also, Take out of their proper *Tables*, the *Anomalies* belonging to 4 *Years* and 7 *Years*; and write them in their proper *Columns* against their respective *Times*: As 359.96 deg. for the ⊖, and 48.86 deg. for ♄; which set down against 4 *Years*: Do the like for 7 *Years*: And for the *Month* of *April*, which must be taken out of the *Table* of *Months* in the *Leap-Years*, because 1712 will be *Leap-Year*: And for the 15 *Days*, take the *Anomalies* answering to them out of the Table of *Anomalies* for *Days*.

Thirdly, all the *Anomalies* being thus exscribed out of the *Tables*, and set orderly one under another; you must add them together, and you shall find the Sum of the ⊖ *Anomaly* to be 1047.45 deg. And the *Anomaly* of ♄ to be 222.76 deg. But, because the Sum of the *Earths Anomalies* is above 360, you must take that Number, as oft as you can (which is twice, *viz.* 720 deg.) and the Remainder 327.45 deg. is the *Anomaly* of the ⊖ *Reduced*: But, the *Anomaly* of ♄ being less than 360 deg. needs no such *Reduction*. And thus have you computed the *Anomalies* of the ⊖ and ♄, for the 16th Day of *May*, in the Year 1712 at Noon.

CHAP. IV.

How the Degrees *of* Anomaly *of the* Earth, *or any of the* Planets *(found as before) in their Orbits; are to be* Reduced *to* Degrees *in the* Ecliptick.

FOR the performance hereof, the following *Table* is subservient, for it will readily (by inspection only) answer your desire.

A TABLE, shewing what Point of the *Ecliptick* answers to any Degrees of Anomaly, that the *Earth*, or any of the other *Planets* is found to have in its own *Orbit*.

Anomalies in their Orbits.

Anomaliæ Degrees.	⊖ S.	D.	♄ S.	D.	♃ S.	D.	♂ S.	D.	♀ S.	D.	☿ S.	D.
0	♑	7. 00	♐	27. 50	♒	7. 82	♍	0: 35	♒	2: 82	♐	14: 97
10		16. 65	♑	6. 44		16. 92		8: 70		12: 69		21: 64
20		26. 32		15. 42		26. 04		17: 09		22: 59		28: 37
30	♒	6. 00		24. 44	♏	5. 20		25: 54	♓	2: 42	♑	5: 20
40		15. 70	♒	3. 50		14. 42	♎	0: 10		12: 19		12: 19
50		25. 45		12. 72		23. 75		12: 80		22: 23		19: 39
60	♓	5. 24		22. 05	♐	3. 20		21: 69	♈	2: 14		26: 84
70		15. 09	♓	1. 50		12. 74	♏	0: 80		12: 07	♒	4: 64
80		25. 00		11. 07		22. 44		10: 15		22: 03		12: 85
90	♈	4. 94		21. 0	♑	2. 30		19: 80	♉	2: 01		21: 55
100		15. 00	♈	1. 00		12. 34		29: 75		12: 07	♓	0: 85
110		25. 05		11. 22		22. 05	♐	10: 02		22: 07		10: 85
120	♉	5. 20		21. 65	♒	2. 89		20: 64	♊	2: 14		21: 65
130		15. 40	♉	2. 26		13. 42	♑	1: 60		12: 20	♈	3: 35
140		25. 65		13. 07		24. 10		12: 87		22: 30		16: 00
150	♊	6. 00		24. 01	♓	4. 90		24: 48	♋	2. 42		29: 62
160		16. 27	♊	5. 12		15. 82	♒	6: 27		12: 55	♉	14: 18
170		26. 64		16. 29		26. 80		18: 26		22: 68		29: 37
180	♋	7. 00		27. 50	♈	7. 32	♓	0: 35	♌	2: 82	♊	15: 00
190		17. 37	♋	8. 72		18. 84		12: 44		13: 00	♋	0: 55
00		27. 73	[illegible]	19. 88		29. 82		24: 42		23: 09		15: 76
210	♌	8. 05	♌	0. 99	♉	10. 74	♈	6: 24	♍	3: 23	♌	0: 29
220		18. 35		11. 94		21. 54		17: 82		13: 34		13: 90
230		28. 60		22. 74	♊	2. 22		29: 10		23: 46		26: 55
240	♍	8. 80	♍	3. 55		12. 75	♉	10: 07	♎	3: 50	♍	8: 25
250		19. 00		13. 79		23. 9		20: 69		13: 57		19: 05
266		29. 04		24. 00	♎	3. 30	♊	1: 09		23: 60		29: 05
270	♎	9. 07	♋	4. 02		13. 33		10: 90	♏	3: 62	♎	8: 35
280		19. 02		13. 84		23. 20		20: 55		13: 60		17: 05
290		28. 92		23. 49	♌	2. 90		29: 90		23: 57		25: 26
300	♏	8. 76	♏	2. 95		22. 45	♋	9: 01	♐	3: 50	♏	3: 07
310		18. 55		12. 29		21: 89		15: 90		13: 42		10: 52
320		28. 30		21. 49	♍	1: 23		26: 60		23: 32		17: 73
330	♐	8. 02	♐	0. 34		10: 44	♌	5: 17	♑	3: 23		24: 72
340		17. 69		9. 60		19: 66		13: 62		13: 09	♐	1: 52
350		27. 35		18. 57		28. 43		22: 0		23: 60		8: 26
350	[illegible]	7. 00	♑	27. 50	♎	7: 82	♍	0: 35	♒	2: 82		14: 95

The

The Uſe of the *Table*.

The *Table* conſiſts of Seven *Columns*: In the firſt, towards the Left-hand, is placed the Degrees of *Anomaly*, from 1 to 360, by every 10th Degree. And in the other Six *Columns* noted at the Head with ⊕, ♄, ♃, ♂, ♀, ☿, you have the Degrees, and 100 Parts of the *Ecliptick* which anſwer thereunto. So that if you ſeek the Degrees of *Anomaly* of any *Planet* (found as by the former *Table*) in the firſt *Column*, you ſhall find againſt thoſe Degrees of *Anomaly*, the *Sign*, *Degree*, and 100 *Part* of the *Ecliptick* anſwerable thereunto.

So in the foregoing *Example* of the *Earth* ⊕ and *Saturn* ♄, where the *Anomaly* of the *Earth* was found to be 328, and of *Saturn* 223 (neareſt)——Now, if you look for theſe *Numbers* in the firſt *Column* of this *Table*, you ſhall find, that againſt 330, (which is the neareſt Number to 328) in the *Earths Column*, ♐ 8.02 de. And that Degree of the *Ecliptick* anſwers to 330 deg. of the *Earths Anomaly* in its *Orbit*——Alſo, againſt 220 in the firſt *Column* (which is the neareſt to 223, the *Anomaly* of *Saturn*) you ſhall find in the *Column* for *Saturn*, ♌ 11.93 deg. And that Degree of the *Ecliptick* anſwers to 320 deg. of *Saturns Anomaly* in his *Orbit*: And ſo of any other.

But in the Uſe of this *Table* you are to *note*, That there being but every *Tenth Degree* of *Anomaly* in the firſt *Column*, the *Degrees* in the other *Columns* for the *Planets* give thoſe Signs and Degrees of the *Ecliptick*, of every *Tenth Degree* of *Anomaly*: And therefore you muſt make Proportion (which how to perform, we ſuppoſe to be known) and therefore in our Example.

					D. M.
The Anomaly of	the Earth	being	328	the Ecliptick Degrees anſwering, are	15.52 ♐
	Saturn		223		15.02 ♌

CHAP.

CHAP. V.

The Use of the Theories, and by them to find the Places *of the* Earth *and* Planets, *both in* Longitude *and* Latitude, *&c. for any time proposed. As for* MARS *and the* EARTH, *on the* 18*th of* October 1705.

THE *Orbits* of the EARTH and MARS being described in the *Theory*, as is before shewed how to do; *First*, Collect the equal *Anomalies* for the *Earth* and *Mars*, for the Time proposed, and their *Reductions* to the *Ecliptick*, as is here done.

The Equal Anomalies *of the* Earth *and* Mars, *for the* 18*th Day of* October 1705 *at Noon.*

	♁ Anomaly.	♂ Anomaly.
Æpoche for Years 1700	194. 26	218. 08
Years Compleat 4	359. 96	45. 59
Month of *September* Compleat	269. 07	143. 06
Days Compleat 17	16. 76	8. 91
The Sum	840. 05	415. 64
Circle Subftr.	720	360.
The Equal-Anomalies	120. 05	55. 64
Their Reduction to the Ecliptick	5. 12 ♉	17. 68 ♎

Secondly, The *Equal Anomaly* of the *Earth*, reduced to the *Ecliptick*, being in 5. 12 de. of *Taurus*, lay a Ruler to the Centre ⊙, and 5.12 deg. ♉, (as at *c*) and where it crosseth the *Orbit* of the *Earth*, make ♁, for that is the Place of the *Earth* in his *Orbit* at that time.

Thirdly, The *Equal Anomaly* of MARS, reduced to the *Ecliptick Circle*, being in 17.68 deg. of *Libra*: Lay a Ruler to the Centre

Centre ☉ and 17.68 deg ♎ (as at *d*,) and where it crosses the *Orbit* of MARS, make ♂; for that is the place of MARS in his *Orbit* at that time.

Fourthly, Draw the Right-Lines, ☉ ⊖, ☉ ♂ and ♂ ⊖, constituting the Right-lined Triangle ☉ ♂ ⊖.

The *Theory* thus finished, I come now to shew,

I. To find the *Sun's true Place* in the *Ecliptick*, for the time proposed

Lay a Ruler upon the Line ☉ ⊖, and it will cut the *Ecliptick Circle* in b ☉ b, in 5. 51 deg. of *Scorpio*; and that is the *Sun's Longitude* (or *Place* in the *Ecliptick*) at the time proposed.

II. To find the *Longitude* (or true *Place* in the *Ecliptick*) of MARS for the same time.

Through the Point ☉ in the Centre, draw a Right-Line parallel to the Line ♂ ⊖, as the short Line *a* ♂ *a*, which will cross the *Ecliptick Circle* in 25.00 deg of *Libra* ♎, where make the Character of MARS; for that is his true *Place* of *Longitude* in the *Ecliptick*, for the time proposed.

Mark now the Triangle ☉ ♂ ⊖, made upon the *Theory*, by the Centre ☉, and the Point of ⊖ and ☉ in their *Orbits*: For in that Triangle,

The Side	☉ ♂	is the distance of	♂	from the	Sun.
	☉ ⊖		⊖		Sun.
	⊖ ♂		♂		Earth.

All which are to be *Measured* upon *Proper Scales* for the several *Planets*. And

The Angle	♂ ☉ ⊖	is the Angle at the	Sun	or of	*Commitation.*
	♂ ⊖ ☉		Earth		*Elongation.*
	⊖ ♂ ⊖		Planet		*Paral. of the Orb.*

These *Angles* may be Measured (very near the Truth) by help of your *Scale* of *Chords*; but more accurately (when the *Sides* are known) by *Trigonometrical Calculation*: Of which hereafter.

But

But for the measuring of the *Sides*, *Proper Scales* must be made for each *Planet*, and the *Earth*. And, *Fig.* LXXVI.

The *Proportions* between the *Length* of the *Aphelial Lines* of the several *Planets*, (taken from the *Center* of the *Theories*, to the *Aphelial Points*) and the *Scales* by which to measure their *Distances* from the *Sun* or *Earth*; are these exhibited in this *Table*, *viz.*

			Parts.
For	Saturn	as 100 to	85. 63
	Jupiter		92. 87
	Mars		56. 73
	Earth		69. 38

Which Numbers will make *Proper Scales* for measuring the *Distances* in each *Theory*. And the *Scale* for the *Earth*, will serve, also, for the *Theories* of *Venus* and *Mercury*.

And note farther, That all *Distances* measured upon these *Proper Scales*, are

In	Saturn	to be *multiplied* by	400
	Jupiter		200
	Mars		100
	Venus		50
	Mercury		50

And these *Products* shall *Reduce* the *Distances* measured upon the *Proper Scales*, into *Semidiameters* of the *Earth*.

III. To find the *Distance* of MARS from the *Sun* and from the *Earth*, for the Time proposed.

Take in your *Compasses* the length of the *Aphelial Line Ap.* ♂, ⊙, and opening the *Sector* in the Points 56.73 and 56.73. The *Sector*, as so opened, shall be the *Scale* whereby to measure the *Distances* in the *Theory* of MARS. Wherefore, if you take in your Compasses (out of the Theory) of *Mars*.

Fig. LXXVI.

					Parts.
The Line	☉ ♂	And apply it to	☉ ♂	to contain	55. 50
	☉ ⊖	the *Sect.* so open'd,	☉ ⊖		34. 75
	⊖ ♂	you shall find	⊖ ♂		88. 20

Which Parts multiplied by 100, will give

For	☉ ♂	3550	*Semidiameter* of the *Earth.*
	☉ ⊖	3475	
	⊖ ♂	8820	

IV. To find the *Angle* of *Commination*, or *Angle* at the *Sun*) ♂ ☉ ⊖.

Lay a Ruler to ☉ the Centre, and the Point ⊖ in its *Orbit*; and it will cut the *Ecliptick Circle* in the Point *c*: Also lay a Ruler upon ☉ the Centre, and the Point *Mars* in its *Orbit*, and it will cut the *Ecliptick* in the Point *d*; then will the Degrees of the *Ecliptick*, intercepted between *c* and *d*, be the quantity of the *Angle* ♂ ☉ ⊖, the *Angle* at the *Sun* (or of *Commination*) and will be found to be 162.00 deg.

V. To find the *Angles* of *Elongation* and *Parallax* of the *Orb* (or the *Angles* at the *Earth* and *Planet*) ☉ ⊖ ♂ and ☉ ⊖ ♂.

These (as I said before) may be measured upon the *Theory* by the assistance of the *Scale* of *Chords*: But the *Angle* of *Commination* and the *distance* of the *Earth* and *Mars* being known, they may more exactly be found by *Trigonometrical Calculation*: Thus;

As the Sum of the Sides ♂ ☉ and ⊖ ☉. 9025	3.954447
Is to their Difference 2075.	3.317018
So is the Tang. of half the Angles at ♂ and ⊖, 9. 00 d.	9.199712
	12.516730
To the Tangent of 2.09 deg.	8.562283

Which added to 9.00 deg. gives 11.09 deg. for the *Angle* at the *Earth* (or of *Elongation*) ☉ ⊖ ♂ —— And substracted from 9.00 deg. leaves 6.91 deg. for the *Angle* at *Mars* (or of the *Parallax* of the *Orb*.

VI. To

365

Fig.
LXXVI.

ust
⊙;
he
⊖
gle
ill.
en.

ty

7
5
2

e
as,

8
3
2
5
7

ac

364
Fig.
LXXVI.

Fig. LXXVI.

VI. To find the Latitude of MARS at the fame time.

Before the *Latitude* can be found, the *Argument of Latitude* muft be had: To the attaining whereof, Lay a Ruler to the Centre ⊙, and the Point ☊ in the *Planets Orbit*; fo will the Ruler cut the *Ecliptick* in the Point *e*: And a Ruler laid from the Centre ⊙ to ♂, will cut the Ecliptick in the Point *d*: So that the *Angle* *d* ⊙ *e* is the *Argument of Latitude*; and in this our *Example* will be found (by meafuring the Degrees of the *Ecliptick Circle* between *d* and *e*) to be 149.75 deg.

The *Argument of Latitude* thus attained, the true *Latitude* may be found by the following Canon.

As the Radius, Sine 90 de.	10.
Is to the the Sine of the *Arg. of Latit.* (149.75 d.)	9.7022357
So is the Tang. of *Mars Greateft Incl.* 1.84 de.	8.5068445
To the Tangent of 0.92 deg.	18.2090802

Which 0.92 Degrees is the *Latitude* of MARS feen at the *Sun*; and is *North*, becaufe the *Inclination* of MARS was *Northward*.

But to find the *Latitude* as feen at the *Earth*

The Proportion is,

As the *Diftance* ♂ ⊕, 8820	3.945468
Is to the *Diftance* ⊙ ♂, 5550	3.744293
So is the Tang. of the *Latit.* feen at the *Sun* 0.92 de.	8.205702
	11.949995
To the Tangent of 0.58 deg.	8.004527

Which is the *Latitude* of *Mars* feen from the *Earth*.

The

Figure LXXVII.

The Equal Motions of the Anomalies of the Earth, and the other Five Planets, ♄, ♃, ♂, ♀, ☿, For the 18th Day of *October*, at Noon, *Anno* 1705.

	♁	♄	♃	♂	♀	☿
Æpocha for Years 1700	194:26	83:88	128:88	218:08	248:42	245:58
Motion in Years Compleat 4	359:96	48:86	121:40	45:59	180:69	218:86
September Compleat	269.07	9:13	22:68	143:06	77:38	37:20
Days Compleat 17	16:79	0:57	1:41	8:91	27,24	69:57
The Sums	840:08	142:44	274:37	415:64	533:73	571:21
Circles Subst.	720:			360:	360:	360:
The Equal Anomalies	120:08	142:44	274:37	55:64	173:73	211:21
Their Reduction to the Eclipt.	5:12 ♉	15:19 ♉	17:70 ♋	17:83 ♎	26:41 ♋	16:10 ♌
The Planets Longitudes	6:00 ♓	17:00 ♌	29:00 ♋	25:00 ♎	27:00 ♍	17:50 ♎
Distances from the Sun	44:30	76:25	88:00	55:50	51:08	24:50
Distances from the Earth		67:51	86:00	88:20	78:03	77:02
Latitude		2:56 S	0:25 N	0:57 N	1:42 N	2:06

FINIS.

Lightning Source UK Ltd.
Milton Keynes UK
UKHW030114231118
332791UK00011B/849/P